INDUSTRIAL AUTOMATION AND ROBOTICS

Second Edition

INDUSTRIAL AUTOMATION AND ROBOTICS

Second Edition

JEAN RIESCHER WESTCOTT
A.K. GUPTA
S.K. ARORA

MERCURY LEARNING AND INFORMATION
Boston, Massachusetts

Publisher: David Pallai
MERCURY LEARNING AND INFORMATION
121 High Street, 3rd Floor
Boston, MA 02110
info@merclearning.com
www.merclearning.com
800-232-0223

J. R. Westcott / A. K. Gupta / S. K. Arora. *Industrial Automation and Robotics, Second Edition.*
ISBN: 978-1-68392-961-1

The material in Chapter 18 and the videos on the companion files appear and were adapted from *Real-Time Embedded Components and Systems with Linux and RTOS* by S.Siewert and J. Pratt. Mercury Learning and Information, 2016. ISBN: 978-1-942270-04-1. Chapter 19 appeared in "Unmanned Aerial Vehicles," by P.K. Garg, © Copyright 2021. Mercury Learning and Information. All Rights Reserved.

Library of Congress Control Number: 2023944997

232425321   Printed on acid-free paper in the United States of America

Our titles are available for adoption, license, or bulk purchase by institutions, corporations, etc. For additional information, please contact the Customer Service Dept. at 800-232-0223(toll free).

All of our titles are available in digital format at *academiccourseware.com* and other digital vendors. *Companion files for this title are available by contacting info@merclearning.com.* The sole obligation of MERCURY LEARNING AND INFORMATION to the purchaser is to replace the disc, based on defective materials or faulty workmanship, but not based on the operation or functionality of the product.

CONTENTS

PREFACE

The purpose of this book is to present an introduction to the multidisciplinary field of automation and robotics for industrial applications. The book initially covers the important concepts of hydraulics and pneumatics and how they are used for automation in an industrial setting. It then moves to a discussion of circuits and using them in hydraulic, pneumatic, and fluidic design. The latter part of the book deals with electric and electronic controls in automation, robotics, robotic programming, and applications of robotics in industry. New chapters on unmanned aerial vehicles and the promise of AI in industrial automation have been added. Companion files are available with applications and videos.

Companion Files

Companion files (videos, lab projects, and figures from the book) for this title are available by contacting the publisher at *info@merclearning.com*.

Acknowledgments

The material in Chapter 18 and the videos on the companion files appear in and were adapted from *Real-Time Embedded Components and Systems with Linux and RTOS* by S.Siewert and J. Pratt. Mercury Learning and Information, 2016. ISBN: 978-1-942270-04-1. I would like to thank Jen Blaney of Mercury Learning for her professional and patient assistance with this product, Sean Westcott for his input and support, and to dedicate this project to the truest friend, Sandy.

Jean Riescher Westcott
October 2023

CHAPTER 1

AUTOMATION

INTRODUCTION

The word *automation* comes from the Greek word *"automatos,"* meaning self-acting. The word automation was coined in the mid-1940s by the U.S. automobile industry to indicate the automatic handling of parts between production machines, together with their continuous processing at the machines. The advances in computers and control systems have extended the definition of automation. By the middle of the 20th century, automation had existed for many years on a small scale, using mechanical devices to automate the production of simply shaped items. However the concept only became truly practical with the addition of the computer, whose flexibility allowed it to drive almost any sort of task.

DEFINITION OF AUTOMATION

Automation can generally be defined as the process of following a pre-determined sequence of operations with little or no human labor, using specialized equipment and devices that perform and control manufacturing processes. Automation in its full sense, is achieved through the use of a variety of devices, sensors, actuators, techniques, and equipment that are capable of observing the manufacturing process, making decisions concerning the changes that need to be made in the operation, and controlling all aspects of it.

OR

Automation is the process in industry where various production operations are converted from a manual process, to an automated or mechanized process.

Example*:* Let's assume that a worker is operating a metal lathe. The worker collects the stock, already cut to size, from a bin. He then places it in the lathe chuck, and moves the various hand-wheels on the machine to create a component; a bolt could be such an item. When finished the worker begins the process again to make another item. This would be a manual process. If this process were automated, the worker would place long lengths of bar into the feed mechanism of an automatic lathe. The lathe mechanisms feed the material into the chuck, turn the piece to the correct shape and size, and cut it off the bar before beginning another item. This is an example of an automated machine in a manufacturing process.

Automation is a step beyond mechanization, where human operators are provided with machinery to help them in their jobs. Industrial robotics are said to be the most visible part of automation. Modern automated processes are mostly controlled by computer programs, which through the action of sensors and actuators, monitor progress and control the sequences of events until the process is complete. Decisions made by the computer ensure that the process is completed accurately and quickly.

Many people fear that automation will result in layoffs and unemployment; they believe that its evils considerably outweigh its benefits. Basically, automation does take over jobs performed by workers; but automation does not need to bring about unemployment, as some people fear, *for three very positive reasons:*

First, in terms of the numbers of workers required to produce a product, the reduction is a temporary displacement which can be offset by the demands of a broadening market, as well as the creation of new industries. It still takes many workers to build, service, and operate any automatic machine.

Second, automation does not happen overnight; it is an evolutionary process. Manual, direct-labor work will be progressively transformed into work, which will be cleaner, easier, safer, and more rewarding to the worker, who, through the process of automation itself, will be trained for the more skillful accomplishments required in the better jobs of the future.

Third, and most important, automation is the necessary solution to a predicted shortage of labor. It is designed to do the work of people who are not there; it is a solution to a problem, not a cause.

Automation is a technology dealing with the application of mechatronics and computers for production of goods and services. Manufacturing automation deals with the production of goods. It includes:

- Automatic machine tools to process parts.
- Automatic assembly machines.
- Industrial robots.
- Automatic material handling.
- Automated storage and retrieval systems.
- Automatic inspection systems.
- Feedback control systems.
- Computer systems for automatically transforming designs into parts.
- Computer systems for planning and decision making to support manufacturing.

The decision to automate a new or existing facility requires the following considerations to be taken into account:

- Type of product manufactured.
- Quantity and the rate of production required.
- Particular phase of the manufacturing operation to be automated.
- Level of skill in the available workforce.
- Reliability and maintenance problems that may be associated with automated systems.
- Economics.

MECHANIZATION VS AUTOMATION

Mechanization refers to the use of powered machinery to help a human operator in some task. The use of hand-powered tools is not an example of mechanization. The term is most often used in industry. The addition of powered machine tools; such as the steam powered lathe dramatically reduced the amount of time needed to carry out various tasks, and improves productivity. Today very little construction of any sort is carried out with hand tools. Automation and mechanization are often confused with each other, though it should not be too hard to keep them apart. Mechanization saves the use of human muscles; automation saves the use of human judgment. Mechanization displaces physical labor, whereas automation displaces mental labor.

Mechanization is the replacement of human power by machine power. Mechanization often replaces craftwork and creates jobs for unskilled labor. It also only affects one or two industries at a time. Mechanization moves slowly and the job displacement is short-term. Mechanization is what occurred during the industrial revolution. *Automation* is the replacement of human thinking with computers and machines. Automation tends to create jobs for skilled workers at the expense of unskilled and semi-skilled workers. It affects many industries at the same time, moving rapidly. It also creates longer-term job displacement and has been more characteristic since the 1950s.

ADVANTAGES OF AUTOMATION

Manufacturing companies in virtually every industry are achieving rapid increases in productivity by taking advantage of automation technologies. When one thinks of automation in manufacturing, robots usually come to mind. The automotive industry was the early adopter of robotics, using these automated machines for material handling, processing operations, and assembly and inspection. Automation can be applied to manufacturing of all types. The advantages of automation are:

- Increase in productivity.
- Reduction in production costs.
- Minimization of human fatigue.
- Less floor area required.
- Reduced maintenance requirements.

- Better working conditions for workers.
- Effective control over production process.
- Improvement in quality of products.
- Reduction in accidents and hence safety for workers.
- Uniform components are produced.

GOALS OF AUTOMATION

Automation has certain primary goals as listed below:

- Integrate various aspects of manufacturing operations so as to improve the product quality and uniformity, minimize cycle times and effort, and thus reduce labor costs.
- Improve productivity by reducing manufacturing costs through better control of production. Parts are loaded, fed, and unloaded on machines more efficiently. Machines are used more effectively and production is organized more efficiently.
- Improve quality by employing more repeatable processes.
- Reduce human involvement, boredom, and possibility of human error.
- Reduce workpiece damage caused by manual handling of parts.
- Raise the level of safety for personnel, especially under hazardous working conditions.
- Economize on floor space in the manufacturing plant by arranging the machines, material movement, and related equipment more efficiently.

SOCIAL ISSUES OF AUTOMATION

Automation has contributed to modern industry in many ways. Automation raises several important social issues. Among them is *automation's impact on employment/unemployment*. Automation leads to fuller employment. When automation was first introduced, it caused widespread fear. It was thought that the displacement of human workers by computerized systems would lead to unemployment (this also happened with mechanization, centuries earlier). In fact the opposite was true, the freeing up of the labor force allowed more people to enter information jobs, which are typically higher paying. One odd side effect of this shift is that "unskilled labor" now pays very well in most industrialized nations, because fewer people are available to fill such jobs leading to supply and demand issues.

Some argue the reverse, at least in the long term. First, automation has only just begun and short-term conditions might partially obscure its long-term impact. For instance many manufacturing jobs left the United States during the early 1990s, but a massive up scaling of IT jobs at the same time offset this as a whole. Currently, for manufacturing companies, the purpose of automation has shifted from increasing productivity and reducing costs to increasing quality and flexibility in the manufacturing process.

Another important social issue of automation is *better working conditions*. The automated plants needs controlled temperature, humidity, and dust free environment for proper functioning of automated machines. Thus the workers get a very good environment to work in.

Automation leads to *safety* of workers. By automating the loading and unloading operations, the chances of accidents to the workers get reduced.

Workers expect an increase in *standard of living* with the help of automation. Standards of living go up with the increase in productivity, and automation is the sure method of increasing productivity. The cost of color TVs, washing machines, and stereos has declined, thus enabling a large number of households to buy these products.

LOW COST AUTOMATION

Low cost automation (LCA) is a technology that creates some degree of automation around the existing equipment, tools, methods, people, etc. by using standard components available in the market with low investment, so that the payback period is short.

The benefits of low cost automation are numerous. It not only simplifies the process, but also reduces the manual content without changing the basic set up. Major advantages of low cost automation are low investment, increased labor productivity, smaller batch size, better utilization of the material, and process consistency leading to less rejections.

A wide range of activities such as loading, feeding, clamping, machining, welding, forming, gauging, assembly, and packing can be subjected to low cost automation systems adoption. Besides, low cost automation can be very useful for process industries manufacturing chemicals, oils, or pharmaceuticals. Many operations in food processing industries, which need to be carried out under totally hygienic conditions, can also be rendered easy through low cost automation systems.

A wide variety of systems (mechanical, hydraulic, pneumatic, electrical, and electronics) are available for deployment in LCA systems. However, each has its own advantages as well as limitations. For uncomplicated situations, one can build a simple LCA device using any of the above systems, through a rapid techno-economic evaluation. However, in most of the practical applications, hybrid systems are used because that can allow use of the advantages of different devices, while simultaneously minimizing individual disadvantages.

Issues in Low Cost Automation

1. **Assessment of the Current Productivity Level:** There are some simple procedures for this. Work sampling (activity sampling) is one of them. It needs no equipment and little time to collect the data. If the data is processed, considerable information will come out about the current productivity level.

2. **PMTS:** Predetermined Motion and Time Studies is a very useful tool to check whether an existing manual operation is correctly pasted. If the time taken is more than desirable, PMTS will help in identifying it and improving it.

3. **Design for Automation and Assembly:** When components are made and assembled manually one may not have thought about the complexity of automation. For example, putting together half a dozen nuts and bolts is very easy in a manual assembly but very complex for an automatic system.

TYPES OF AUTOMATION

Fixed Automation (Hard Automation)

Fixed automation refers to the use of special purpose equipment to automate a fixed sequence of processing or assembly operations. It is typically associated with high production rates and it is relatively difficult to accommodate changes in the product design. This is also called *hard automation*. For example, GE manufactures approximately 2 billion light bulbs per year and uses fairly specialized, high-speed automation equipment. Fixed automation makes sense only when product designs are stable and product life cycles are long. Machines used in hard-automation applications are usually built on the *building block*, or *modular principle*. They are generally called *transfer machines*, and consist of the following two major components: *powerhead production units* and *transfer mechanisms*.

Advantages

- Maximum efficiency.
- Low unit cost.
- Automated material handling —fast and efficient movement of parts.
- Very little waste in production.

Disadvantages

- Large initial investment.
- Inflexible in accommodating product variety.

Programmable Automation

In programmable automation, the equipment is designed to accommodate a specific class of product changes and the processing or assembly operations can be changed by modifying the control program. It is particularly suited to "batch production," or the manufacture of a product in medium lot sizes (generally at regular intervals). The example of this kind of automation is the CNC lathe that produces a specific product in a certain product class according to the "input program." In programmable automation, reconfiguring the system for a new product is time consuming because it involves reprogramming and set up for the machines, and new fixtures and tools. Examples include numerically controlled machines, industrial robots, etc.

Advantages

- Flexibility to deal with variations and changes in product.
- Low unit cost for large batches.

Disadvantages

- New product requires long set up time.
- High unit cost relative to fixed automation.

Flexible Automation (Soft Automation)

In flexible automation, the equipment is designed to manufacture a variety of products or parts and very little time is spent on changing from one product to another. Thus, a flexible manufacturing system can be used to manufacture various combinations of products according to any specified schedule. With a flexible automation system, it is possible to quickly incorporate changes in the product (which may be redesigned in reaction to changing market conditions and to consumer feedback) or to quickly introduce a new product line. For example, Honda is widely credited with using flexible automation technology to introduce 113 changes to its line of motorcycle products in the 1970s. Flexible automation gives the manufacturer the ability to produce multiple products cheaply in combination than separately.

Advantages

- Flexibility to deal with product design variations.
- Customized products.

Disadvantages

- Large initial investment.
- High unit cost relative to fixed or programmable automation.

CURRENT EMPHASIS IN AUTOMATION

Currently, for manufacturing companies, the purpose of automation has shifted from increasing productivity and reducing costs, to broader issues, such as increasing quality and flexibility in the manufacturing process. The old focus on using automation simply to increase productivity and reduce costs was short-sighted, because it is also necessary to provide a skilled workforce who can make repairs and manage the machinery. Moreover, the initial costs of automation were high and often could not be recovered by the time entirely new manufacturing processes replaced the old. (Japan's "robot junkyards" were once world famous in the manufacturing industry.)

Automation is now often applied primarily to increase quality in the manufacturing process, where automation can increase quality substantially. For example, automobile and truck pistons used to be installed into engines manually. This is rapidly being transitioned to automated machine installation, because the error rate for manual installment was around 1–1.5%, but is 0.00001% with automation. Hazardous operations, such as oil refining, the manufacturing of industrial chemicals, and all forms of metal working, were always early contenders for automation.

Another major shift in automation is the increased emphasis on flexibility and convertibility in the manufacturing process. Manufacturers are increasingly demanding the ability to easily switch from one manufacturing product to other without having to completely rebuild the production lines.

REASONS FOR AUTOMATION

1. Shortage of labor

2. High cost of labor

3. Increased productivity: Higher production output per hour of labor input is possible with automation than with manual operations. Productivity is the single most important factor in determining a nation's standard of living. If the value of output per hour goes up, the overall income levels go up.

4. Competition: The ultimate goal of a company is to increase profits. However, there are other goals that are harder to measure. Automation may result in lower prices, superior products, better labor relations, and a better company image.

5. Safety: Automation allows the employee to assume a supervisory role instead of being directly involved in the manufacturing task. For example, die casting is hot and dangerous and the work pieces are often very heavy. Welding, spray painting, and other operations can be a health hazard. Machines can do these jobs more precisely and achieve better quality products.

6. Reducing manufacturing lead-time: Automation allows the manufacturer to respond quickly to the consumers needs. Second, flexible automation also allows companies to handle frequent design modifications.

7. Lower costs: In addition to cutting labor costs, automation may decrease the scrap rate and thus reduce the cost of raw materials. It also enables just-in-time manufacturing which in turn allows the manufacturer to reduce the in-process inventory. It is possible to improve the quality of the product at lower cost.

REASONS FOR NO AUTOMATION

1. Labor resistance: People look at robots and manufacturing automation as a cause of unemployment. In reality, the use of robots increases productivity, makes the firm more competitive, and preserves jobs. But some jobs are lost. For example, Fiat reduced its work force from 138,000 to 72,000 in nine years by investing in robots. GM's highly automated plant built in collaboration with Toyota in Fremont, California employs 3,100 workers in contrast to 5,100 at a comparable older GM plant.

2. Cost of upgraded labor: The routine monotonous tasks are the easiest to automate. The tasks that are difficult to automate are ones that require skill. Thus, manufacturing labor must be upgraded.

3. Initial investment: Cash flow considerations may make an investment in automation difficult even if the estimated rate of return is high.

ISSUES FOR AUTOMATION IN FACTORY OPERATIONS

- Task is too difficult to automate.
- Short product lifecycle.
- Customized product.
- Fluctuating demand.
- Reduce risk of product failure.
- Cheap manual labor.

STRATEGIES FOR AUTOMATION

- Specialization of operations.
- Combined operations.
- Simultaneous operations.
- Integration of operations.
- Increased flexibility.
- Improved material handling and storage.
- On-line inspection.
- Process control and optimization.
- Plant operations control.
- Computer Integrated Manufacturing.

EXERCISES

1. Differentiate between mechanization and automation.

2. Identify some of the major reasons for automation.

3. List the levels of automation.

4. Discuss the concept of low cost automation with the help of suitable examples.

5. What are the types of automation that can be used in a production system? Compare them for their features and drawbacks.

6. Discuss the various levels of automation.

7. Write short notes on "low cost automation."

8. Identify major socio-economic considerations favoring automation.

9. State the advantages of automating production operations.

10. List the strategies for automation.

11. Compare hard automation with soft automation.

12. List the advantages of flexible automation.

13. List at least four reasons why automation is required in industry.

BASIC LAWS AND PRINCIPLES

FLUID PROPERTIES

Force

A force is a push or a pull, or more generally anything that can change an object's speed or direction of motion. The International System of Units (SI) unit used to measure force is the Newton (symbol N).

$$F = ma$$

where F stands for force in Newton, m stands for mass in Kg and a represents acceleration expressed as meters divided by seconds squared m/s^2.

Pressure

Pressure is the ratio of force to the area over which the force acts.

Mathematically, it can be expressed as:

$$p = \frac{F}{A}$$

where p is pressure, F is force, and A represents area. Pressure is usually expressed in Newton per square meter, given the name Pascal, and traditionally, it was expressed in pounds force per square inch (PSI).

Atmospheric Pressure

Atmospheric pressure is defined as the pressure due to the weight of the atmosphere (air and water vapor) on the earth's surface. Atmospheric pressure is determined by a mercury column barometer, that is why it is sometimes called as barometric pressure. The average atmospheric pressure at sea level has been defined as 1.01325 bars, or 14.696 pounds per square inch absolute (PSIA).

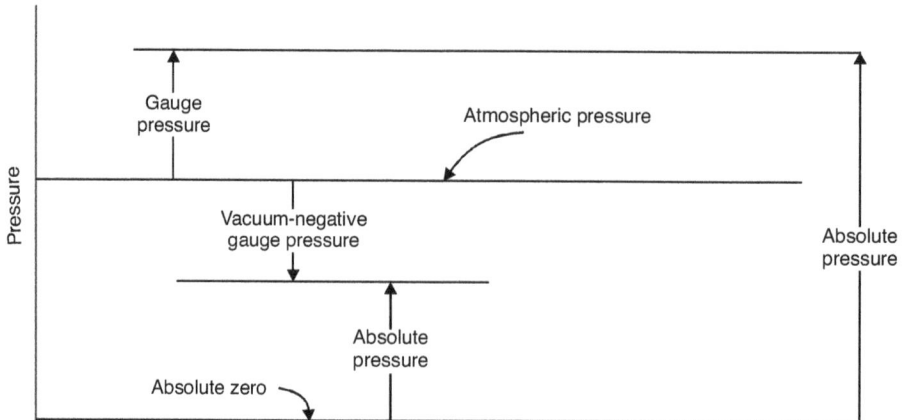

FIGURE 2.1 Pressure Relationship.

Absolute Pressure

Absolute pressure can be given as gauge pressure plus barometric or atmospheric pressure. Absolute pressure is referenced against absolute zero pressure, or a complete vacuum. The units of absolute pressure are followed by suffix "a," such as psia. If we hold an absolute pressure instrument in the open air, the reading should be well above zero, in the range of 14.7 to 12 psia.

Gauge and Vacuum Pressure

Gauge pressure is referenced against the atmospheric pressure at the measurement point. The units of gauge pressure are followed by a "g," such as psig. A gauge pressure instrument should always read zero when exposed to atmospheric pressure. Similarly, when the pressure falls below atmospheric, it is called vacuum pressure, sometimes it is also called negative gauge pressure.

Based upon the above discussions, the following equations can be derived:

$$P_{abs} = P_{atm} + P_{gauge}$$
$$P_{abs} = P_{atm} - P_{vacuum}$$

Where
P_{abs} = Absolute pressure

P_{atm} = Atmospheric pressure

P_{gauge} = Gauge pressure

P_{vacuum} = Vacuum pressure ($-$ ve gauge pressure).

Conversion of various units of pressure in Pascal

Unit	Symbol	No. of pascals
Bar	bar	1 x 10^5 Pa
Millibar	mbar	100 Pa
Hectopascal	hPa	100 Pa
conventional mm of Hg	mmHg	133.322 Pa
conventional inch of Hg	in Hg	3,386.39… Pa
Torr	torr	101325/760 ≈ 133.322 Pa
pound-force per square inch	lbf/in^2	6,894.76 ≈ 6895 Pa

Pascal's Law

Blaise Pascal formulated this basic law in the mid-17th century. His law states that pressure in a confined fluid is transmitted undiminished in every direction and acts with equal force on equal areas and at right angles to container walls. Hydraulic brakes, lifts, presses, syringe pistons, etc. work on the principle of Pascal's law.

According to Pascal's law, inside the pipes of a confined system pressure is uniform at all points. Mathematically,

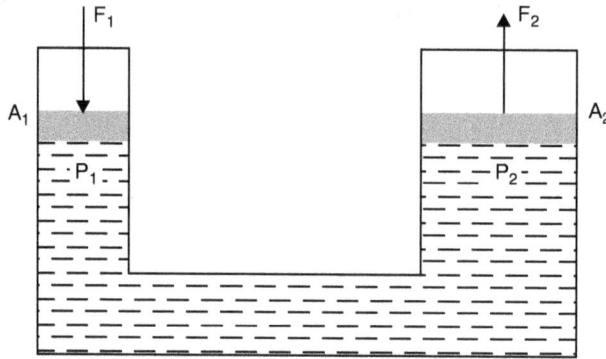

FIGURE 2.2 Pascal's Law Illustrated.

$$\frac{F_1}{A_1} = P_1 = P_2 = \frac{F_2}{A_2}$$

$$F_2 = \frac{A_2}{A_1} \cdot F_1$$

From the above expression, $P_1 = P_2$, therefore F_2 is greater than F_1 because A_2 is greater than A_1. This means that, in order to obtain a greater output force, it is enough to have suitably sized surfaces available.

Flow and Flow Rate

The volume of a substance passing a point per unit time is called flow and the volume of water, a pump or a compressor can move during a given amount of time is called, "flow rate."

Volumetric Flow Rate

It is the volume of the fluid flowing through a cross section per unit time. Air related flows are usually expressed in cubic feet per minute (CFM) and for liquid-based fluids, they are expressed as liters or gallons per minute (LPM or GPM) or cubic meters per second, etc.

$$\text{Volumetric Flow Rate} = \text{Area} \times \text{Velocity}$$

Mass Flow Rate

Volumetric flow rate times density, i.e., pounds per hour or kilograms per minute.

$$\text{Mass Flow Rate} = \text{Area} \times \text{Velocity} \times \text{Density}$$

Conversion of various flow rate units into m³/s

Unit	Symbol	No. of m³/s
Liters/second	l/s	10^{-3} m³/s
Gallons/second	gps	0.003788 m³/s
cubic feet/min	cfm	4.719×10^{-4} m³/s

Bernoulli's Equation

It states that, for a non-viscous, incompressible fluid in steady flow, the sum of pressure, potential, and kinetic energies per unit volume is constant at any point. Mathematically, it can be expressed as:

$$\frac{\rho \cdot v_1^2}{2} \quad p_1 = \frac{\rho \cdot v_2^2}{2} + p_2 = \text{Constant}$$

Where

$$g = \text{gravity}$$
$$v = \text{flow speed}$$
$$h = \text{height}$$
$$p = \text{pressure}$$
$$\rho = \text{density}$$

Bernoulli's principle states that in fluid flow, an increase in velocity occurs simultaneously with a decrease in pressure. It is named for the Dutch/Swiss mathematician/scientist Daniel Bernoulli; this phenomenon can be seen in airplane lift, a carburetor, the flow of air around the ball, etc.

Venturi Effect

A fluid passing through smoothly varying constrictions is subject to changes in velocity and pressure, as described by Bernoulli's principle. In case of fluid or airflow through a tube or pipe with a constriction in it, the fluid must speed up in the restriction, reducing its pressure, and producing a partial vacuum.

As shown in the Fig. 2.3 fluid density $= (\rho)$, area $= (A)$, and velocity $= (V)$. Let the properties of fluid at entrance and exit be (ρ_1, A_1, V_1) and at constriction be (ρ_2, A_2, V_2). There is a drop in pressure at the constriction as shown by the height of the column and it is due to conservation of energy. The fluid experiences a gain in kinetic energy and a drop in pressure as it enters the constriction; this effect is called *Venturi effect, it is named after the Italian physicist Giovanni Battista Venturi.*

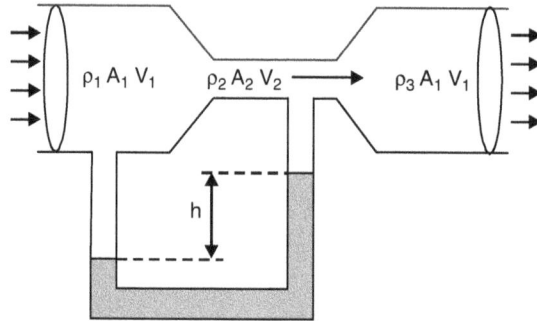

FIGURE 2.3 Venturi's Law Illustrated.

Continuity Equation

It is simply a mathematical expression of the principle of conservation of mass. Mass is neither created nor destroyed. For a steady flow, it states that:

$$\text{Mass flow rate in} = \text{mass flow rate out}$$
$$\rho_1 \, A_1 \, V_1 = \rho_2 \, A_2 \, V_2 \;\; (\rho_1 = \rho_2)$$
$$A_1 \, V_1 = A_2 \, V_2$$

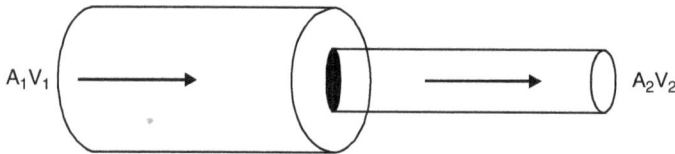

The "continuity equation" is a direct consequence of the rather trivial fact that what goes into the pipe must come out. This has the important consequence that as the area of the hole decreases, the velocity of the fluid must increase, in order to keep the flow rate constant.

Specific Weight, Density, and Specific Gravity

(a) Specific Weight or Weight Density

The weight per unit volume of a substance. Usually it is expressed in N/m^3 or lbs/ft^3. Mathematically,

$$\rho = \frac{w}{V}$$

Where
$$\rho = \text{density}(N/m^3)$$
$$w = \text{weight}(N)$$
$$V = \text{volume}(m^3)$$

(b) Density

Density is defined as the ratio of the mass of an object to its volume; usually it is expressed in kg/m^3 or g/cm^3. Mathematically,

$$\rho = \frac{m}{V}$$

Where
$$\rho = \text{density}(kg/m^3)$$
$$m = \text{weight}(kg)$$
$$V = \text{volume}(m^3)$$

(c) Specific Gravity

The ratio of the density (or specific weight) of a substance to the density (or specific weight) of a standard fluid is called *Specific gravity* or *Relative density*. The usual standard of comparison for solids and liquids is water at 4°C at atmospheric pressure. Gases are commonly compared to dry air, under standard conditions (0°C and atmospheric pressure).

Specific gravity is not expressed in *units*, as it is purely a ratio. Mathematically,

$$SG = \frac{m_0}{V \; \rho_{H_2O}}$$

$$SG = \text{specific gravity}$$

Where
$$m_0 = \text{mass weight of oil to be compared(kg)}$$
$$V = \text{volume of oil}(m^3)$$
$$\rho_{H_2O} = \text{density of water}(1000 \text{ kg/m}^3)$$

Compressibility and Bulk Modulus

Compressibility is the measure of change in volume of substance when pressure is exerted on it. Liquids are incompressible fluids. For each atmosphere increase in pressure, the volume of water would decrease 46.4 parts per million. The hydraulic brake systems used in most cars operate on the principle that there is essentially no change in the volume of the brake fluid when pressure is applied to this liquid.

On the other hand, the volume of the gases can be readily changed by exerting an external pressure on the gas. An internal combustion engine provides a good example of the ease with which gases can be compressed.

The compressibility is the reciprocal of the bulk modulus. Compressibility is denoted by "k" and is expressed mathematically as:

$$k = \frac{1}{b}$$

Where B is called the bulks modulus of elasticity and is defined as the ratio of change in pressure to volumetric strain (change in volume/original volume) over a fluid element. It is expressed as follows:

$$B = -\frac{dP}{V_{change}/V_{initial}}$$

Where

$B = $ Bulks modulus

$V_{initial} = $ Original volume before application of force.

$V_{change} = $ Net change in volume after the application of force.

$-$ Ve sign indicates that volume decreasesas pressure increases.

Viscosity and Viscosity Index

Viscosity is the measure of the internal friction of a fluid or its resistance to flow. A hydraulic fluid that is too viscous usually causes high-pressure drop, sluggish operation, low-mechanical efficiency, and high-power consumption. Low-viscosity fluids permit efficient low-drag operation, but tend to increase wear, reduce volumetric efficiency, and promote leakage.

Viscosity index is an arbitrary scale, which indicates how the viscosity of a fluid varies with changes in temperature. The higher the viscosity index, the lower the viscosity changes with respect to temperature and vice versa. Ideally, the fluid should have the same viscosity at very low temperatures as well as at high temperatures. In reality, this cannot be achieved. This change is common to all fluids. Heating tend to make fluids thinner and cooling makes them thicker.

Gas Laws

(a) Boyle's Law

English scientist Robert Boyle investigated the relationship between the volume of a dry ideal gas and its pressure. It states that at constant temperature, the pressure is inversely proportional to the volume of a definite amount of gas. Mathematically,

$$P_1V_1 = P_2V_2$$

therefore $\qquad PV = C$

Where
$P_1 = $ Initial Pressure
$V_1 = $ Initial Volume
$P_2 = $ Final Pressure
$V_2 = $ Final Volume.

(b) Charle's Law

French scientist Jacques Charles experimented with gas under constant pressure and his observations have been formalized into Charle's law.

The volume of a gas at constant pressure is directly proportional to the absolute temperature. Mathematically, it can be expressed as:

$$V_1 / T_1 = V_2 / T_2$$

therefore $\qquad V / T = C$

Where
$V_1 = $ Initial Volume
$T_1 = $ Initial Temperature
$T_1 = $ Final Volume
$T_2 = $ Final Temperature.

(c) Gay-Lussac's Law

French scientist Joseph Gay-Lussac investigated the relationship between the pressure of a gas and its temperature. It states that the pressure of a gas at constant volume is directly proportional to the absolute temperature. The mathematical statement is as follows:

$$P_1 / T_1 = P_2 / T_2$$

therefore $\quad\quad P / T = C$

Where $\quad\quad\quad P_1 = $ Initial Pressure

$\quad\quad\quad\quad\quad T_1 = $ Initial Temperature

$\quad\quad\quad\quad\quad P_2 = $ Final Pressure

$\quad\quad\quad\quad\quad T_2 = $ Final Temperature.

(d) Combined Gas Laws

Any two of the three gas laws of Boyle, Charles, or Gay-Lussac can be combined, hence the name, combined gas law. In short, this combined gas law is used when it is difficult to keep either the temperature or pressure constant:

$$P_1 V_1 T_2 = P_2 V_2 T_1$$

This relationship can be used to predict pressure, volume, and temperature relationships where any five of the six variables are known.

Moisture in the Air

(a) Humidity

Humidity is the concentration of water vapor in the air. The concentration can be expressed as *specific humidity*, *absolute humidity*, or *relative humidity*. A device used to measure humidity is called a *hygrometer*.

(*i*) *Specific humidity.* Is defined as mass of water vapor present per kg of dry air. It is expressed in g/kg of dry air. Humidity is measured by means of a hygrometer.

(*ii*) *Absolute humidity.* Is expressed as the mass of water vapor contained in a given volume of air. The hotter the air is, the more water it can contain. It may be measured in grams of vapor/cubic meter of air. Absolute humidity finds greatest application in ventilation and air-conditioning problems.

(*iii*) *Relative humidity.* Is defined as the ratio (usually expressed as a percentage) of the mass of water in a given volume of moist air divided by the maximum mass of water that can be held by that same volume of air (saturated air) at a given temperature. We can compare how much water vapor is present in the air to how much water vapor would be in the air if the air were saturated. A reading of 100 percent relative humidity means that the air is totally saturated with water vapor and cannot hold any more.

Example: If 20 grams of water vapor were present in each m³ of dry air, and the air would be saturated with 30 grams of water vapor per m³ of dry air, the relative humidity would be 20/30 = 66.66%.

Practically, "relative humidity" is the amount of moisture in the air at a certain temperature. It is called "relative" because it is being compared to the maximum amount of moisture that could be in the air at the same temperature.

(b) Dew Point Temperature and Holding Capacity of Air

Air present at certain temperatures could consume a certain quantity of water in it, likewise when this temperature is attained, air becomes completely saturated. If the air is further cooled then water will start condensing out of it. Dew point is the temperature at which water vapor begins to condense out of the air. Dew points can be defined and specified for ambient air or for compressed air. Dew point normally occurs when a mass of air has a relative humidity of 100%. This temperature can be recorded by a thermometer.

(c) Atmospheric Dew Point

Atmospheric dew point is the value of the temperature at which moisture present in the air begins to condense at atmospheric pressure, *i.e.*, at 1.01325 bar. Atmospheric dew point is not at all important for pneumatics, as pressure is always more than atmospheric pressure in a pneumatic line.

(d) Pressure Dew Point

Pressure dew point is the value of the temperature at which moisture present in the air begins to condense at pressures more than the atmospheric pressure. As pressures encountered in pneumatics are generally more than atmospheric so it is of great importance. Obviously, at higher pressures, the water present in air condenses at higher temperatures in comparison to atmospheric pressure, so dew point should be kept very low so as to ensure the least amount of moisture in the pneumatic line (as moisture is the biggest enemy in pneumatics).

Energy, Work, and Power

Energy is the ability to do work and is expressed in foot pound (ft lb) or Newton meter (Nm). The three forms of energy are potential, kinetic, and heat. Work measures accomplishments; it requires motion to make a force do work. Power is the rate of doing work or the rate of energy transfer.

Potential Energy

Potential energy is energy due to position. An object has potential energy in proportion to its vertical distance above the earth's surface. For example, water held back by a dam represents potential energy because until it is released, the water does not work. In hydraulics, potential energy is a static factor. When force is applied to a confined liquid, potential energy is present because of the static pressure of the liquid. Potential energy of a moving liquid can be reduced by the heat energy released. Potential energy can also be reduced in a moving liquid when it transforms into kinetic energy. A moving liquid can, therefore, perform work as a result of its static pressure and its momentum.

Kinetic Energy

Kinetic energy is the energy a body possesses because of its motion. The greater the speed, the greater the kinetic energy. When water is released from a dam, it rushes out at a high velocity jet, representing energy of motion—kinetic energy. The amount of kinetic energy in a moving liquid is directly proportional to the square of its velocity. Pressure caused by kinetic energy may be called *velocity pressure*.

Heat Energy and Friction

Heat energy is the energy a body possesses because of its heat. Kinetic energy and heat energy are dynamic factors. Pascal's Law dealt with static pressure and did not include the friction factor. Friction is the resistance to relative motion between two bodies. When liquid flows in a hydraulic circuit, friction produces heat. This causes some of the kinetic energy to be lost in the form of heat energy. Although friction cannot be eliminated entirely, it can be controlled to some extent. The three main causes of excessive friction in hydraulic systems are:

(i) Extremely long lines.

(ii) Numerous bends and fittings or improper bends.

(iii) Excessive velocity from using undersized lines.

EXERCISES

1. What are the differences between a liquid and a gas?

2. What do you mean by pressure dew point? Does it have any importance in pneumatics?

3. State the importance of gas laws.

4. Define the terms specific weight, density, and specific gravity.

5. What is the effect of temperature on viscosity of fluids?

6. State continuity equation.

7. Differentiate between the terms viscosity and viscosity index.

8. What is the difference between pressure and force?

9. Explain venturi effect. Give the name of important pneumatic equipment, which uses this principle.

10. State Bernoulli's equation.

11. What is meant by the term bulk modulus?

12. State and prove Pascal's law.

13. What is the relationship between atmospheric, absolute, and vacuum pressure?

14. Differentiate between absolute and relative humidity.

BASIC PNEUMATIC AND HYDRAULIC SYSTEMS

INTRODUCTION TO FLUID POWER

A fluid power system transmits and controls energy through the use of pressurized fluid. The term *fluid power* applies to both hydraulics and pneumatics. With hydraulics, that fluid is a liquid such as oil or water. With pneumatics, the fluid is typically compressed air or inert gas. Hydraulics uses oil or liquid as the medium that cannot be compressed and pneumatics, which involves gases, uses air or gas as the medium that can be compressed. It is a term, which was created to collect the generation, control, and application of smooth, effective power of pumped or compressed fluids (either liquids or gases). This power is used to provide force and motion to various mechanisms. This force and motion may be in the form of push, pull, rotate, regulate, or drive.

Fluid power is one of three commonly used methods of transmitting power in an industry; the others are electrical and mechanical power transmission. Electric power transmission uses an electric current flowing through a wire to transmit power. Mechanical power transmission uses gears, pulleys, chains, etc. to transmit power. Fluid power's motive force comes from the principle that pressure applied to a confined fluid is transferred uniformly and undiminished to every portion of the fluid and to the walls of the container that holds the fluid. A surface such as a cylinder piston will move if the difference in force across the piston is larger than the total load plus frictional forces. The resulting net force can then accelerate the load proportionately to the ratio of the force divided by the mass.

Fluid power encompasses most applications that use liquids or gases to transmit power in the form of mechanical work, pressure, and/or volume in the system. This definition includes all systems that rely on pumps and compressors to transmit specific volumes and pressures of liquids or gases within a closed system. Fluid power is used in the steering, brake system, and automatic transmissions of cars and trucks. In addition to the automotive industry, fluid power is used to control airplanes and spacecraft, harvest crops, mine coal, drive machine tools, and process food. Fluid power can be effectively combined with other technologies through the use of sensors, transducers, and microprocessors.

BASIC ELEMENTS OF FLUID POWER SYSTEM

The basic elements of fluid power system are:

- Power device: Pump or Compressor
- Control valves
- Actuators: Cylinders or Motors

Figure 3.1 shows the basic elements of a fluid power system connected by fluid power lines. These elements are discussed in detail in next chapters.

ADVANTAGES AND DISADVANTAGES OF FLUID POWER

Advantages

There are few advantages, which make fluid power so popular. These are listed below:

Fluid Power Line   Fluid Power Line

| Power Device | | Control Valves | | Actuator |

Pump
or
Compressor

Cylinder
or
Motor

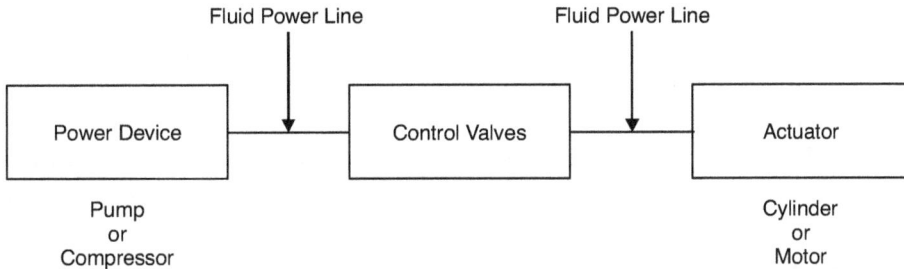

FIGURE 3.1 Elements of Fluid Power System.

- **No need of intermediate equipment:** They eliminate the need for complicated systems of gears, cams, and levers. Motion can be transmitted without the slack inherent in the use of solid machine parts.

- **Less wear and tear:** The fluids used are not subject to breakage as are mechanical parts, and the mechanisms are not subjected to great wear.

- **Multi-function control:** A single hydraulic pump or air compressor can provide power and control for numerous machines or machine functions when combined with fluid power manifolds and valves.

- **Constant force or torque:** This is a unique fluid power attribute.

- **Flexibility:** Hydraulic components can be located with considerable flexibility. Pipes and hoses instead of mechanical elements virtually eliminate location problems.

- **Comparatively small pressure losses:** The different parts of a fluid power system can be conveniently located at widely separated points, because the forces generated are rapidly transmitted over considerable distances with small loss. These forces can be conveyed up and down or around corners with small loss in efficiency and without complicated mechanisms.

- **Multiplication and variation of force:** Very large forces can be controlled by much smaller ones and can be transmitted through comparatively small lines and orifices. Linear or rotary force can be multiplied from a fraction of an ounce to several hundred tons of output.

- **Accurate and easy to control:** We can start, stop, accelerate, decelerate, reverse, or position large forces with great accuracy.

- **High horsepower and low weight:** Pneumatic components are compact and lightweight.

- **Smoothness:** Fluid systems are smooth in operation. Vibration is kept to a minimum.

- **Overload protection:** In case of an overload, an automatic release of pressure can be guaranteed; automatic valves guard the system against a breakdown from overloading so that the system is protected against breakdown or strain.
- **Wide variety of motions:** Fluid power systems can provide widely variable motions in both rotary and straight-line transmission of power.
- **Low speed torque:** Unlike electric motors, air or hydraulic motors can produce large amounts of torque (twisting force) while operating at low speeds. Some hydraulic and air motors can even maintain torque at zero speed without overheating.
- **Less human intervention:** The need for control by hand can be minimized.
- **Low operating costs:** Fluid power systems are economical to operate their high efficiency with minimum friction loss keeps the cost of a power transmission at a minimum.
- **Safety in hazardous environments:** Fluid power can be used in mines, chemical plants, near explosives, and in paint applications because it is inherently spark-free and can tolerate high temperatures.
- **Better force control:** Force control is much easier with fluid systems than for electric motors. Fluid actuators, either hydraulic or pneumatic, are well suited to walking robots because they are high force, low speed actuators. They provide much higher force densities than electric systems.
- **Simpler design:** In most cases, a few pre-engineered components will replace complicated mechanical linkages.

Disadvantages

The main disadvantage of a fluid system is maintaining the precision parts when they are exposed to bad climates and dirty atmospheres. Protection against rust, corrosion, dirt, oil deterioration, and other adverse environmental conditions is very important.

APPLICATIONS OF FLUID POWER

Mobile

Fluid power is used to transport, excavate, and lift materials, as well as control or power mobile equipment. End use industries include construction, agriculture, marine, and the military. Applications include backhoes, graders, tractors, truck brakes and suspensions, spreaders, and highway maintenance vehicles.

Industrial

Fluid power is used to provide power transmission and motion control for the machines of industry. End use industries range from plastics to paper production. Applications include metal working equipment, controllers, automated manipulators, material handling, and assembly equipment.

Aerospace

Fluid power is used for both commercial and military aircraft, spacecraft, and related support equipment. Applications include landing gear, brakes, flight controls, motor controls, and cargo loading equipment.

PNEUMATICS VS. HYDRAULICS

Fluid power can be broadly divided into two fields: *pneumatics and hydraulics*. Both pneumatics and hydraulics are applications of fluid power.

Pneumatic systems use compressed gas such as air or nitrogen to perform work processes whereas hydraulic systems use liquids such as oil and water to perform work processes. Pneumatic systems are open systems, exhausting the compressed air to the atmosphere after use whereas hydraulic systems are closed systems, recirculating the oil or water after use.

A fluid power system uses hydraulics or pneumatics to deliver extremely powerful pushing and pulling forces to machinery. Much of the factory equipment used to lift and move large components is powered by hydraulics. For example, forklift trucks, opencast mining equipment, and multi-purpose agricultural spraying equipment all use hydraulic systems to operate their lifting arms. Pneumatics, on the other hand, are used for a variety of purposes that include delivering power to tools like jack hammers, air guns, and complex industrial equipment for conveying, separating, and air handling goods.

The extensive use of hydraulics and pneumatics to transmit power is due to the fact that properly constructed fluid power systems possess a number of favorable characteristics. They eliminate the need for complicated systems of gears, cams, and levers. The fluids used are not subject to breakage as are mechanical parts, and the mechanisms are not subjected to great wear.

Some points of difference between hydraulics and pneumatics are shown in Table 3.1.

TABLE 3.1 Difference between Hydraulics and Pneumatics.

	Pneumatics	*Hydraulics*
Pressure level	5–10 bar	Up to 200 bar
Actuating forces	Pneumatic actuators can produce only low or medium size forces.	Hydraulic actuators are suitable for very high loads.
Element cost	Pneumatic elements such as cylinders and valves are less costly as compared to hydraulic elements.	Hydraulic elements can cost from 5 to 10 times more than similar size pneumatic elements.
Transmission lines	Transmission lines in pneumatics are made up of inexpensive flexible plastic tubing. Only single line is needed in pneumatics to simply exhaust the air into atmosphere.	Transmission lines in hydraulics are made up of metal tubing with expensive fittings to withstand high working pressure and to avoid leaks. Also return lines are needed in hydraulics to return the oil from each cylinder back to reservoir.
Stability	Low stability because air is compressible.	High stability because oil is incompressible.
Speed control	Difficult to control the speed of pneumatic cylinders or motors.	Easy to control the speed.

ADVANTAGES AND DISADVANTAGES OF PNEUMATICS

Advantages

- Pneumatic systems are clean because they use compressed air. If a pneumatic system develops a leak, it will be air that escapes and not oil.
- Pneumatic systems are cheaper to run than other systems.
- Inherently modulating actuators and sensors.
- Explosion proof components.
- High efficiency, example, a relatively small compressor can fill a large storage tank to meet intermittent high demands for compressed air.
- Ease of design and implementation.

- High reliability, mainly because of fewer moving parts.
- Compressed gas can be stored, allowing the use of machines when electrical power is lost.
- Easy installation and maintenance.

Disadvantages

- Low accuracy and control limitation because of compressibility.
- Noise pollution.
- Leakage of air can be of concern.
- Need for a compressor producing clean and dry air.
- Cost of air piping.
- Need for regular component calibration.

ADVANTAGES AND DISADVANTAGES OF HYDRAULICS

Advantages

- Through the use of simple devices, an operator can readily start, stop, speed up, slow down, and control large forces with very close and precise tolerance.
- High power output from a compact actuator.
- Hydraulic power systems can multiply forces simply and efficiently from a fraction of an ounce to several hundred tons of output.
- Force can be transmitted over distances and around corners with small losses of efficiency.
- There is no need for complex systems of gears, cams, or levers to obtain a large mechanical advantage.
- Extreme flexibility of approach and control. Control of a wide range of speed and forces is easily possible.
- Safety and reliability.
- Hydraulic systems are smooth and quiet in operation. Vibration is kept to a minimum.

Disadvantages

- Hydraulic systems are expensive.
- System components must be engineered to minimize or preclude fluid leakage.

- Protection against rust, corrosion, dirt, oil deterioration, and other adverse environment is very important.
- Maintenance of precision parts when they are exposed to bad climates and dirty atmospheres.
- Fire hazard if leak occurs.
- Adequate oil filtration must be maintained.

APPLICATIONS OF PNEUMATICS

- Operation of heavy or hot doors
- Lifting and moving in slab moulding machines
- Spray painting
- Bottling and filling machines
- Component and material conveyor transfer
- Unloading of hoppers in building, mining, and chemical industry
- Air separation and vacuum lifting of thin sheets
- Dental drills

APPLICATIONS OF HYDRAULICS

- Machine tool industry
- Plastic processing machines
- Hydraulic presses
- Construction machinery
- Lifting and transporting
- Agricultural machinery

BASIC PNEUMATIC SYSTEM

Pneumatic systems use pressurized air to make things move. A basic pneumatic system consists of an air generating unit and an air-consuming unit. Air compressed in a compressor is not ready for use. The air has to be filtered, the moisture present in air has to be dried, and, for different applications in the plant, the pressure of air has to be varied. Several other treatments are given to the air before it reaches finally to the actuator. Figure 3.2 gives an overview of a basic pneumatic system. Some accessories are added for economical and efficient operation of system.

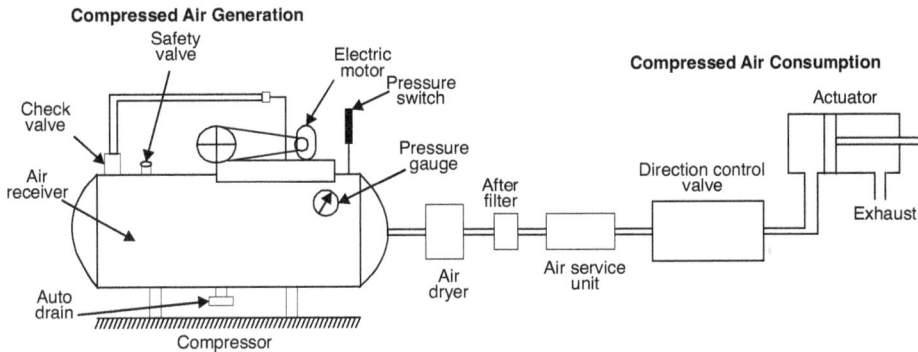

FIGURE 3.2 Basic Pneumatic System.

A typical pneumatic power system includes the following components:

1. Compressor
2. Electric motor
3. Air receiver
4. Pressure switch
5. Safety valve
6. Auto drain
7. Check vlve
8. Pressure gauge
9. Air dryer
10. After filter
11. Air service unit
12. Direction control valve
13. Pneumatic actuator

1. Compressor: A device, which converts mechanical force and motion into pneumatic fluid power, is called a compressor. Every compressed-air system begins with a compressor, as it is the source of airflow for all the downstream equipment and processes.

2. Electric Motor: An electric motor is used to drive the compressor.

3. Air Receiver: An air receiver is a container in which air is stored under pressure.

4. Pressure Switch: A pressure switch is used to maintain the required pressure in the receiver. It adjusts the *high pressure limit* and *low pressure limit* in the receiver. The compressor is automatically turned off when the pressure is about to exceed the high limit and it is also automatically turned on when the pressure is about to fall below the low limit.

5. Safety Valve: The function of the safety valve is to release extra pressure if the pressure inside the receiver tends to exceed the safe pressure limit of the receiver.

6. **Auto Drain:** Air condenses to give out water in the receiver and a device called auto drain directs this water out.

7. **Check Valve:** The valve enables flow in one direction and blocks flow in a counter direction is called check valve. Once compressed air enters the receiver via check valve, it is not allowed to go back even when the compressor is stopped.

8. **Pressure Gauge:** The pressure gauge tells us the pressure inside the compressor receiver.

9. **Air Dryer:** A device for reducing the moisture content of the working compressed air.

10. **After Filter:** A filter that follows the compressed air dryer and usually used for the protection of downstream equipment from desiccant dust, etc. is called an after filter. The name filter refers to a device whose primary function is the removal of insoluble contaminants from a liquid or a gas with the help of porous media.

11. **Air Service Unit:** Filter, regulator, and lubricator combined in one device is popularly known as an air service unit or F.R.L unit. Its purpose is to supply air to other successive applications in the line. It provides clean air at the required pressure with lubricant added to it to increase the life of equipment.

12. **Direction Control Valve:** Directional-control valves are devices used to change the flow direction of fluid within a pneumatic/hydraulic circuit. They control compressed-air flow to cylinders, rotary actuators, grippers, and other mechanisms in packaging, handling, assembly, and countless other applications. These valves can be actuated either manually or electrically.

13. **Pneumatic Actuator:** A device in which power is transferred from one pressurized medium to another without intensification. Pneumatic actuators are normally used to control processes requiring quick and accurate response, as they do not require a large amount of motive force. They may be reciprocating cylinders, rotating motors, or may be robot end effectors.

A few more components can be added to the system; first, an air intake filter and second, an air intercooler (if a multistage compressor is used). The function of the former is to prevent the entry of vast quantities of dust and dirt along with air and the latter is to cool the air again to room temperature after it is discharged from the low-pressure compressor.

BASIC HYDRAULIC SYSTEM

Figure 3.3 gives an overview of a basic hydraulic system.

A typical hydraulic power system includes the following components:

1. Electric motor
2. Hydraulic pump
3. Strainers and filters
4. Pressure gauge
5. Pressure relief valves
6. Check valve
7. Direction control valve
8. Hydraulic actuator
9. Reservoir

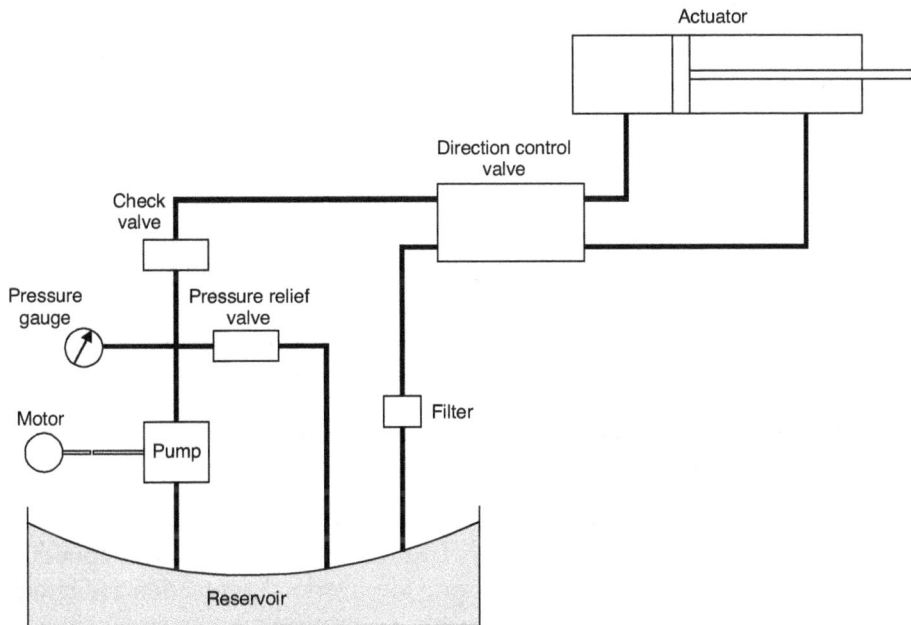

FIGURE 3.3 Basic Hydraulic System.

1. Electric Motor: An electric motor is used to drive the pump.

2. Hydraulic Pump: Hydraulic pumps convert mechanical energy from a prime mover (engine or electric motor) into hydraulic (pressure) energy.

The pressure energy is used then to operate an actuator. Pumps push on a hydraulic fluid and create flow.

3. Strainers and Filters: To keep hydraulic components performing correctly, the hydraulic liquid must be kept as clean as possible. Foreign matter and tiny metal particles from normal wear of valves, pumps, and other components are going to enter a system. Strainers, filters, and magnetic plugs are used to remove foreign particles from a hydraulic liquid and are effective as safeguards against contamination.

Strainers: A strainer is the primary filtering system that removes large particles of foreign matter from a hydraulic liquid. Even though its screening action is not as good as a filter's, a strainer offers less resistance to flow.

Filters: A filter removes small foreign particles from a hydraulic fluid and is most effective as a safeguard against contaminates. They are classified as full flow or proportional flow:

(*a*) *Full-Flow Filter*: In a full-flow filter, all the fluid entering a unit passes through a filtering element. Although a full-flow type provides a more positive filtering action, it offers greater resistance to flow, particularly when it becomes dirty.

(*b*) *Proportional-Flow Filters*: This filter operates on the venturi principle in which a tube has a narrowing throat (venturi) to increase the velocity of fluid flowing through it. Flow through a venturi throat causes a pressure drop at the narrowest point. This pressure decrease causes a sucking action that draws a portion of a liquid down around a cartridge through a filter element and up into a venturi throat.

4. Pressure Gauge: A pressure gauge tells us the pressure of fluid going into the valve.

5. Pressure Relief Valves: Relief valves are the most common type of pressure-control valves. The relief valves function may vary, depending on a system's needs. They can provide overload protection for circuit components or limit the force or torque exerted by a linear actuator or rotary motor. The internal design of all relief valves is similar. The valves consist of two sections: a body section containing a piston that is retained on its seat by a spring(s), depending on the model, and a cover or pilot-valve section that hydraulically controls a body piston's movement. The adjusting screw adjusts this control within the range of the valves. Valves that provide emergency overload protection do not operate as often because other valve types are used to load and unload a pump. However, relief valves should be cleaned regularly by reducing their pressure adjustments to flush out any possible sludge deposits that may accumulate. Operating under reduced pressure will clean out sludge deposits and ensure that the valves operate properly after the pressure is adjusted to its prescribed setting.

6. **Check Valve:** Check valves are the most commonly used in fluid-powered systems. They allow flow in one direction and prevent flow in the other direction. They may be installed independently in a line, or they may be incorporated as an integral part of a sequence, counterbalance, or pressure-reducing valve. The valve element may be a sleeve, cone, ball, poppet, piston, spool, or disc. Force of the moving fluid opens a check valve; backflow, a spring, or gravity closes the valve.

7. **Direction Control Valve:** Directional-control valves are devices used to change the flow direction of fluid within a pneumatic/hydraulic circuit. They control compressed-air flow to cylinders, rotary actuators, grippers, and other mechanisms in packaging, handling, assembly, and countless other applications.

8. **Hydraulic Actuator:** A hydraulic actuator receives pressure energy and converts it to mechanical force and motion. An actuator can be linear or rotary. A linear actuator gives force and motion outputs in a straight line. It is more commonly called a cylinder but is also referred to as a ram, reciprocating motor, or linear motor. A rotary actuator produces torque and rotating motion. It is more commonly called a hydraulic motor or motor.

9. **Reservoir:** A reservoir stores a liquid that is not being used in a hydraulic system. It has many other important functions as well.

- It also allows gases to expel and foreign matter to settle out from a liquid.
- It functions as a cooler.
- It functions as a "coarse strainer," providing sedimentation of impurities.
- It functions as an air and water separator.
- It functions as a foundation for pumps, etc.

A properly constructed reservoir should be able to dissipate heat from the oil, separate air from the oil, and settle out contaminates that are in it. It should be high and narrow rather than shallow and broad. The oil level should be as high as possible above the opening to a pump's suction line. This prevents the vacuum at the line opening from causing a vortex or whirlpool effect, which would mean that a system is probably taking in air. Most mobile equipment reservoirs are located above the pumps. This creates a flooded-pump-inlet condition. This condition reduces the possibility of pump cavitation; a condition where all the available space is not filled and often metal parts will erode. Most reservoirs are vented to the atmosphere. A vent opening allows air to leave or enter the space above the oil as the level of the oil goes up or down. This maintains a constant atmospheric pressure above the oil.

HYDRAULIC SYSTEM DESIGN

Each component in the system must be compatible with and form an integral part of the system. Example, an inadequate size filter on the inlet of a pump can cause cavitation and subsequent damage to the pump. Each component in the system has a maximum rated speed, torque, or pressure. Loading the system beyond the specifications increases the possibility of failure.

All lines must be of proper size and free of restrictive bends. An undersized or restricted line results in a pressure drop in the line itself. Some components must be mounted in a specific position with respect to other components or lines. The housing of an in-line pump, for example, must remain filled with fluid to provide lubrication.

Systems should be designed for correct operating pressures. The correct operating pressure is the lowest pressure, which will allow adequate performance of the system function and still remain below the maximum rating of the components and machine. Always set and check pressures with a gauge.

FLUIDS USED IN HYDRAULICS

The oil in a hydraulic system must first and foremost transfer energy, but the moving parts in components must also be lubricated to reduce friction and consequent heat generation. Additionally, the oil must lead dirt particles and friction heat away from the system and protect against corrosion. Oil requirements include:

- Good lubricating properties.
- Good wear properties.

▪ Suitable viscosity.

▪ Good corrosion inhibitor.

▪ Good anti-aeration properties.

▪ Reliable air separation.

▪ Good water separation motor oil.

Liquids being used include mineral oil, water, phosphate ester, water-based ethylene glycol compounds, and silicone fluids. The three most common types of hydraulic liquids are petroleum-based, synthetic fire-resistant, and water-based fire-resistant.

Petroleum-Based Fluids

The most common hydraulic fluids used are the petroleum-based oils. These fluids contain additives to protect the fluid from oxidation (antioxidant), to protect system metals from corrosion (anticorrosion), to reduce tendency of the fluid to foam (foam suppressant), and to improve viscosity.

Synthetic Fire-Resistant Fluids

Petroleum based oils contain most of the desired properties of a hydraulic liquid. However, they are flammable under normal conditions and can become explosive when subjected to high pressures and a source of flame or high temperatures. Nonflammable synthetic liquids have been developed for use in hydraulic systems where fire hazards exist.

Water-Based Fire-Resistant Fluids

The most widely used water-based hydraulic fluids may be classified as water-glycol mixtures and water-synthetic base mixtures. The water-glycol mixture contains additives to protect it from oxidation, corrosion, and biological growth and to enhance its load-carrying capacity. Fire resistance of the water mixture fluids depends on the vaporization and smothering effect of steam generated from the water. The water in water-based fluids is constantly being driven off while the system is operating. Therefore frequent checks to maintain the correct ratio of water are important.

EXERCISES

1. Define fluid power.

2. What are the different fluid power elements? Discuss.

3. What is the difference between fluid power and hydraulics and pneumatics?

4. Discuss the basic elements of an automated system.

5. What are the advantages and disadvantages of fluid power?

6. Distinguish between hydraulic and pneumatic systems.

7. What are the properties of pneumatic energy that make it a fit for the engineering applications?

8. What are the advantages of hydraulics over pneumatics?

9. Discuss the application of hydraulics in automation.

10. Compare the different features of a hydraulic system with those of a pneumatic system.

11. Draw a sketch of a hydraulic system and explain its components.

12. Draw a sketch of a pneumatic system and explain its components.

13. List the applications of hydraulics and pneumatics.

14. Discuss the various types of fluid used in hydraulics.

15. What are the reasons for pressure drop in hydraulics?

CHAPTER 4

PUMPS AND COMPRESSORS

INTRODUCTION

The power source is the key element in a fluid-power system. In a pneumatic system, the power source is an air compressor, while in a hydraulic system, it is a pump. These normally are driven by an electric motor or internal combustion engine. Storage devices are used along with most systems so that they can be made to work more efficiently. In hydraulic systems, the storage device is an accumulator; in pneumatic systems, it is a tank or receiver. However, most pneumatic systems are used with a receiver. One may define the fluid power generator as a means of converting the mechanical energy of a motor or engine into potential energy of the fluid.

A *pump* is the heart of the hydraulic system. It is a mechanical device used to move liquids under pressure or to raise them from a lower level to a higher level. This is done by creating suction at the inlet and high pressure at the outlet. Pumps are used throughout the plant to move fluids from one place to another. Every system from fuel oil to water use pumps. They play a vital role in supplying fluids at the correct flow rate and pressure to components downstream of the pumps. The selection of the pump class and type for a certain application is influenced by system requirements, system layout, fluid characteristics, intended life, energy cost, code requirements, and materials of construction.

A *compressor* is a device that takes in air at atmospheric pressure and delivers it at a pressure higher than atmospheric. Every compressed air system begins with a compressor, which is a continuous source of airflow for all the downstream equipment and processes. The main parameters of any air compressor are capacity and its pressure; capacity does the work and the pressure affects the rate at which work is done.

PUMPS VS. COMPRESSORS

The purpose of both the compressor and the pump is to raise the pressure of fluid. The main difference between a pump and a compressor is that the fluid delivered by compressor, *i.e.*, air, is compressed and under pressure at the time it is delivered, even if there is no load on the system. Most devices used to compress air are very similar in concept and perhaps even in hardware to hydraulic pumps, and selection considerations are similar. The only other substantive difference is that most hydraulic systems are powered by a single pump that is actually a part of the system, whereas most pneumatic systems are often powered by a single compressor, which is almost a utility in the plant like water or electric service. Hydraulic pumps deliver high-pressure fluid flow to the pump outlet. Hydraulic pumps are powered by mechanical energy sources to pressurize fluid. A hydraulic pump, when powered by pressurized fluid, can rotate in a reverse direction and act as a motor.

POSITIVE DISPLACEMENT VS. NON POSITIVE DISPLACEMENT DEVICES

Positive displacement devices are also called hydrostatic devices. A positive displacement device is one that displaces (delivers) the same amount of fluid for each rotating cycle of the pumping element. Constant delivery during each cycle is possible because of the close tolerance fit between the pumping element and the pump case. They use a mechanism that seals fluid in a chamber and forces it out by reducing the volume of the chamber. This action increases the fluid's pressure. Positive displacement devices can be of either *fixed or variable displacement*. The output of a fixed displacement device remains constant during each pumping cycle and at a given pump speed. The output of a variable displacement device can be changed by altering the geometry of the displacement chamber. Generally, these devices are used for low volume and high lift.

Non positive displacement devices are those where fluid is compressed by the dynamic action of rotating vanes or impellers imparting velocity and pressure to the fluids. These devices are also called as *Rotodynamic* or *hydrodynamic devices*.

CLASSIFICATION OF HYDRAULIC PUMPS

All pumps may be classified as either positive displacement or non positive displacement pumps.

Most pumps used in hydraulic systems are positive displacement. Figure 4.1 shows the classification of pumps.

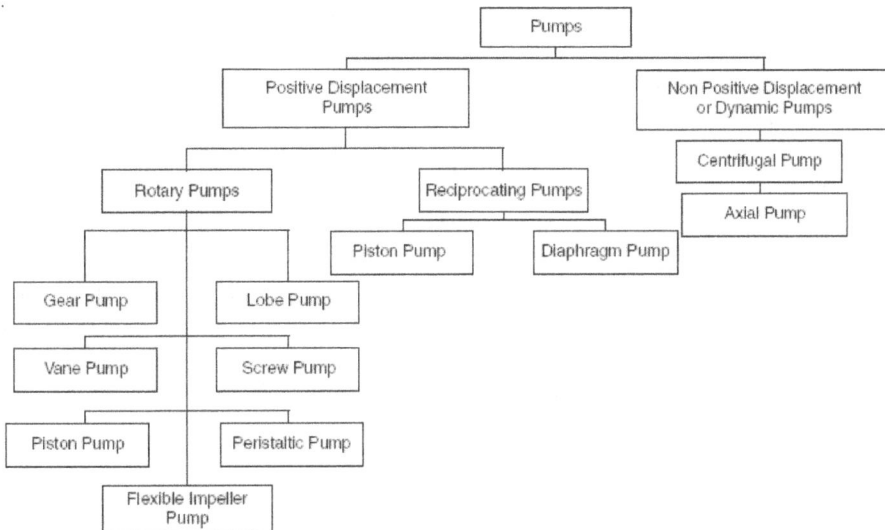

FIGURE 4.1 Classification of Pumps.

POSITIVE DISPLACEMENT PUMPS

The word displacement refers to how much fluid a pump can move in a single rotation. Positive displacement pumps displace a known quantity of liquid with each revolution of the pumping elements. This is done by trapping fluid between the pumping elements and a stationary casing. Pumping element designs include gears, lobes, rotary pistons, vanes, and screws. Positive displacement pumps are found in a wide range of applications such as chemical processing, liquid delivery, marine, pharmaceutical, as well as food, dairy, and beverage processing. These pumps are used when higher head increases are required. Their versatility and popularity are due to their relatively compact design, high-viscosity performance, continuous flow regardless of differential pressure, and ability to handle high differential pressure. Positive displacement pumps can be classified as:

- Rotary pumps.
- Reciprocating pumps.
- Metering pumps.

ROTARY PUMPS

Rotary pumps operate in a circular motion and displace a constant amount of liquid with each revolution of the pump shaft. In general, this is accomplished by pumping elements (e.g., gears, lobes, vanes, screws) moving in such a way as to expand volumes to allow liquid to enter the pump. These volumes are then contained by the pump geometry until the pumping elements move in such a way as to reduce the volumes and force liquid out of the pump. Flow from rotary pumps is relatively unaffected by differential pressure and is smooth and continuous. Rotary pumps have very tight internal clearances which minimize the amount of liquid that slips back from discharge to the suction side of the pump. Because of this, they are very efficient. These pumps work well with a wide range of viscosities, particularly high viscosities. They can handle almost any liquid that does not contain hard and abrasive solids, including viscous liquids. A few important designs of rotary pumps are listed and explained below:

- Gear pump
- Lobe pump
- Vane pump

- Peristaltic pump
- Screw pump
- Piston pump
- Flexible impeller pump

Gear Pump

A *gear pump* works on the principle that when a pair of gears is being driven by an external means, fluid moving into the pump as a partial vacuum is formed and then discharged though the outlet by the meshing of the gears. They are one of the most common types of pumps for hydraulic fluid power applications. Gear pumps are *fixed displacement*, meaning they pump a constant amount of fluid for each revolution. Gear pumps are built to operate in the most punishing applications, particularly those with large amounts of debris and extremely high temperatures, such as steel mills, foundries, and mining operations. There are two main variations of gear pumps:

1. External gear pumps
2. Internal gear pumps

External gear pumps use two external spur gears and *internal gear pumps* use an external and an internal spur gear.

External Gear Pump

A typical external-gear pump is shown in Figure 4.2. These pumps come with a straight spur, helical, or herringbone gear. Straight spur gears are easiest to cut and are the most widely used.

Helical and herringbone gears run more quietly, but cost more.

External gear pumps are similar in pumping action to internal gear pumps in that two gears come into and out of mesh to produce flow. The gear pump consists of pump housing that contains a pair of meshed gears. The external gear pump uses two identical gears rotating against each other, one gear is driven by a motor called drive gear and it in turn drives the other gear called driven gear. Each gear is supported by a shaft with bearings on both sides of the gear.

Figure 4.3 shows the working of an external gear pump. The gear pump works by creating a partial vacuum at the pump inlet. As the gears mesh, a partial vacuum is created at the pump inlet as shown in Figure 4.3 *(a)*.

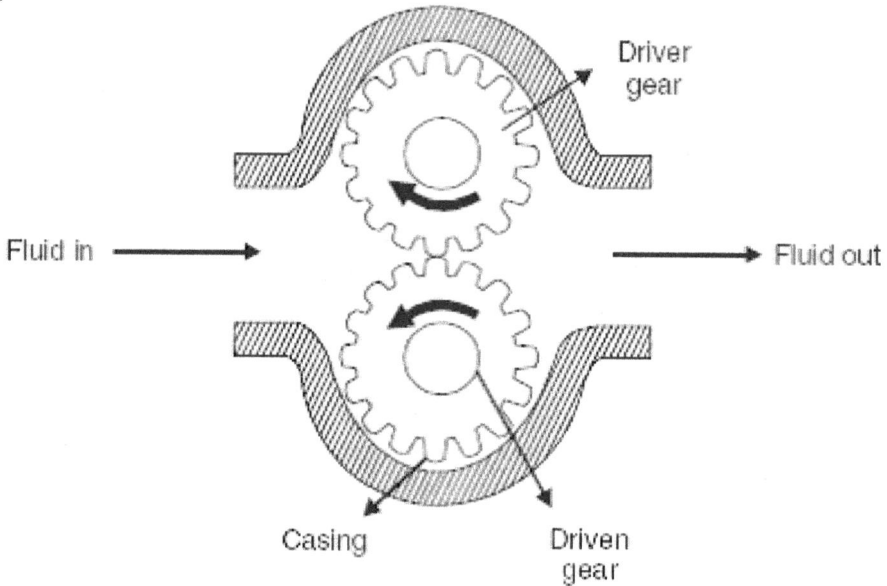

FIGURE 4.2 External Gear Pump.

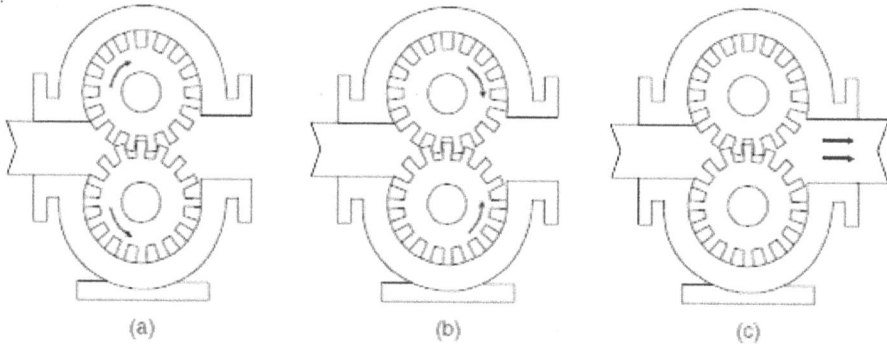

FIGURE 4.3 Working of External Gear Pump.

This allows the atmospheric pressure to push oil from the reservoir into the pump. The oil is carried into the cavity and is trapped by the gear teeth as they rotate. (Refer to Figure 4.3 (b).) As the gears come out of the mesh, they create expanding volume on the inlet side of the pump. Finally, the meshing of the gears forces oil through the outlet port under pressure as shown in Figure 4.3 (c).

Advantages

- High speed
- High pressure
- No overhung bearing loads
- Relatively quiet operation
- Design accommodates a wide variety of materials

Disadvantages

- No solids allowed
- Fixed end clearances

Applications

- Various fuel oils and lube oils.
- Chemical additive and polymer metering.
- Chemical mixing and blending (double pump).
- Industrial and mobile hydraulic applications (log splitters, lifts, etc.).
- Acids and caustic (stainless steel or composite construction).
- Low volume transfer or application.

Internal Gear Pump

The working principle of the internal gear pump is similar to that of the external gear pump. Figure 4.4 shows an internal gear pump. Here

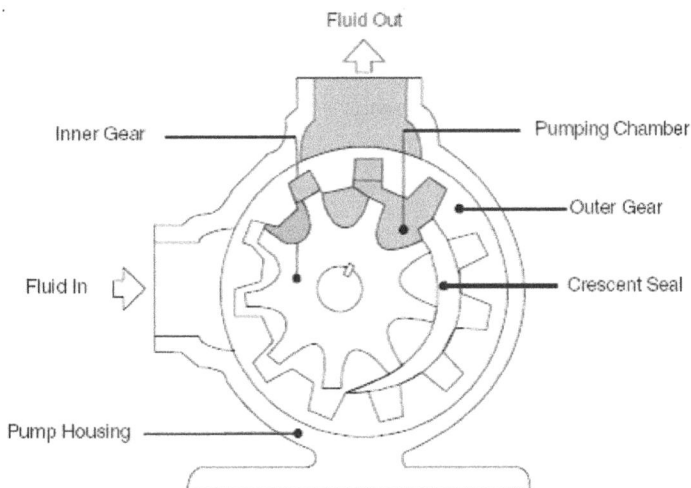

FIGURE 4.4 Internal Gear Pump.

an inner gear meshes onto the inside of a larger gear. This is unlikely the external gear pump where both gears mesh on their outer surface. Pumping chambers are formed between the inner gear teeth and the outer gear teeth and oil is forced out of pump outlet where the teeth mesh. The crescent seal ensures the separation of the inlet and outlet port. The crescent seal prevents liquid from flowing backwards from the outlet.

The rotation of the internal gear by a shaft causes the external gear to rotate, because the two are in mesh. Everything in the chamber rotates except the crescent, causing a liquid to be trapped in the gear spaces as they pass the crescent. Fluid is carried from an inlet to the discharge, where it is forced out of a pump by the gears meshing. As fluid is carried away from the inlet side of a pump, the pressure is diminished, and fluid is forced in from the supply source. The size of the crescent that separates the internal and external gears determines the volume delivery of this pump. A small crescent allows more volume of a fluid per revolution than a larger crescent.

Advantages

- Only two moving parts
- Excellent for high-viscosity liquids
- Constant and even discharge regardless of pressure conditions
- Operates well in either direction
- Easy to maintain
- Flexible design offers application customization

Disadvantages

- Usually requires moderate speeds
- Medium pressure limitations
- One bearing runs in the product pumped
- Overhung load on shaft bearing
- Expensive to manufacture and maintain

Applications

Common internal gear pump applications include, but are not limited to:

- All varieties of fuel oil and lube oil.
- Resins and polymers.
- Alcohols and solvents.

- Asphalt, bitumen, and tar.
- Food products such as corn syrup, chocolate, and peanut butter.
- Paint, inks, and pigments.
- Soaps and surfactants.

Gerotor Pump. Gerotor pump is one of the most common types of internal gear pumps. (Refer to Figure 4.5.) It consists of an internal gear inside an external gear. The internal gear is keyed to the drive shaft and has one fewer tooth than an external gear. During rotation each tooth of the internal gear is in constant contact with the external gear. Due to the extra tooth, the outer gear rotates slowly. Space between the rotating teeth increases during the first half of each turn thus taking fluid in. In the last half, the space decreases thus forcing fluid out through the outlet port.

Lobe Pump

The lobe pump is similar to an external gear pump in operation, *i.e.*, the fluid flows around the interior of the casing. It differs from the external gear pump in the way the gears are driven. In a gear pump, one gear drives the other; in a lobe pump, both rotors are driven through suitable drive

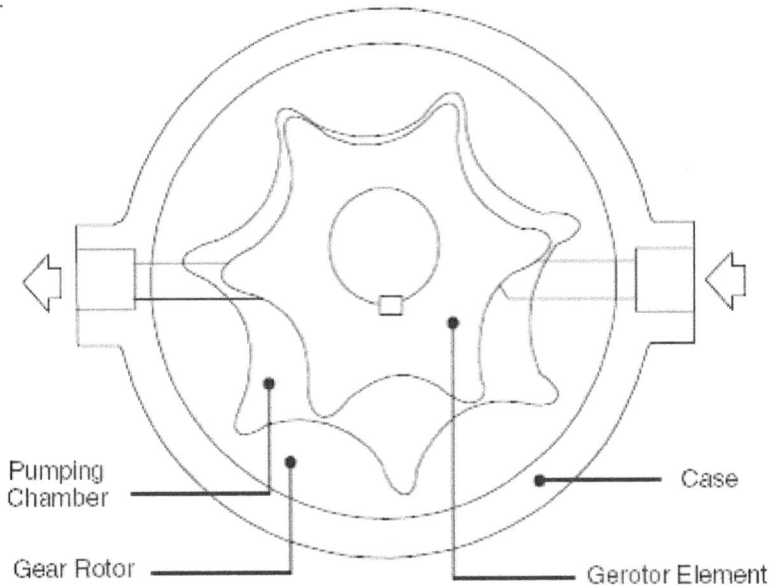

FIGURE 4.5 Gerotor Pump.

gears outside of the pump casing chamber. Gears are replaced by impellers which have two or three lobe shaped teeth. These teeth are wider and more rounded than the teeth of a regular gear pump. As a result, displacement is higher. Unlike external gear pumps, the pumping gear elements or lobes do not make contact. Figure 4.6 illustrates a three-lobe pump with two impellers. The number of lobes will determine the amount of pulsation from the pump output. The greater the number of lobes, the more constant is the discharge from the pump.

As the rotors start to rotate, an expanding cavity is formed by the rotation of the lobes, which creates a vacuum at the inlet port, drawing fluid in the pumping chamber. Fluid travels around the interior of the casing in the pockets between the lobes and the casing. It does not pass between the lobes. Finally, the fluid is forced or pressed out of discharge port of the pump in continuous smooth flow and pressure is generated by the meshing of the lobe rotors. The lobed impellers are easier to replace and tend to wear less with abrasives than the gears in the gear pump.

Advantages

- Pass medium solids
- No metal-to-metal contact

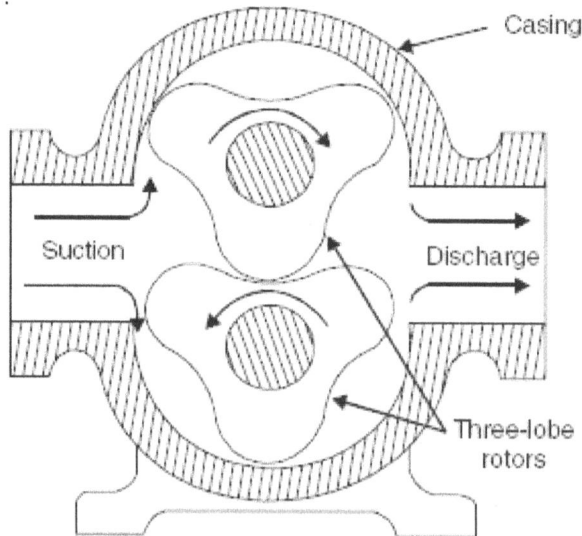

FIGURE 4.6 Lobe Pump with Two Impellers.

 ▪ Long-term dry run (with lubrication to seals)
 ▪ Non-pulsating discharge

   Disadvantages

 ▪ Requires timing gears
 ▪ Costly
 ▪ Requires two seals
 ▪ Reduced lift with thin liquids

   Applications

   Common rotary lobe pump applications include, but are not limited to:

 ▪ Polymers
 ▪ Paper coatings
 ▪ Soaps and surfactants
 ▪ Paints and dyes
 ▪ Rubber and adhesives
 ▪ Pharmaceuticals
 ▪ Food applications

   They are popular in these diverse industries because they offer superb sanitary qualities, high efficiency, reliability, corrosion resistance, and good clean-in-place characteristics.

Vane Pump

   A rotary vane pump is a positive displacement pump that consists of vanes mounted to a rotor that rotates inside of a cavity. Vane pumps are available in a number of vane configurations including sliding vane (*left*), flexible vane, swinging vane, rolling vane, and external vane. Each type of vane pump offers unique advantages. For example, external vane pumps can handle large solids. Flexible vane pumps, on the other hand, can only handle small solids but create good vacuum. Sliding vane pumps can run dry for short periods of time and handle small amounts of vapor.

   Sliding vane pumps operate quite differently from gear and lobe types. The most simple vane pump consists of a circular rotor rotating inside a larger circular cam ring. (Refer to Figure 4.7). The centers of these two circles are offset, causing eccentricity. The vanes are fitted into the rotor slots and the rotor is mounted on the drive shaft. The rotor rotates inside the cam ring by a prime mover. The vanes forced out of their slots contact and follow the inner surface of the housing. This occurs due to the centrifugal force which pushes the vanes out of their slots and holds them in place creating a seal against the cam ring. There is an increase in volume from the point where the cam ring surface is closest to the rotor. Due to an increase in volume, a partial vacuum is created and oil is pushed into the inlet from the reservoir. As the fluid crosses the rotor centerline, the chambers become progressively smaller as the cam ring forces the vanes into their slots. As the volume decreases, fluid is forced out of the pump outlet.

Advantages

- Simple in construction
- High efficiency and low cost
- Handles thin liquids at relatively higher pressures
- Sometimes preferred for solvents, LPG
- Can run dry for short periods
- Develops good vacuum

Disadvantages

- Complex housing and many parts
- Not suitable for high pressures

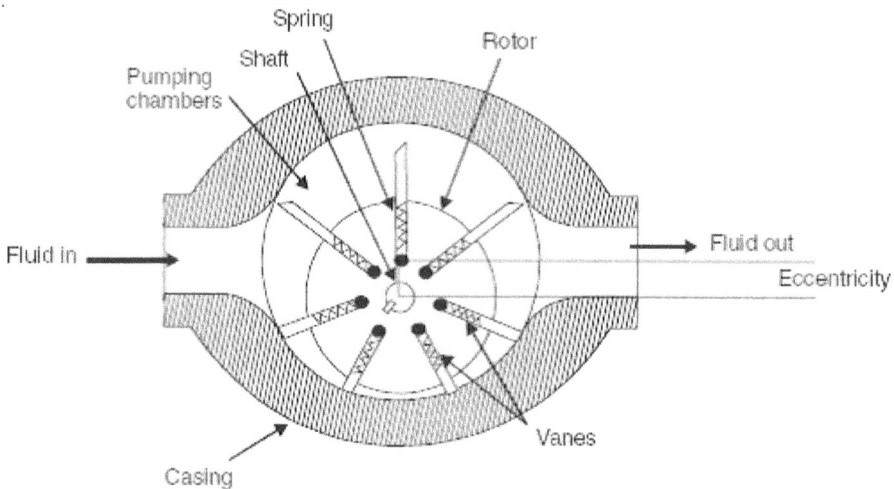

FIGURE 4.7 Vane Pump.

- Not suitable for high viscosity
- Not good with abrasives

    Applications

- Aerosol and Propellants.
- Aviation Service - Fuel Transfer, Deicing.
- Auto Industry - Fuels, Lubes, Refrigeration Coolants.
- Bulk Transfer of LPG and NH_3.
- LPG Cylinder Filling.
- Used in spray painting equipment.

Peristaltic Pump

A peristaltic pump is a type of positive displacement pump used for pumping a variety of fluids. (Refer to Figure 4.8.) This pump is based on an electrometric tube through which the fluid is forced. A flexible tube passes through the fixed casing of the pump. A rotor with rollers attached to it moves and presses against the flexible tube. The tube is compressed at a number of points in contact with the rollers. This squeezing action produces an even flow of fluid. The fluid is moved through the tube with each rotating motion. For proper action, it is important that the tubing is flexible enough to allow the rollers to squeeze it until it is completely closed. The pumped fluid is completely isolated from the moving parts, which permits

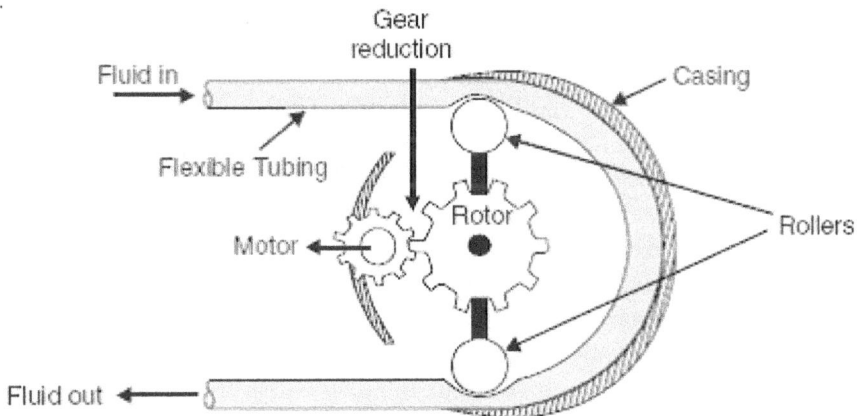

FIGURE 4.8 Peristaltic Pump.

pumping of corrosive substances. The flow rate of the pump is related directly to the diameter of the tube and the speed of rotation of the drive.

Advantages

- Hydraulically operated speed variations to meet all job demands.
- Enables pumping of heavy, fibered, thick, and abrasive materials at high volumes.
- The pump's mechanical maintenance is limited to hose wear. Pumped material does not come in contact with any moving parts.
- Easy to sterilize and clean the inside surfaces of the pump.
- Replacement of the hose is quick and easy.
- Dry running can occur without causing damage to the pump.
- Low cost of maintenance.
- Inexpensive to manufacture because there are no moving parts in contact with the fluid.

Applications

Some common applications include pumping aggressive chemicals, high solid slurries, and other materials where isolation of the product from the environment, and the environment from the product, are critical.

Screw Pump

Screw pumps are a unique type of rotary positive displacement pump in which the flow through the pumping element is truly axial. Screw pumps consist of helical screws which revolve in a fixed casing. As the screw rotates in the casing, a cavity created between the screw and the casing progresses towards the discharge side of the pump. This movement creates a partial vacuum which draws liquid into the pump. The shape of the casing at the discharge end is such that the cavity becomes closed. Screw pumps produce a constant discharge with negligible pulsation. They have an exceptionally long life expectancy. Significant disadvantages of screw pumps are that they are bulky and heavy. Application of screw pumps in agriculture is limited to food processing and hydraulic systems for machine tools. Screw pumps can be of *twin screw* or *multiple screw* types.

Twin-Screw Pumps

In twin-screw pumps, timing gears are used to control the relative motion of the screws. In pumps with more than two screws, a single central screw causes the complimentary rotation of the adjacent screws. Twin-screw pumps have the following characteristics:

- Applicable for all types of liquid, from low to high viscosity, chemically neutral or aggressive.
- Screws do not touch. Therefore low shear rate, low emulsion, low friction/less torque.
- High reliability and efficiency.
- Quiet operation.
- High self-priming property.
- Low pulsation flow.
- Compact and space-saving design.
- Long life.

Multi-Screw Pumps

In multi-screw pumps, the fluid is transferred under the action of a number of screws meshed together in a casing provided with channels to suit the screws. These pumps are used in the chemical process industry and in the oil industry for applications on oil rigs. They are used for pumping fuel oil, lubrication oil, seawater, paints, etc. Multiple screw pumps have the following characteristics:

- Output is smooth.
- Produce a constant discharge with negligible pulsation.
- A wide range of fluid viscosities can be handled.
- The pump is self-priming especially when the screws are wetted.
- The volumetric and mechanical efficiency is good.
- The pumps are quiet operating.
- The pumps have a high level of reliability.

Piston Pumps

A rotary piston pump has a rotary motion rather than a back and forth motion as in the reciprocating piston pumps. There are two basic designs of rotary piston pumps available:

- Radial piston pump
- Axial piston pump

In a radial design, the number of pistons and cylinders are arranged radially around the rotor hub, while in the axial design they are located in a parallel position with respect to the rotor shaft.

Piston pumps operate at higher efficiencies than gear and vane pumps and are used for high-pressure applications with hydraulic oil or fire-resistant fluids. The piston pumps are extensively used for power transfer applications in the off-shore power transmission, agricultural, aerospace, and construction industries, etc.

Construction

The pump consists of a block with a number of symmetrically arranged cylindrical pistons around a common center line. The pistons are caused to reciprocate in and out under the action of a separate fixed or rotating plate (axial pistons) or an eccentric bearing ring (radial pump). Each piston is interfaced with the inlet and outlet port via a special valve arrangement so that as it moves out of its cylinder, it draws fluid in and as it moves back, it pushes the fluid out.

Rotary piston pumps include a number of variations which works on a similar principle:

- Radial piston pump
- Swash plate piston pump
- Wobble plate pump
- Bent axis piston pump

Radial Piston Pump

A radial piston pump shown in Figure 4.9 is the simplest design of a piston pump. It consists of a rotating cylinder containing equally spaced radial pistons arranged radial around the cylinder center line. A spring pushes the pistons against the inner surface of an encircling stationary ring mounted eccentric to the cylinder. The pistons draw in fluid during half a revolution and drive fluid out during the other half. The greater the ring eccentricity, the longer the pistons stroke and the more fluid they transfer.

Swash Plate Pump

Swash plate pumps shown in Figure 4.10 have a rotating cylinder containing parallel pistons arranged radially around the cylinder center line. A spring pushes the pistons against a stationary swash plate located at one end of the cylinder, which sits at an angle to the cylinder. The pistons draw in

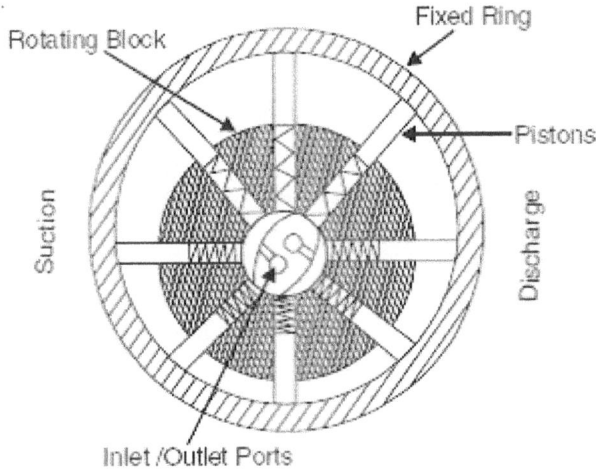

FIGURE 4.9 Radial Piston Pump.

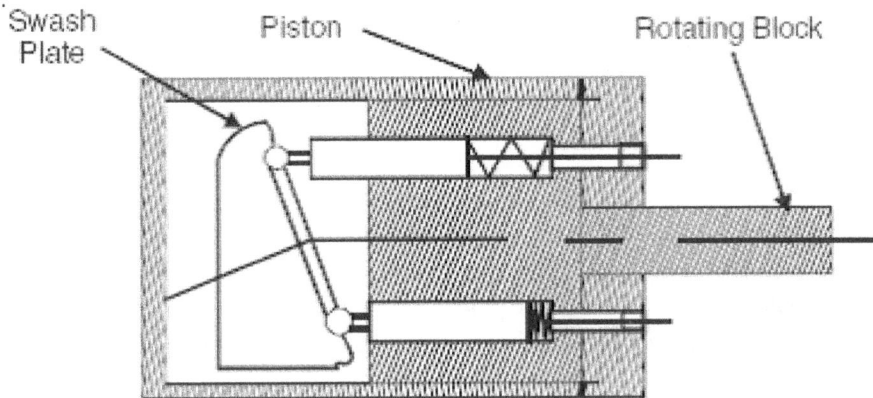

FIGURE 4.10 Swash Plate Pump.

fluid during half a revolution and drive fluid out during the other half. The greater the swash plate angle relative to the cylinder center line, the longer the pistons stroke and the more fluid they transfer.

Wobble Plate Pump

This pump includes a stationary piston block containing a number parallel pistons arranged radially around the block center (at least five). (Refer to Figure 4.11.) The end of each piston is forced against a rotating wobble

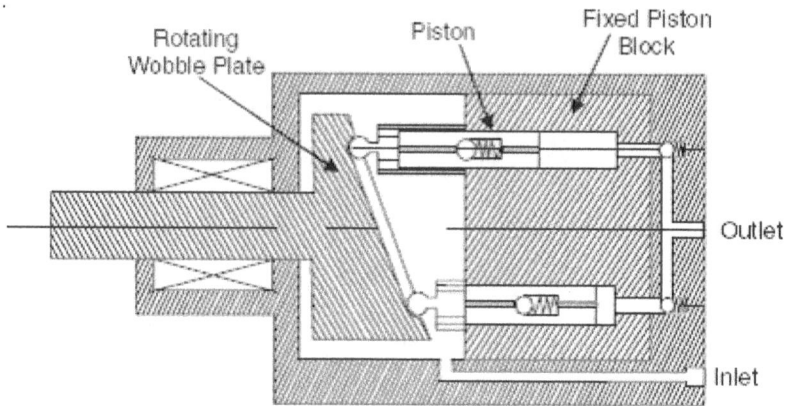

FIGURE 4.11 Wobble Plate Pump.

vplate by springs. The wobble plate is shaped with varying thickness around its center line and thus, as it rotates, it causes the pistons to reciprocate at a fixed stroke. The pistons draw in fluid from the cavity during half a revolution and drive fluid out at the rear of the pump during the other half. The fluid flow is controlled using non-return valves for each piston.

Bent Axis Pump

Another design of piston pumps is the bent axis piston pump. It consists of a cylinder block placed at an angle offset from the drive shaft. The piston rods are attached to the shaft flange by ball joints. (Refer to Figure 4.12.) The cylinder block is splined to the shaft and this shaft is joined to the drive shaft by a universal link. This link ensures that the drive shaft and the cylinder block are aligned properly and rotate together. When the drive shaft rotates, the pistons are forced in and out of their bores because the distance between the cylinder block and the shaft flange varies. Due to the reciprocating action of the pistons, the fluid is pushed in through the inlet port by atmospheric pressure and discharge from the outlet port.

Flexible Impeller Pump

Flexible impeller pumps consist of flexible bladed impellers which are placed eccentrically in casings. The impeller blades unfold as they pass the suction port, creating a partial vacuum which causes liquid to flow into the pump. As the rotor moves, the blades bend due to the eccentric placement

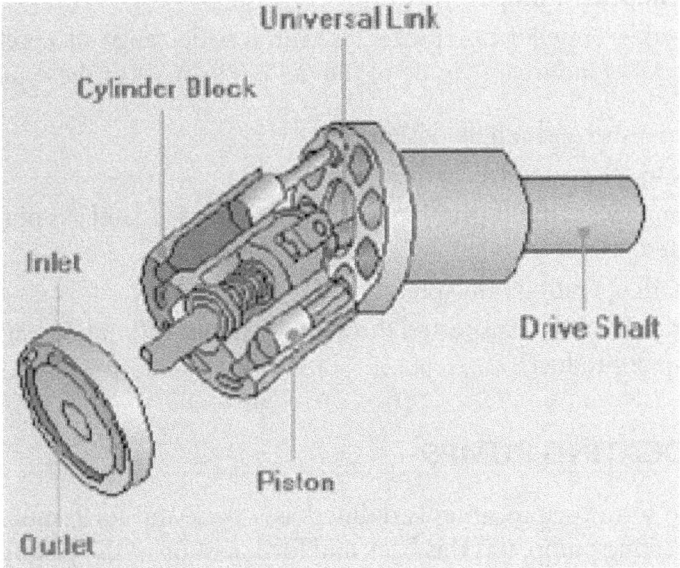

FIGURE 4.12 Bent Axis Pump.

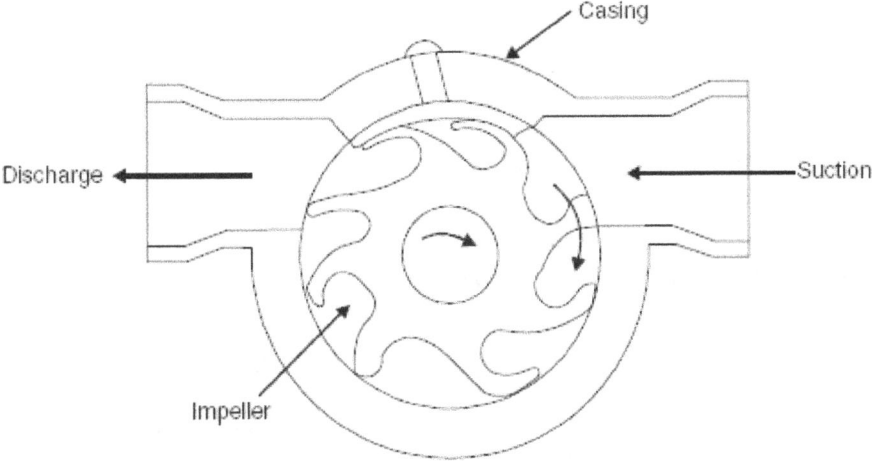

FIGURE 4.13 Flexible Impeller Pump.

of the rotor, resulting in a squeezing action on the liquid and increased pressure. (Refer to Figure 4.13.)

Benefits of Impeller Pumps

Flexible impeller pumps are used in a wide range of applications in most types of industries. Some of the unique features of these pumps are:

- Inexpensive and simple design.
- Easy to select, service, and use.
- Extremely good self-priming capacity displaces air and gas positively or aerated products.
- Smooth operating principle.
- Hard/soft solids transported through the pump without damage to pump or product.

RECIPROCATING PUMPS

The term reciprocating is defined as a back and forth motion. In the reciprocating pump, it is this back and forth motion of the piston inside the cylinder that provides the flow of fluid. Reciprocating pumps, like rotary pumps, operate on the positive principle, *i.e.*, each stroke delivers a definite volume of fluid to the system. Reciprocating pumps are generally very efficient and are suitable for very high heads at low flows.

Types of Reciprocating Pumps

Reciprocating pumps are classified as:

- Piston pump/plunger pump
- Diaphragm pump

This classification is based on the type of reciprocating element used to transfer energy to the fluid. These types of pumps operate by using a reciprocating piston or diaphragm. The liquid enters a pumping chamber via an inlet valve and is pushed out via an outlet valve by the action of the piston or diaphragm. *Reciprocating pumps* operate with high efficiencies. These are suitable for very high heads at low flows and can handle varying pressures at a constant speed.

Piston Pump/Plunger Pump

Piston pumps/plunger pumps are reciprocating pumps that use a plunger or piston to move fluid through a cylindrical chamber. In piston pumps, a piston, which is attached to a mechanical linkage, transforms the rotary motion of a drive wheel into the reciprocating motion of the piston.

A single-acting piston pump is shown in Figure 4.14. During suction stroke, as shown in Figure 4.14 (*a*), liquid enters the cylinder through the suction check valve. During compression stroke, as shown in Figure 4.14 (*b*), the liquid is forced into the discharge line through the discharge check valve. This action is similar to the action of a piston in the cylinder of an automobile engine.

The flow rate of a simple piston pump is not constant because there is no flow on the intake stroke and the flow varies from zero to the maximum and back to zero on each compression stroke. Pulsation of flow can be reduced by using a double-action piston pump as shown in Figure 4.15, where the volume on both sides of the piston is used for pumping liquid. In this case, each suction stroke is a companion compression stroke for the opposite side of the pump and the liquid is pumped on both strokes.

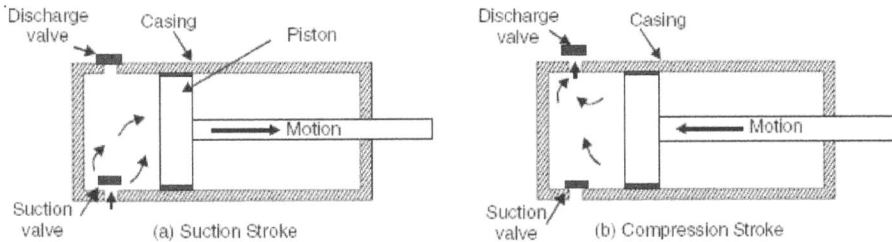

FIGURE 4.14 Single-Acting Piston Pump.

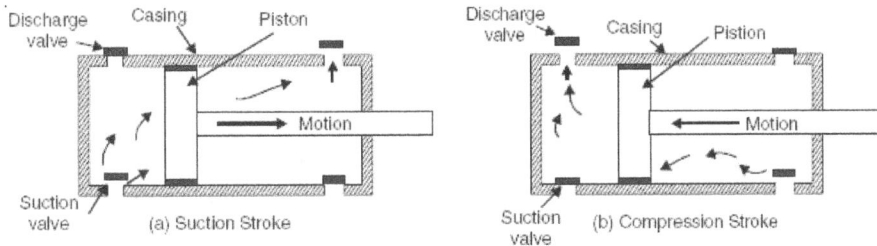

FIGURE 4.15 Double-Acting Piston Pump.

Characteristics of Piston Pumps

■ These pumps can create high pressures.
■ They deliver a constant flow rate independent of the discharge pressure.
■ They can run dry.
■ They have a long life.
■ These pumps are bulky and heavy.

Diaphragm Pump

Diaphragm pumps are common industrial pumps that use positive displacement principle to move liquids. Sometimes this type of pump is also called a membrane pump. These devices typically include a single diaphragm and chamber, as well as suction and discharge check valves to prevent backflow. Pistons are either coupled to the diaphragm, or used to force hydraulic oil to drive the diaphragm. The operation of a diaphragm pump is similar to that of a piston pump. The pulsating motion is transmitted to the diaphragm through a fluid or a mechanical drive, and then through the diaphragm to the pumped liquid. A schematic of this type of pump is shown in Figure 4.16.

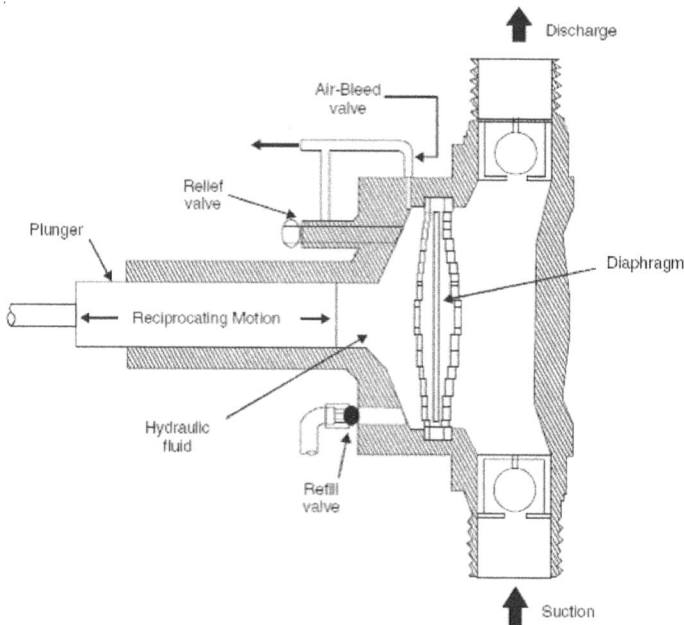

FIGURE 4.16 Diaphragm Pump.

Characteristics of Diaphragm Pump

- The pumped liquid does not come into contact with most of the working parts of the pump. Hence, diaphragm pumps are suitable for pumping corrosive liquids.
- Diaphragm pumps can run dry for extended periods of time without damage.
- Most diaphragm pumps can be adjusted for calibration when running.
- Diaphragm pumps are highly reliable.
- These pumps can handle abrasive materials such as acids, chemicals, concrete, coolants, combustible or corrosive materials, and effluents.
- Disadvantages of these pumps include limited head and capacity range, and the necessity of check valves in the suction and discharging nozzles.
- These pumps are used in a variety of industries such as aerospace or defense, agriculture or horticulture, automotive, brewery or distillery, construction, cryogenic, dairy, or flood control applications, medical, pharmaceutical, and biotechnology applications; power generation; the pulp and paper industries.

METERING PUMP

A metering pump is used to pump liquids at adjustable flow rates that are precise when averaged over time. Delivery of fluids in precise adjustable flow rates is sometimes called metering. The term *metering pump* is based on the application or use rather than the exact kind of pump used. They are also called proportioning or controlled-volume pumps.

Metering pumps are available in either a diaphragm or piston/plunger, and are designed for clean service as dirty liquid can easily clog the valves and nozzle connections. A piston/plunger type metering pump is shown in Figure 4.17. Many metering pumps are piston-driven. Piston pumps are positive displacement pumps which can be designed to pump at practically constant flow rates. In piston driven metering pumps, there is a piston/plunger, which goes in and out of a correspondingly shaped chamber in the pump head. The inlet and outlet lines are joined to the piston chamber. There are two check valves attached to the pump head, one at the inlet and the other at the outlet. The inlet valve allows flow from the inlet line to the piston chamber, but not in the reverse direction. The outlet valve allows flow from the chamber to the outlet line, but not in reverse. The motor repeatedly moves the piston into and out of the piston chamber, causing

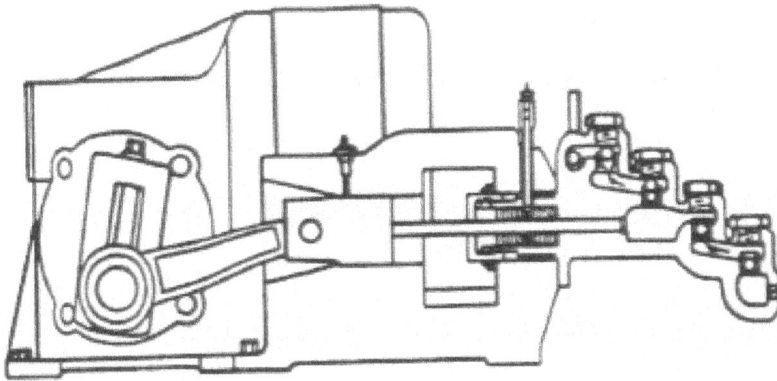

FIGURE 4.17 Plunger Type Metering Pump.

the volume of the chamber to repeatedly become smaller and larger. When the piston moves out, a vacuum is created. Low pressure in the chamber causes liquid to enter and fill the chamber through the inlet check valve, but higher pressure at the outlet causes the outlet valve to shut. Then when the piston moves in, it pressurizes the liquid in the chamber. High pressure in the chamber causes the inlet valve to shut and forces the outlet valve to open, forcing liquid out at the outlet.

DYNAMIC/NON POSITIVE DISPLACEMENT PUMPS

Hydrodynamic or non positive displacement pumps are low pressure, high volume pumps. These pumps operate by developing a high liquid velocity and converting the velocity to pressure in a diffusing flow passage. Dynamic pumps usually have lower efficiencies than positive displacement pumps, but also have lower maintenance requirements. Dynamic pumps are also able to operate at fairly high speeds and high fluid flow rates. These pumps are large in size. Dynamic pumps are classified as:

- Centrifugal pumps
- Axial pumps

CENTRIFUGAL PUMPS

Centrifugal pumps differ from rotary pumps in that they rely on kinetic energy rather than mechanical means to move fluid. A centrifugal pump is a rotodynamic pump that uses a rotating impeller to increase the velocity of a fluid. Fluid enters the pump at the center of a rotating impeller and gains energy as it moves to the outer diameter of the impeller. Fluid is forced out of the pump by the energy it obtains from the rotating impeller. Typically the volute shape of the pump casing (increasing in volume), or the diffuser vanes (which serve to slow the fluid, converting to kinetic energy in to flow work) are responsible for the energy conversion. The energy conversion results in an increased pressure on the downstream side of the pump, causing flow. Centrifugal pumps can transfer large volumes of liquid but efficiency and flow decrease rapidly as pressure and/or viscosity increases.

There are several kinds of centrifugal pumps. Every centrifugal pump has two characteristics that are same; each has an impeller that forces the liquid being pumped into a rotary motion, and each has a casing, which directs the liquid to the impeller. The liquid leaves the impeller as the impeller rotates. The liquid leaves with a higher velocity and pressure than it had when it entered. There is a conversion of some of the velocity to pressure that takes place before the liquid leaves the pump; this partial conversion takes place in the pump casing. A centrifugal pump is shown in Figure 4.18.

Centrifugal pumps are classified into three general categories:

Radial flow: a centrifugal pump in which the pressure is developed wholly by centrifugal force.

Mixed flow: a centrifugal pump in which the pressure is developed partly by centrifugal force and partly by the lift of the vanes of the impeller on the liquid.

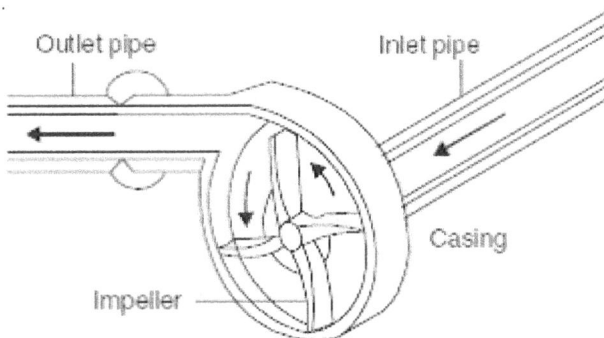

FIGURE 4.18 Centrifugal Pump.

Axial flow: a centrifugal pump in which the pressure is developed by the propelling or lifting action of the vanes of the impeller on the liquid.

Advantages

Some of the advantages of centrifugal pumps are smooth flow through the pump and uniform pressure in the discharge pipe, low cost, and an operating speed that allows for a direct connection to steam turbines and electric motors. They are widely used in chemical processing plants and oil refineries.

Disadvantages

Primary disadvantages are higher cost and limited flow at higher pressure.

PUMP SELECTION PARAMETERS

The process involved in the selection of a suitable pump for a given application depends on many parameters. These factors include:

- Cost
- Pressure ripple and noise
- Suction performance
- Contaminant sensitivity
- Speed
- Weight
- Fixed or variable displacement
- Maximum pressure and flow, or power
- Fluid type.

COMPARISON OF POSITIVE AND NON POSITIVE DISPLACEMENT PUMPS

Parameter	Centrifugal Pumps	Reciprocating Pumps	Rotary Pumps
Optimum Flow and Pressure Applications	Medium/High Capacity, Low/Medium Pressure	Low Capacity, High Pressure	Low/Medium Capacity, Low/Medium Pressure
Maximum Flow Rate	6,300+ l/s	630+ l/s	630+ l/s

Parameter	Centrifugal Pumps	Reciprocating Pumps	Rotary Pumps
Low Flow Rate Capability	No	Yes	Yes
Maximum Pressure	400+ bar	6,500+ bar	270+ bar
Requires Relief Valve	No	Yes	Yes
Smooth or Pulsating Flow	Smooth	Pulsating	Smooth
Variable or Constant Flow	Variable	Constant	Constant
Self-priming	No	Yes	Yes
Space Considerations	Requires Less Space	Requires More Space	Requires Less Space
Costs	Lower Initial, Lower Maintenance, Higher Power	Higher Initial, Higher Maintenance, Lower Power	Lower Initial, Lower Maintenance, Lower Power
Fluid Handling	Suitable for a wide range including clean, clear, non-abrasive fluids to fluids with abrasive, high-solid content. Not suitable for high viscosity fluids. Lower tolerance for entrained gases.	Suitable for clean, clear, non-abrasive fluids. Specially fitted pumps suitable for abrasive-slurry service. Suitable for high viscosity fluids. Higher tolerance for entrained gases.	Requires clean, clear, non-abrasive fluid due to close tolerances. Optimum performance with high viscosity fluids. Higher tolerance for entrained gases.

AIR COMPRESSORS

Every compressed air system begins with a compressor, the source of airflow for all the equipment and processes. Air compressors are utilized to raise the pressure of a volume of air. Air compressors are versatile mechanical tools that use one or numerous pistons to pump compressed air into a defined space. An air compressor is defined as a component that takes in air at atmospheric pressure and delivers it at a higher pressure.

Compressed Air

Compressed air is widely used throughout industry and is considered one of the most useful and clean industrial utilities. It is simple to use, but complicated and costly to create. Compressed air is air that is condensed and contained at a pressure that is greater than the atmosphere. The process takes a given mass of air, which occupies a given volume of space, and reduces it into a smaller space. In that space, greater air mass produces greater pressure. Compressed air is used in many different manufacturing operations.

Applications of Compressed Air

Compressed air is used in almost every industry like automotive, construction, universities, hospitals, mining, agriculture, food and beverage, consumer goods, pharmaceutical, electronics, and more.

- For maintenance work, plants can use air-operated drills, screwdrivers, and wrenches.
- Painting can be done using paint-spraying systems.
- Sprinkler systems are controlled by air pressure, which keeps water from entering the pipes until heat breaks the seal and releases the pressure.
- Pneumatic tools are used on a production line.
- Used in the foundry for cleaning large castings, and to remove weld scale, rust, and paint in other industries.
- Grinding, wire brushing, polishing, sanding, shot blasting, and buffing are performed efficiently with compressed air in the automotive, aircraft, rail car, locomotive, vessel shops, shipbuilding, other heavy machinery, and other industries.

TYPES OF AIR COMPRESSORS

Air compressors are of two types:

- Positive displacement air compressors
- Non positive/dynamic air compressors

In the positive displacement type, a given quantity of air or gas is trapped in a compression chamber and the volume it occupies is mechanically reduced, causing a corresponding rise in pressure prior to discharge. At constant speed, the air flow remains essentially constant with variations in discharge pressure.

Dynamic compressors impart velocity energy to continuously flowing air or gas by means of impellers rotating at very high speeds. The velocity energy is changed into pressure energy both by the impellers and the discharge volutes or diffusers.

The compressors are further segmented into several compressor types as shown in Figure 4.19.

The function of all of them is to draw in air from the atmosphere and produce air at substantially higher pressures.

POSITIVE DISPLACEMENT COMPRESSORS

Positive displacement compressors decrease the volume and increase the pressure of a quantity of air by mechanical means. These are classified as:

- Rotary compressors
- Reciprocating compressors

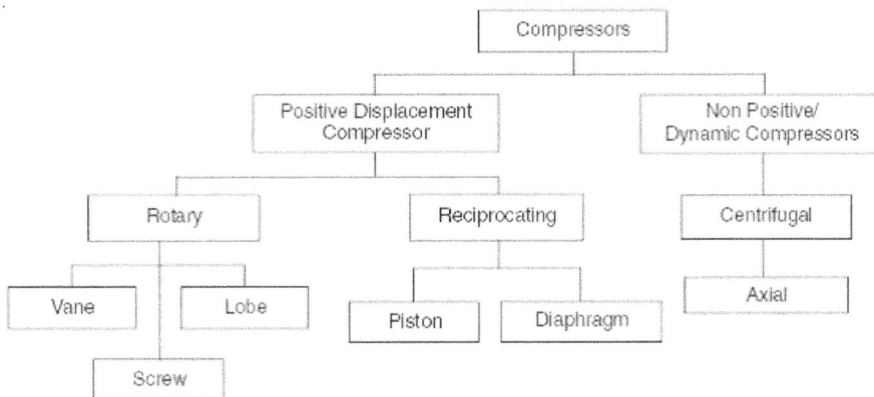

FIGURE 4.19 Classification of Compressors.

ROTARY COMPRESSORS

Rotary air compressors are *positive displacement* compressors. Rotary compressors have rotors that give a continuous pulsation free discharge. They operate at high speeds. Their capital costs are low. They are compact in size, have low weight, and are easy to maintain. These units are basically oil cooled (with air cooled or water cooled oil coolers) where the oil seals the internal clearances. Because the cooling takes place right inside the compressor, the working parts never experience extreme operating temperatures. Rotary compressors are classified into three general groups:

- Screw compressor
- Lobe compressor
- Vane compressor

Screw Compressor

Rotary screw compressors are positive displacement compressors. Compression is achieved via the meshing of two helically cut rotor profiles. One rotor is cut as a male profile, and the other as a female profile. These

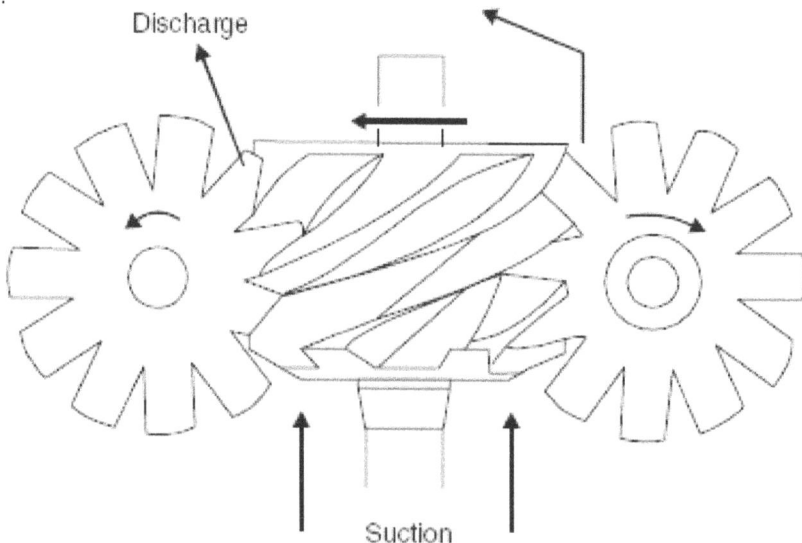

FIGURE 4.20 Single-Screw Compressor.

two rotors spin in opposite directions. The rotary screw compressor can be *single screw* or *twin screw*.

A *single-screw* compressor uses a single main screw rotor meshing with two gate rotors with matching teeth. The main screw is driven by the prime mover, typically an electric motor. (Refer to Figure 4.20.)

A *twin-screw* compressor consists of two intermeshing screws or rotors, which trap gas between the rotors and the compressor case. (Refer to Figure 4.21.) The motor drives the male rotor, which in turn drives the female rotor. Both rotors are encased in a housing provided with air inlet and outlet ports. Air is drawn through the inlet port into the voids between the rotors. As the rotors move, the volume of trapped air is successively reduced and compressed by the rotors coming into mesh.

These compressors are available as dry or wet (oil-flooded) screw types. In the dry-screw type, the rotors run inside of a stator without a lubricant (or coolant). The heat of compression is removed outside of the compressor, limiting it to a single-stage operation. In the oil-flooded screw type compressor, the lubricant is injected into the air, which is trapped inside of the stator. In this case, the lubricant is used for cooling, sealing, and lubrication. The air is removed from the compressed gas-lubricant mixture in a separator.

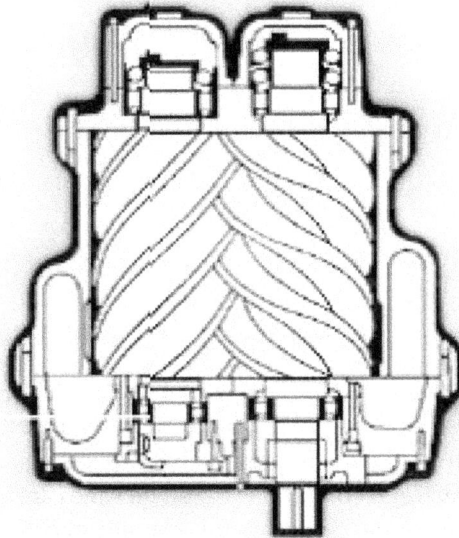

FIGURE 4.21 Twin-Screw compressor.

Because of simple design and few wearing parts, rotary screw air compressors are easy to maintain, operate, and provide great installation flexibility. Advantages of the rotary screw compressor include smooth, pulse-free air output in a compact size with high output volume over a long life.

Lobe Compressor

A schematic diagram for a rotary lobe compressor is provided in Figure 4.22. In this type of compressor, the rotors do not touch and a certain amount of slip exists. The slip increases as the output pressure increases. The principle of operation is analogous to the rotary screw compressor, except that with the lobe compressor, the mating lobes are not typically lubricated for air service. As the lobe impellers rotate, gas is trapped between the lobe impellers and the compressor case, where the gas is pressurized through the rotation of lobes and then discharged. The bearings and timing gears are lubricated using a pressurized lubricating system or sump.

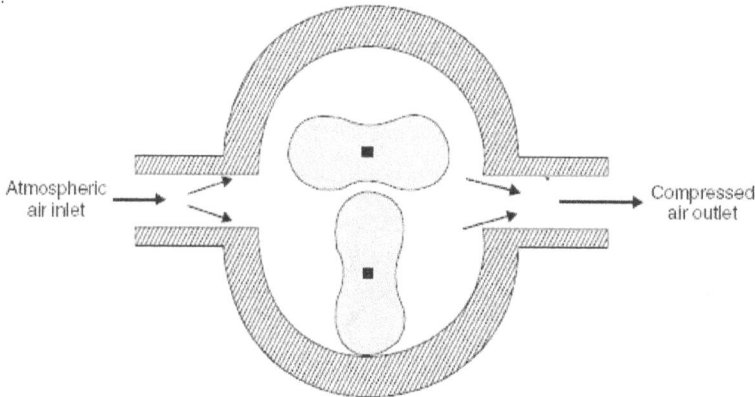

FIGURE 4.22 Lobe Compressors.

Vane Compressor

A rotary vane compressor is schematically illustrated in Figure 4.23. Rotary vane compressors consist of a rotor which is mounted offset in a larger housing which can be of circular or a more complex shape. The rotor has slots along its length, each slot contains a vane. The vanes are thrown outwards by centrifugal force when the compressor is running and the vanes move in and out of the slot because the rotor is eccentric to the casing. The vanes sweep the cylinder, sucking air in on one side and ejecting it on the other.

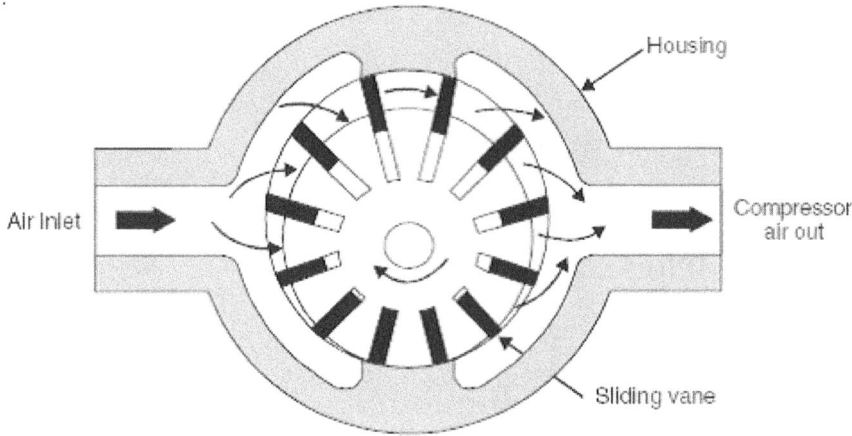

FIGURE 4.23 Vane Compressor.

RECIPROCATING COMPRESSORS

Reciprocating compressors were the first of the modern air compressor designs. Reciprocating air compressors are *positive displacement* machines in which pistons are driven by a crankshaft. This means they are taking in successive volumes of air that is confined within a closed space and elevating this air to a higher pressure. The reciprocating air compressor accomplishes this by using a piston within a cylinder as the compressing and displacing element. As the piston enters the down stroke, air is drawn into the cylinder from the atmosphere through an air inlet valve. During the up stroke, the piston compresses the air and forces it through a discharge control valve and out of the compressor. These are available in single-acting and double-acting configurations. The reciprocating air compressor is considered single acting when the compressing is accomplished using only one side of the piston. A compressor using both sides of the piston is considered double acting. Reciprocating compressors are of two types:

- Piston compressors
- Diaphragm compressors

    Advantages

- Good for small applications.
- Cheap and simple to operate.
- Operates over a wide range of pressures.

Disadvantages

■ Noisy.

■ Maintenance problems.

■ Not good for small applications.

■ Oil free air units are expensive.

PISTON COMPRESSORS

Piston compressors are of two types:

Single-Stage Piston Compressor

The single-stage reciprocating compressor has a piston that moves downward during the suction stroke, expanding the air in the cylinder as shown in Figure 4.24. The expanding air causes pressure in the cylinder to drop. When the pressure falls below the pressure on the other side of the inlet valve, the valve opens and allows air in until the pressure equalizes across the inlet valve. The piston bottoms out and then begins a compression stroke. The upward movement of the piston compresses the air in the

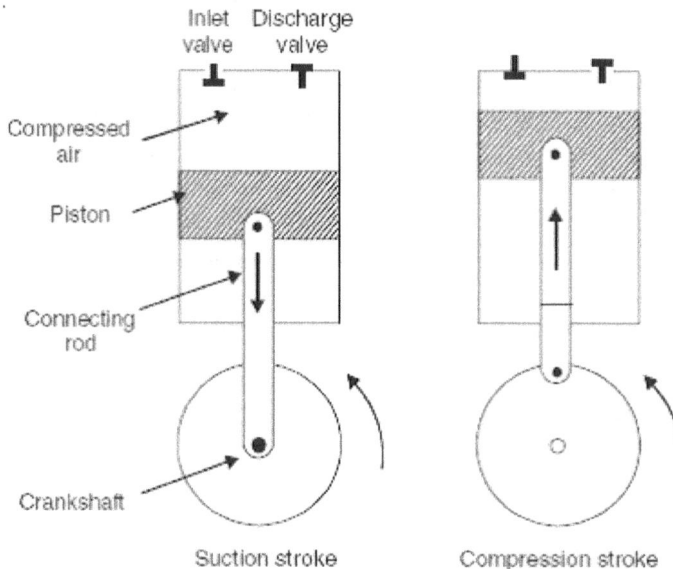

FIGURE 4.24 Single-Stage Piston Compressor.

cylinder, causing the pressure across the inlet valve to equalize and the inlet valve to reset. The piston continues to compress air during the remainder of the upward stroke until the cylinder pressure is great enough to open the discharge valve against the valve spring pressure. Once the discharge valve is open, the air compressed in the cylinder is discharged until the piston completes the stroke.

Two-Stage Piston Compressor

To avoid an excessive rise in temperature, multistage compressors are provided with intercoolers. These heat exchangers remove the heat of compression from the gas and reduce its temperature to approximately the temperature existing at the compressor intake. Such cooling reduces the volume of gas going to the high pressure cylinders, reduces the power required for compression, and keeps the temperature within safe operating limits.

These compressors can generate high pressures than single-stage compressors. The most common type is two-stage or double-acting compressors. This style of reciprocating air compressor utilizes a double-acting piston (compression takes place on both sides of the piston), piston rod, crosshead, connecting rod and crankshaft. (Refer to Figure 4.25.) Double-acting

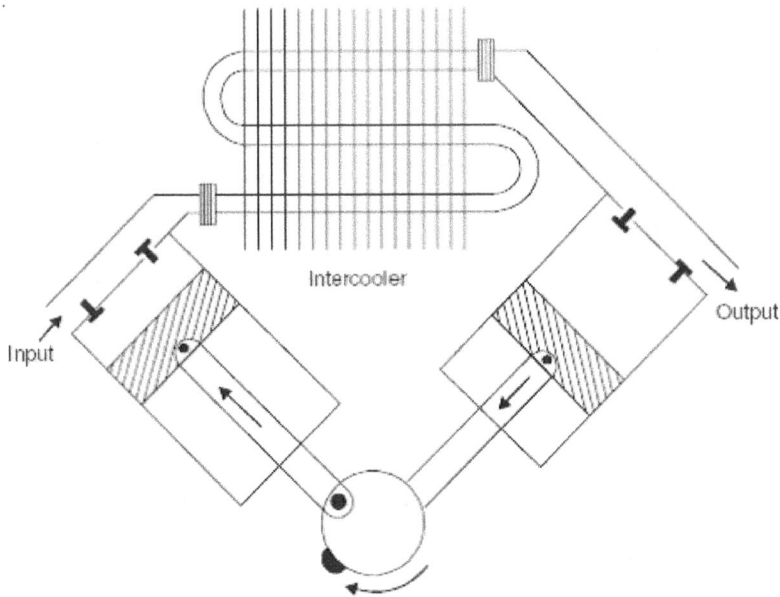

FIGURE 4.25 Two-Stage Reciprocating Compressor.

compressors are available in single- and multi-cylinder and single- and multi-stage configurations. Most double-acting air compressors are available either water-cooled or air-cooled. Lubrication of double-acting compressors is accomplished with a positive pressure oil pump. The oil required to both lubricate and protect the compressor is circulated via the oil pump to the cylinders and the crankshaft bearings. Discharge air temperatures in these machines may often exceed 300°C in some of the applications. On the whole, the reciprocating compressors run hotter than other types of compressors.

The compressed air from the first cylinder enters the second-stage cylinder at greatly reduced temperature after passing through the intercooler, thus improving efficiency as compared to that of a single-stage unit.

Advantages

■ Highest efficiency.
■ Longest service life.
■ Field serviceability.

Disadvantages

■ Highest initial cost.
■ High installation cost.
■ High maintenance cost

DIAPHRAGM COMPRESSOR

A diaphragm compressor (also known as a membrane compressor) is a variant of the conventional reciprocating compressor. The compression of gas occurs by the movement of a flexible membrane, instead of an intake element. The back and forth movement of the membrane is driven by a rod and a crankshaft mechanism. (Refer to Figure 4.26.) Only the membrane and the compressor box come in touch with the gas being compressed. Diaphragm compressors are used for hydrogen and compressed natural gas (CNG) as well as in a number of other applications.

The oil pressure required to bend the diaphragm is generated by a crank drive with a reciprocating piston. During the compression stroke, the piston pushes oil from the cylinder into the diaphragm head where it flows through the perforated plate to the back side of the diaphragm. The diaphragm is thus forced to bend into the concave diaphragm head cover surface. As the piston moves back, it pulls the diaphragm against the surface of

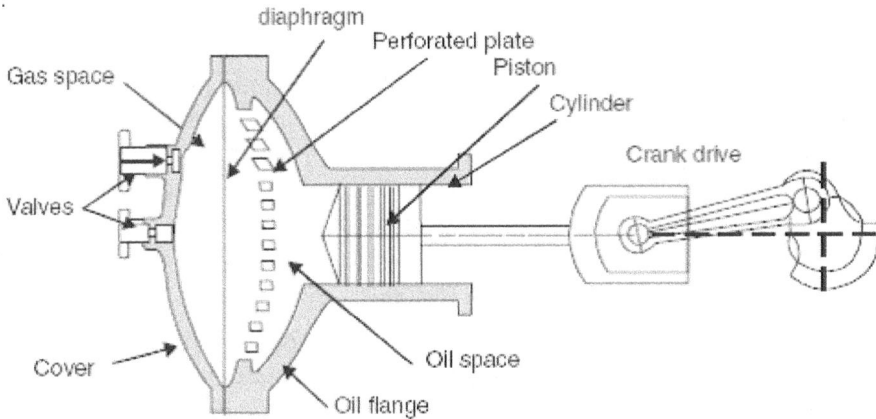

FIGURE 4.26 Diaphragm Compressor.

the perforated plate, which is also concave. So the oscillation frequency of the diaphragm corresponds to the compressor speed.

DYNAMIC COMPRESSORS

These compressors raise the pressure of air or gas by imparting velocity, energy, and converting it to pressure energy. First, rapidly rotating impellers (similar to fans) accelerate the air. Then, the fast flowing air passes through a diffuser section that converts its velocity head into pressure by directing it into a volute. Dynamic compressors include centrifugal and axial types. These types of compressors are widely used in the chemical and petroleum refinery industries for specific services. They are also used in other industries such as the iron and steel industry, etc. Compared to positive displacement type compressors, dynamic compressors are much smaller in size and produce much less vibration.

Centrifugal/Axial Compressors

The results accomplished by centrifugal compressors are the same as by previously explained compressor types, but the centrifugal compressors go about it in an entirely different way. Whereas reciprocating and screw compressors compress air by squeezing the air from a large volume into a smaller one, centrifugal compressors raise pressure by increasing the air's velocity. For this reason, centrifugal compressors are referred to as dynamic compressors.

Centrifugal compressors raise the pressure of air by imparting velocity, using a rotating impeller, and converting it to pressure. Each stage of compression in a centrifugal compressor consists of an impeller which rotates and a stationary inlet and discharge section. The air enters the eye of the impeller, designated D as shown in Figure 4.27. As the impeller rotates, air is thrown against the casing of the compressor. The air becomes compressed as more and more air is thrown out to the casing by the impeller blades. The air is pushed along the path designated A, B, and C. The pressure of the air is increased as it is pushed along this path.

Centrifugal compressors produce high-pressure discharge by converting angular momentum imparted by the rotating impeller (dynamic displacement). In order to do this efficiently, centrifugal compressors rotate at higher speeds than the other types of compressors. These types of compressors are also designed for higher capacity because flow through the compressor is continuous. Compare to other type of compressors, axial flow compressors are mainly used for applications where the head required is low.

Advantages

- High-quality air.
- Moderate efficiency.
- Longer service life.

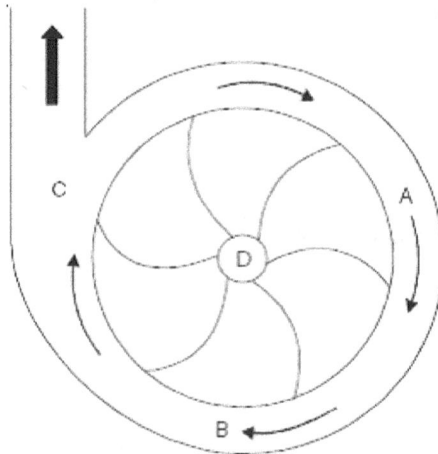

FIGURE 4.27 Centrifugal Compressor.

Disadvantages

- Higher initial cost.
- Must be water-cooled.
- Airflow is sensitive to changes in ambient conditions.

Axial Flow Compressor

The axial flow type air compressor is essentially a large capacity, high speed machine with characteristics quite different from the centrifugal air compressor. Axial flow compressors are used mainly as compressors for gas turbines. The component of an axial flow compressor consists of the rotating element that constructs from a single drum to which are attached several rows of decreasing height blades having airfoil cross sections. Between each rotating blade row is a stationary blade row. All blade angles and areas are designed precisely for a given performance and high efficiency. The efficiency in an axial flow compressor is higher than the centrifugal compressor. The operation of the axial flow compressor is a function of the rotational speed of the blades and the turning of the flow in the rotor. The stationary blades (stator) are used to diffuse the flow and convert the velocity increased in the rotor to a pressure increase. One rotor and one stator make up a stage in a compressor.

COMPARISON OF DIFFERENT COMPRESSORS

Iterm	*Reciprocating*	*Rotary Vane*	*Rotary Screw*	*Centrifugal*
Efficiency at full load	High	Medium-high	High	High
Efficiency at part load	High due to to staging	Poor: below 60% of full load	Poor: below 60% of full load	Poor: below 60% of full load
Efficiency at no load (power as % of full load)	High (10% – 25%)	Medium (30% – 40%)	High-Poor (25% – 60%)	High-Medium (20% – 30%)
Noise level	Noisy	Quiet	Quiet-if enclosed	Quiet
Size	Large	Compact	Compact	Compact
Oil carry over	Moderate	Low-medium	Low	Low
Vibration	High	Almost none	Almost none	Almost none

(Continue)

COMPARISON OF DIFFERENT COMPRESSORS (*Continued*)

Iterm	*Reciprocating*	*Rotary Vane*	*Rotary Screw*	*Centrifugal*
Maintenance	Many wearing parts	Few wearing parts	Very few wearing parts	Sensitive to dust in air
Capacity	Low - high	Low - medium	Low - high	Medium - high
Pressure	Medium - very high	Low - medium	Medium - high	Medium - high

SPECIFICATIONS OF COMPRESSORS

Following parameters are taken into account while selecting a compressor:

- Pressure range (psi)
- Flow rate
- Receiver size (gallons)
- Power supply (may be electric or mechanical)
- Size of installation
- Number of stages (in case of reciprocating)

EXERCISES

1. Classify various types of pumps and compressors.

2. Differentiate between fixed displacement and variable displacement pumps.

3. State at least four points of differences between positive and non positive displacement devices.

4. List the types of rotary pumps.

5. List the characteristics of gear and vane pumps.

6. Name the two basic types of rotary pumps.

7. With the help of a sketch, explain how a vane pump works.

8. Sketch constructional details of a variable capacity axial piston pump, label its components, and explain how it works. Give the standard graphical symbol for such a device.

9. Differentiate between the screw and lobe compressors.

10. Differentiate between external and internal gear pumps with the help of a sketch.

11. With the help of a sketch, explain the construction and working of a peristaltic pump.

12. Compare different types of compressors.

13. Discuss the construction of a compressor.

14. How can reciprocating compressors be classified?

15. What are the properties of a double-acting compressor?

16. What is meant by dynamic compressors?

17. What are the various methods of varying the flow rate of a pump? Illustrate with the help of sketches.

18. List the various parameters for selecting various pumps and compressors.

19. What are the factors to be considered while sizing a compressor?

20. Name any five applications of compressed air.

21. Define screw pump. Explain how it works, the types, and the characteristics.

22. Draw the cross sectional diagram of a two-stage compressor with an intercooler.

23. Sketch explanatory diagrams of any one type of air compressor.

CHAPTER 5

FLUID ACCESSORIES

INTRODUCTION

Fluid Accessories are divided into:

- Pneumatic Accessories
- Hydraulic Accessories.

These are discussed separately in the following paragraphs.

PNEUMATIC ACCESSORIES

AIR RECEIVER

Air receivers are tanks used for compressed air storage and are recommended for all compressed air systems. Using air receivers of unsound or questionable construction can be very dangerous. The vessel should be fitted with a safety valve, pressure gauge, drain, and inspection covers for checking or cleaning inside. Air receivers serve several important purposes:

- Eliminate pulsations from the discharge line.
- Separate some of the moisture; oil and solid particles that might be present from the air as it comes from the compressor or that may be carried over from the after cooler.
- Help reduce dew point and temperature spikes that follow regeneration.
- Offer additional storage capacity made to compensate for surges in compressed air usage.
- Contribute to reduce energy costs by minimizing electric demand charges.

Air receivers are of two types:

1. Wet Air Receivers
2. Dry Air Receivers

1. Wet Air Receivers: Wet receivers provide additional storage capacity and reduce moisture. The large surface area of the air receiver acts as a free cooler, which removes the moisture. Because the moisture is being reduced at this point in the system, the load on filters and dryers will be reduced.

2. Dry Air Receivers: When sudden large air demands occur, dry air receivers should have adequate capacity to minimize a drop in system air pressure. If these pressure drops were not minimized here, the performance of air dryers and filters would be reduced because they would no longer be operating within their original design parameters.

AFTERCOOLER

When free air is drawn into a compressor, all the moisture and dirt floating around in the free air is also drawn into the compressor. Once inside, the air, moisture, and dirt are compressed. The process of compression generates heat because air molecules collide more frequently in a smaller space. Heat and moisture are undesirable in a compressed air system. The aftercooler and the air dryer work to cool the air and remove moisture.

Compressed air discharges from the compressor and enters the aftercooler. *Aftercoolers* are heat exchangers that cool air to within 5-20°F of ambient air temperature immediately after it is discharged from the compressor. The *aftercooler* is a device containing a series of tubes and fins, which uses ambient air to condense out moisture in the compressed air. This condensation process also works to cool the air. Condensed water should be removed from the aftercooler using drain traps. The aftercooler removes approximately 80% of the moisture. However, the remaining 20% is still not acceptable air quality in a manufacturing environment. A large portion of the remaining 20% moisture is removed by an air dryer or a series of air dryers.

Almost all industrial systems require aftercooling. In some systems, aftercoolers are an integral part of the compressor package, while in other systems the aftercooler is a separate piece of equipment. Some systems have both. The aftercooler should be located as close as possible to the discharge of the compressor. Functions of compressed air aftercooler are:

- Cool air discharge from air compressors via the heat exchanger.
- Reduce risk of fire (hot compressed air pipes can be a source of ignition).
- Reduce compressed air moisture level.
- Increase system capacity.
- Protect downstream equipment from excessive heat.

Types of Aftercoolers

There are two basic types of air aftercoolers:

1. Air-cooled
2. Water-cooled

1. Air-Cooled Aftercooler

Air-cooled aftercoolers use ambient air to cool the hot compressed air (Refer to Figure 5.1). The compressed air enters the air-cooled aftercooler. The compressed air travels through either finned tubes or corrugated aluminum sheets of the aftercooler while ambient air is forced over the cooler

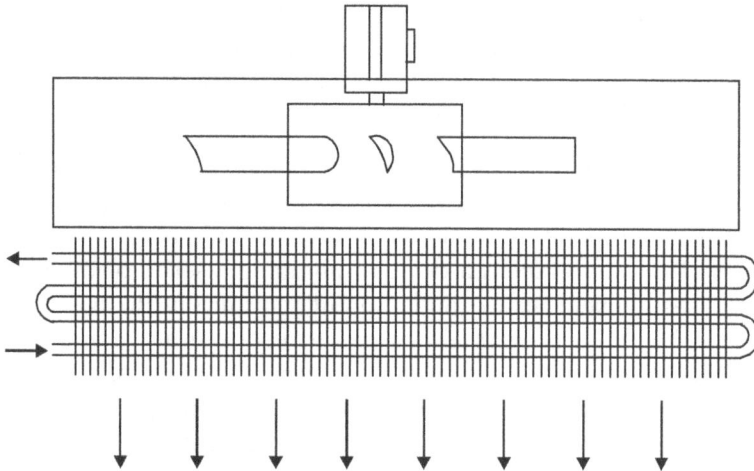

FIGURE 5.1 Air Cooled Aftercooler.

by a motor-driven fan. The cooler, ambient air removes heat from the compressed air.

2. *Water-Cooled Pipe Line Aftercooler*

Refer to Figure 5.2. The pipeline aftercooler consists of a shell with a bundle of tubes fitted inside. Typically the compressed air flows through the tubes in one direction as water flows on the shell side in the opposite direction. Heat from the compressed air is transferred to the water. Water vapor forms as the compressed air cools. The moisture is removed by the moisture separator and drain valve. A modulating valve is recommended to maintain a consistent temperature and reduce water consumption. The disadvantages of a water-cooled aftercooler include high water usage and difficult heat recovery. Advantages of using a water-cooled aftercooler include better heat transfer and no requirement of electricity.

AIR DRYER

Water vapor is contained in high-pressure air. In vapor form, the water does little damage in most components. But if the water is allowed to condense in the system, it can cause great damage. Drying of compressed air is achieved by passing it through dryers. Compressed air dryers reduce the quantity of water vapor, liquid water, hydrocarbon, and hydrocarbon vapor

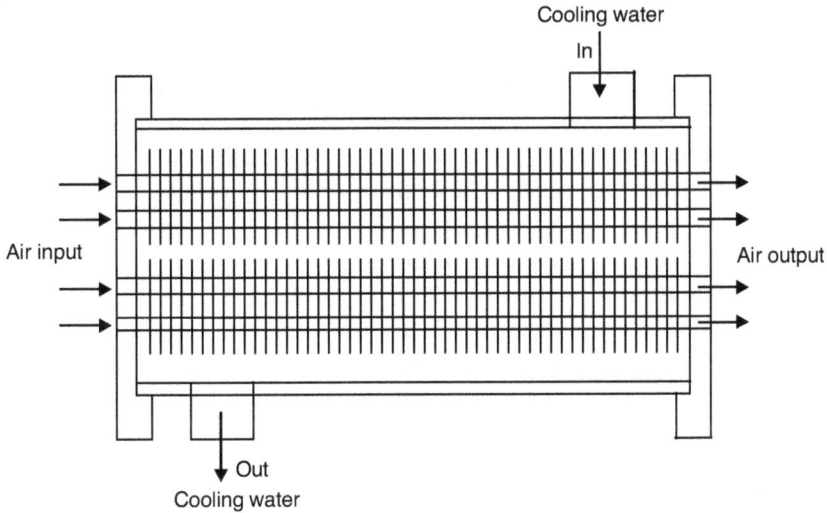

FIGURE 5.2 Water-Cooled Aftercooler.

in compressed air. Moisture in compressed air is harmful. Water damages a compressed air system in several ways:

- *Erosion:* Water mist erodes piping, valves, and other system components.
- *Corrosion:* Mist condenses and combines with salts and acids within the system forming highly corrosive solutions.
- *Microbial Contamination:* Moisture supplies a growth medium for bacteria and mold, which produce acidic waste and can be a health threat.
- *Freezing:* Water can freeze in compressed air lines shutting down the system.

The result is lower productivity, increased maintenance, and higher operating costs. Compressed air is dried to protect the system's piping and process equipment. Dry air also protects against lost product. Most pneumatic equipment has a recommended operating pressure, dryness level, and a maximum operating temperature.

Types of Compressed Air Dryers

The most common types of air dryers are:

1. Refrigerated Air Dryer
2. Desiccant Air Dryer
3. Deliquescent Air Dryer

1. Refrigerated Air Dryers

Refrigerated air dryers are the most economical types of dryer that cool air to 35-50°F, condense, mechanically separate, and discharge the water, and then reheat the air. These utilize a mechanical refrigeration system to cool the compressed air and condense water and hydrocarbon vapor. The saturated, moist air entering is precooled by the exiting cooled air in the precooler/ reheater. Further cooling is provided by refrigerant in the air-to-refrigerant heat exchanger. The separator removes condensed water droplets, oil, and solid particles from the air stream. Condensate is automatically discharged by a condensate drain trap. The compressed air, now free of liquid moisture, is reheated in the air-to-air precooler/reheater and discharged into the compressed air system.

2. Desiccant Air Dryers

Desiccant dryers utilize chemicals beads, called desiccant, to absorb water vapor from compressed air (Refer to Figure 5.3). Silica gel, activated alumina, and molecular sieve are the most common desiccants used. (Silica gel or activated aluminas are the preferred desiccants for compressed air dryers.) A desiccant dries air by absorbing moisture on its surface and holding the water as a mono or bimolecular film. The

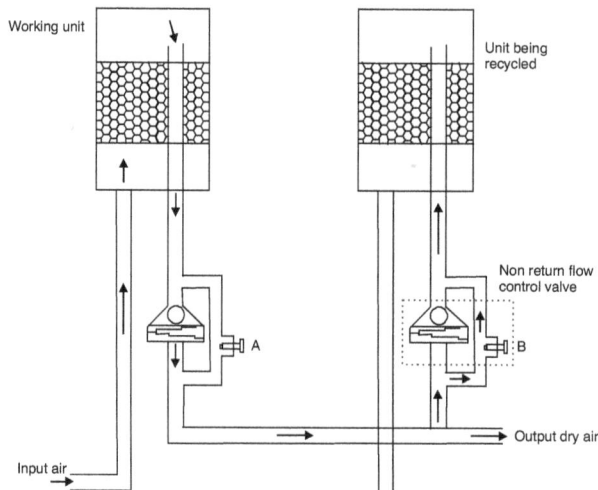

FIGURE 5.3 Desiccant Air Dryer.

method of regeneration, the process of removing absorbed water from the desiccant, is the primary distinguishing feature among the various types of desiccant dryers. Most regenerative desiccant dryers are dual-chamber systems with one chamber on-stream drying the compressed air while the other is off-stream being regenerated. There are three ways to regenerate a desiccant: with air, internal or external heaters, or a heat pump.

3. Deliquescent Air Dryer

Deliquescent dryers are simply large pressure vessels filled with a chemical having an affinity for water (Refer to Figure 5.4). Salt, urea, and calcium chloride are the common chemicals used. As the compressed air passes through the vessel, the salt dissolves in the water vapor and drips to the bottom of the tank where it is drained. The dried air is then discharged through the outlet port at the same temperature at which it entered.

These are the least expensive dryers to purchase and maintain because they have no moving parts and require no power to run, however, they require that the salt be replenished regularly. In addition, the corrosive salt solution can cause drain traps to clog.

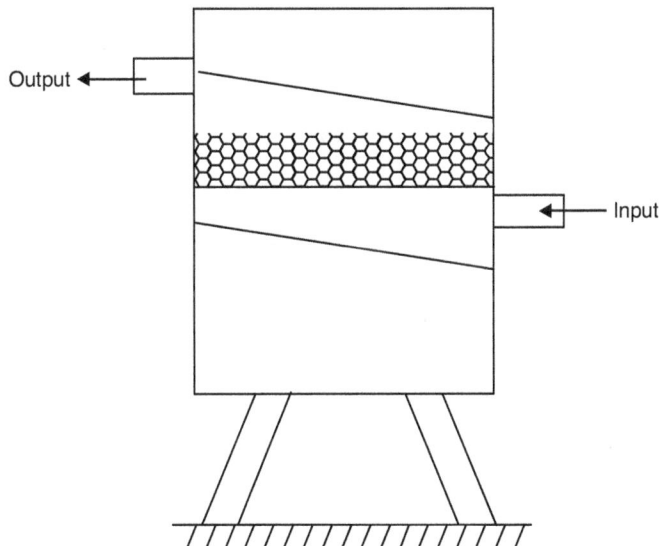

FIGURE 5.4 Deliquescent Air Dryer.

AIR FILTER

It is inevitable that impurities will make their way into the air distribution lines in any system. Pipe scale, rust, moisture, compressor oil, pipe compound, and dirt are some of the contaminants that can damage valve parts and other close fitting parts of downstream devices. In the compressed air system, hard particles assault equipment and piping. The result is damage to the system. Particles in a compressed air system can:

- Plug orifices of sensitive pneumatic instrumentation.
- Wear out seals.
- Erode system components.
- Decrease air dryer capacity.
- Foul heat transfer surfaces.
- Reduce air tool efficiency.
- Damage finished products.

Compressed air contaminants combine inside the air system forming sludge. Sludge clogs pipes and valves causing valves to jam. A filter will remove all foreign matter and allow clean dry air to flow freely. Clean, dry air protects the air system, reduces maintenance costs, and increases finished product yields. It should be installed in the line upstream from all working devices and in such a way that it cannot be bypassed to avoid damage to those devices. The filter capacity should be large enough to handle the required flow of air. In order to properly size a filter for a particular application, the maximum allowable pressure drop that can be caused by the filter should be established.

Construction and Working of Pneumatic Filter

(Refer to Figure 5.5.) Construction of air filter is shown along with its major parts. Air enters from an inlet port and is directed to inside the filter. The compressed air has to pass through the filter element, as there is no other way to pass out. In this course of air passage, the air comes in contact with the element of the filter and finally after removal of all the impurities from air, the compressed air is directed outside by means of an outlet port. Moisture present in the air is blocked by the filter and it can be manually drained out by simply operating the manually operated drain plug.

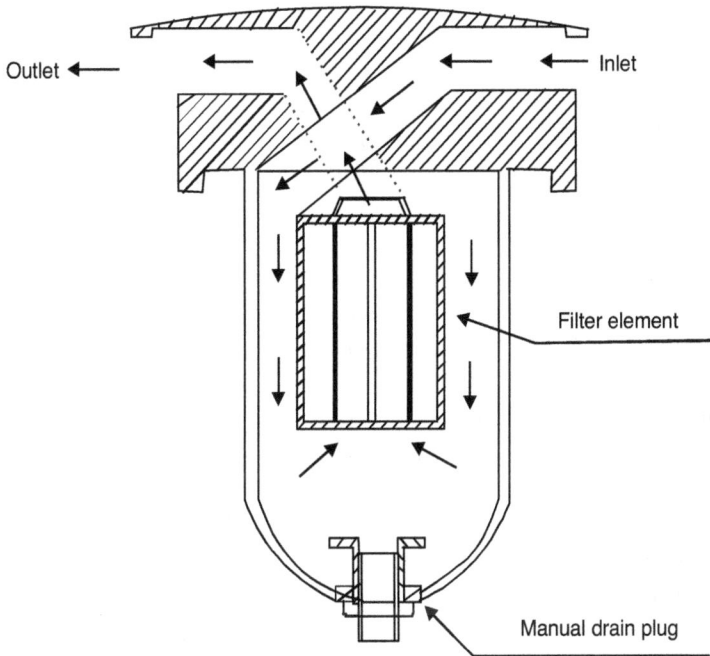

FIGURE 5.5 Pneumatic Filter.

Types of Air Filters

1. General-Purpose Filters

General-purpose filters remove harmful water condensate, pipe scale, dirt, and rust from your compressed air system. This prevents corrosive damage to compressed air equipment and finished products. Typically, general-purpose filters are installed upstream of regulators to prevent valve failure. They are also used as pre-filters to oil removing and coalescing filters to ensure high efficiency and long element life in applications such as paint spraying, instrumentation, and pharmaceutical.

2. Compressor Intake Filter

The first line of defense is the intake air filter, which reduces the bulk contaminant load protecting the compressor from dirt and solids.

3. Coarse Coalesce Filter

A coarse coalescing filter separates large water and oil droplets from the compressed air stream before the coalescing filter. The coarse coalescer also removes large solids.

4. Coalescing Filter

Coalescing filters function in a different way from general-purpose filters. Air flows from inside to outside through a coalescing media. Coalescing by definition means "to come together." It is a continuous process by which small aerosols come in contact with the fibers in the filter media uniting with other collected aerosols and growing to emerge as a droplet on the downstream surface of the media which by its weight is gravitationally drained away. For maximum performance and efficiency, coalescing filters should be preceded by a general-purpose filter.

5. Vapor Filter (Charcoal Filter)

A vapor filter removes organics from the air stream. Organics like taste and odor need to be removed from breathing air systems. In general, industrial applications of vapor compressed air filters remove hydrocarbons and other organic chemical vapors from the air system. Depending on the airflow, vapor filters need to be replaced every few months because of the effectiveness of the activated carbon degrades as it absorbs.

PRESSURE REGULATOR

Pneumatic equipment is designed to operate properly at a certain pressure. Although most equipment will run at pressure higher than recommended. The excess force, torque, and wear can shorten the equipment's life and waste compressed air. A regulator will provide a constant set flow of air pressure at its outlet, thus assessing optimum operation and life of the downstream equipment. An airline regulator is a specialized control valve, which reduces upstream supply pressure level to a specified constant downstream pressure. The size of the regulator is determined by the downstream flow and pressure requirements.

The diagram of a pressure regulator is shown in Figure 5.6. It works on the principle of pressure differences. There are two opposing forces, which maintain a constant output pressure at outlet. Spring force applying downward pressure and the upward force due to fluid pressure act on the downside of diaphragm.

Air at fluctuating pressures is allowed to enter the inlet port. Air is entered to a spacious chamber by applying force against a spring-loaded diaphragm and poppet assembly. The spring force is adjusted with the help

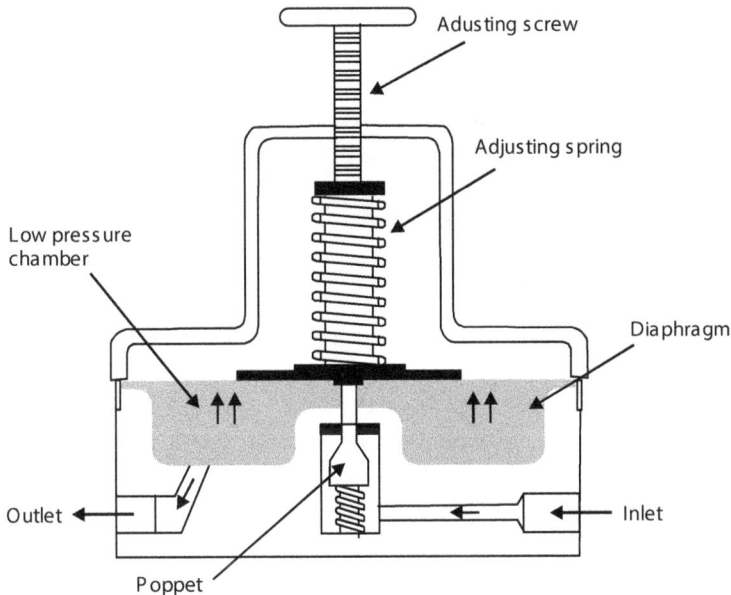

FIGURE 5.6 Pressure Regulator.

of an adjusting screw, depending on how much pressure is required at outlet. If the pressure in the chamber is already high, the entering air has to apply more and more pressure on the poppet and diaphragm, against the force of the spring. On the other hand, if the pressure in the chamber is less than the spring force, the air from an inlet port enters freely and quickly without any resistance and maintains the required pressure at which the spring is adjusted. It is important to add to a pneumatic system a pressure regulator that:

▣ Supplies air at constant pressure regardless of flow variation or upstream pressure.

▣ Helps operate the system more economically by minimizing the amount of pressurized air that is wasted. (This happens when the system operates at pressures higher than needed for the job.)

▣ Helps promote safety by operating the actuator at reduced pressure.

▣ Extends component life because operating at higher-than-recommended pressures increases wear rate and reduces equipment life.

▣ Produces readily controlled variable air pressures where needed.

▣ Increases operating efficiency.

AIR LUBRICATOR

Most pneumatic system components and air tools require oil lubrication for proper operation and long service life. Too little oil can cause excessive wear and premature failure. Too much oil is wasteful and can contaminate, particularly when carried over with the air exhaust. Pneumatic equipment can be lubricated by the use of an airline lubricator. Filtered and regulated air enters the lubricator and is mixed with oil in an aerosol mist. The air is then routed to the operating system.

The working of the lubricator is based on the principle of venturi effect, as shown in Figure 5.7. The compressed air from the compressor, filter, and regulator is passed through the inlet port of lubricator. This high-pressure air has to pass through a converging and diverging passage. While passing from the converging portion, pressure head goes on decreasing and velocity head goes on increasing. When the air reaches the middle portion of nozzle, velocity becomes comparatively very high, due to which suction head is created in the suction pipe and oil is sucked up and mixed in the air passing at high velocity.

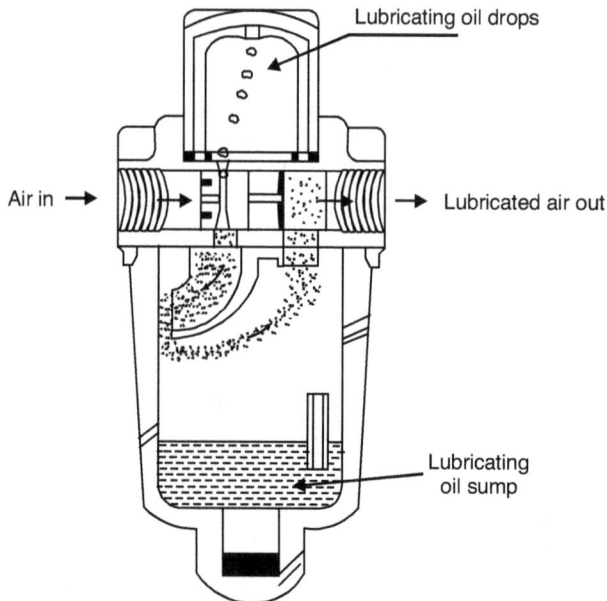

FIGURE 5.7 Air Lubricator.

AIR SERVICE UNIT (F.R.L.)

FRL stands for filter, regulator, and lubricator. Filters, regulators, and lubricators can be combined to ensure optimum compressed air preparation for a specific pneumatic system as shown in Figure 5.8. They form a unit that will prepare the condition of compressed air before delivering it to pneumatic equipment or machinery. This ensures the air supply is clean and dry, the pressure is at the correct level, and fine particles of oil are carried in the air to lubricate the wearing parts between the valves, cylinder, and tools.

PIPELINE LAYOUT

The purpose of the compressor piping system is to deliver compressed air to the points of usage. The compressed air needs to be delivered with enough volume, appropriate quality, and *pressure* to properly power the components that use the compressed air. Compressed air is costly to manufacture. A poorly designed compressed air system can increase energy costs, promote equipment failure, reduce production efficiencies, and increase maintenance requirements. It is generally considered true that any

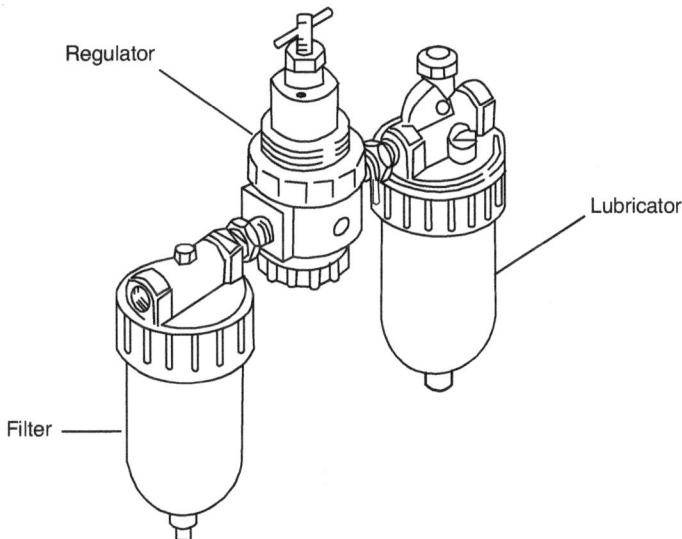

FIGURE 5.8 Air Service Unit (F.R.L.).

additional costs spent improving the compressed air piping system will pay. Discharge piping from a compressor without an integral aftercooler can have very high temperatures. The pipe that is installed here must be able to handle these temperatures. The high temperatures can also cause thermal expansion of the pipe, which can add stress to the pipe.

The layout of the system can also affect the compressed air system. A very efficient compressed air piping system design is a loop design. The loop design allows airflow in two directions to a point of use. This can cut the overall pipe length to a point in half that reduces pressure drop. In many cases, a balance line is also recommended which provides another source of air. Reducing the velocity of the airflow through the compressed air piping system is another benefit of the loop design. In cases where there is a large volume user, an auxiliary receiver can be installed. This reduces the velocity, which reduces the *friction* against the pipe walls and reduces pressure drop. Receivers should be positioned close to the far ends or at points of infrequent heavy use of long distribution lines. Many peak demands for air are short-lived, and storage capacity near these points helps avoid excessive.

Figure 5.9 shows the layout of compressed air pipeline in an industry where compressed air is used in almost all the operations starting from

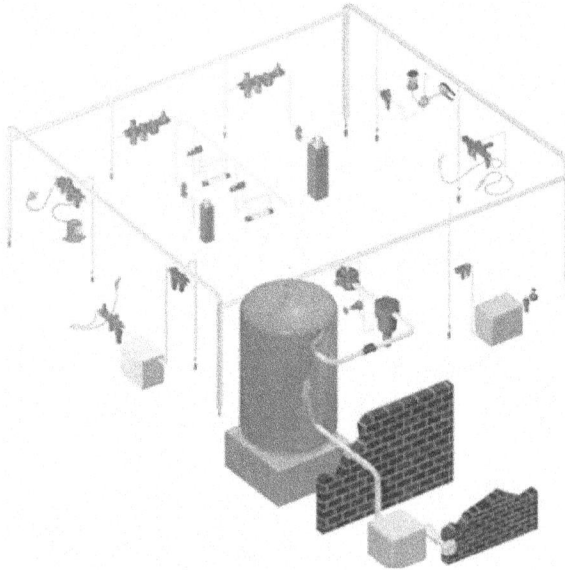

FIGURE 5.9 Pipeline Layout.

running the various types of tools like pneumatic drills or pneumatic hammers to special purpose automatic machines.

Condensate Control

Condensation control must be considered when installing a compressed air piping system. Drip legs should be installed at all low points in the system. A drip leg is an extension of pipe below the airline, which is used to collect condensation in the pipe. At the end of the drip leg, a drain trap should be installed. Preferably an automatic drain will be used.

To eliminate *oil, condensate*, or cooling water (if the water-cooled aftercooler leaks), a low point drain should be installed in the discharge pipe before the aftercooler. Be sure to connect the aftercooler outlet to the separator inlet when connecting the aftercooler and the moisture separator together. If they are not connected properly, it will result in either poor after cooling or poor separation. Another method of controlling the condensation is to take all branch connections from the top of the airline. This eliminates condensation from entering the branch connection and allows the condensation to continue to the low points in the system.

Pressure Drop

Pressure drop in a *compressed air* system is a critical factor. Pressure drop is caused by friction of the compressed air flowing against the inside of the pipe and through valves, tees, elbows, and other components that make up a complete compressed air piping system. Pressure drop can be affected by pipe size, type of pipes used, the number and type of valves, couplings, and bends in the system. Each *header* or main should be furnished with outlets as close as possible to the point of application. This avoids significant *pressure* drops through the hose and allows shorter hose lengths to be used. To avoid carryover of condensed moisture to tools, outlets should be taken from the top of the pipeline. Larger pipe sizes, shorter pipe and hose lengths, smooth wall pipe, long radius swept tees, and long radius elbows all help reduce pressure drop within a compressed air piping system.

Pressure drop can be given by Darcy's equation which can be expressed mathematically as:

$$\Delta p = \lambda \cdot \frac{L}{D} \cdot \frac{\rho V^2}{2}$$

where the pressure loss due to friction Δp is a function of:

the coefficient of laminar, or turbulent flow, λ

the ratio of the length to diameter of the pipe, L/D

the density of the fluid, ρ

the velocity of the flow, V

Piping Material

Common piping materials used in a *compressed air* system include copper, aluminum, stainless steel, and carbon steel. Compressed air piping systems that are 2" or smaller utilize copper, aluminum, or stainless steel. Pipe and fitting connections are typically threaded. Piping systems that are 4" or larger utilize carbon or stainless steel with flanged pipe and fittings. Plastic piping may be used on compressed air systems; however caution must be used because many plastic materials are not compatible with all compressor lubricants. Ultraviolet light (sun light) may also reduce the useful service life of some plastic materials. Installation must follow the manufacturer's instructions.

It is always better to oversize the compressed air piping system you choose to install. This reduces pressure drop, which will pay for itself, and it allows for expansion of the system. Corrosion-resistant piping should be used with any compressed air piping system using oil-free compressors. A non-lubricated system will experience corrosion from the moisture in the warm air, contaminating products, and control systems, if this type of piping is not used.

SEALS

A seal is a ring-shaped component that is designed to prohibit or limit the leakage of fluid from a device. Various types of seals are used in applications that have constantly moving equipment, such as the rotating or reciprocating shafts and cylinders that are an essential part of hydraulic and pneumatic systems.

Classification of Seals

Seals are classified as:

- Based on design requirements
- Based on medium in which seal is used

- Based on material used
- Based on shapes.

1. Based on Design Requirements

    (a) Static Seals: Static seals are placed between surfaces, which do not move. Some form of pressure must be exerted to squeeze the surfaces together and force the seal material into the small imperfections in the surfaces. Some of the static seals are:

- *Gaskets:* Gaskets are cut out of thin sheets of material and placed between mating surfaces, which are then squeezed together by screws or bolts. The materials used are paper, copper, brass, rubber, and so on. Typical applications are between flanges on pipes and flanges on the fluid port of a pump or motor.
- *Rings:* Rings are placed in grooves between the mating surfaces so that they are squeezed when the surfaces are pulled together. The rings are usually circular in section and are then called O-rings. The materials used are natural or synthetic rubbers. Typical applications include flanges, cylinder end caps, and motor bodies.

    (b) Sliding Seals: Sliding seals are mainly used with cylinders to prevent fluid escaping around a piston rod or from passing from one side of a piston to the other. All sliding seals are rings but many different types exist. These may be solid rings such as O-rings or rings with rectangular sections.

    (c) Rotary Seals: Rotary seals are used on pumps and motors to prevent fluid leaking out through the gap between the shaft and the shaft bearing. They are designed with a spring-loaded lip, which presses to the shaft. Oil leaking into the space behind the seal will force the lip even tighter but this space should be drained to prevent the seal from being blown out by pressure.

2. Based on Medium in which Seal is Used

    (a) Hydraulic Seals
    (b) Pneumatic Seals

3. Based on Material Used

    The materials seals are made from generally include rubber or metal, and in some cases, leather or felt. Some of the rubber materials used to create seals include nitrile, silicone, natural rubber, butyl, and styrene butadiene. Additionally, quite a few manufacturers use their own unique materials that they have specially developed for their products. Common seals materials can be categorized as: leather, metal, polymers, elastomers and plastics, asbestos, nylon, etc.

4. Based on Shapes

Various varieties of geometric shapes or configurations of seals are:

1. O-ring seal
2. Quad-ring seal
3. T-ring seal
4. V-ring seal
5. V-cup ring
6. Hat ring

Pneumatic Seals

The purpose of a pneumatic seal is to contain the working fluid (air) within the pneumatic unit and to keep external contamination out. Seals for pneumatic cylinders and valves are required to withstand lower pressures than hydraulic seals. Pneumatic seals are used at higher temperatures than hydraulic seals. The term "pneumatic seals" actually describes a class of seals that are used in applications with either rotary or reciprocating motions. Pneumatic seals are exposed to air with a minimum amount of lubrication. They are often used in pneumatic cylinders and valves and usually are not under high pressure. However, pneumatic seals may be exposed to high operating speeds. Rod seals, piston seals, u-cups, vee-cups, and flange packing are just some of the sealing designs that can be used for a pneumatic seal. As in the case of hydraulic seals sometimes a composite seal is used as a pneumatic seal. Rotary applications need only one pneumatic seal (single acting) because it can seal in the one axial direction the application is moving. However, a reciprocating application will need two pneumatic seals (double acting) one for each of the directions.

HYDRAULIC ACCESSORIES

HYDRAULIC FLUIDS

The term hydraulic fluid generally refers to a liquid used as a power transmitting medium. Choosing a correct fluid for a hydraulic system is important. In a hydraulic system fluid has four basic functions:

1. **Power transmission:** The primary purpose of any hydraulic fluid is to transmit power mechanically throughout a hydraulic power system.

2. **Lubrication:** Hydraulic fluids must provide the lubricating characteristics and qualities necessary to protect all hydraulic system components against friction and wear, rust, oxidation, corrosion, and demulsibility. These protective qualities are usually provided through the use of additives.

3. **Sealing:** Many hydraulic system components, such as control valves, operate with tight clearances where seals are not provided. In these applications, hydraulic fluids must provide the seal between the low-pressure and high-pressure side of valve ports. The amount of leakage will depend on the closeness or the tolerances between adjacent surfaces and the fluid viscosity.

4. **Cooling:** The circulating hydraulic fluid must be capable of removing heat generated throughout the system.

Properties of Hydraulic Fluids

A good hydraulic fluid comprises of:

- **Good Lubricity:** A hydraulic system has various components that contain surfaces that are in close contact and move in relation to each other. A good hydraulic fluid must protect against wear and separate and lubricate such surfaces.

- **Stable Viscosity:** Viscosity is a vital fluid property that varies with temperature and pressure. Fluids having large changes of viscosity with temperature are commonly referred as low viscosity index fluids and those having small changes of viscosity with temperature are known as high viscosity index fluids.

- **Chemical and Physical Stability:** The characteristics of a fluid should remain unchanged during an extended useful life. Because many aspects of stability are chemical in nature, the temperature to which the fluid will be exposed is an important criterion in the selection of a hydraulic fluid.

- **System Compatibility:** The hydraulic fluid should be inert to materials used in or near the hydraulic equipment. If the fluid in any way attacks, destroys, dissolves, or changes parts of the hydraulic system, the system may lose its functional efficiency and may start malfunctioning.

■ **Good Heat Dissipation:** Pressure drops, mechanical friction, fluid friction, leakages, all generate heat. The fluid must carry the generated heat away and readily dissipate it to the atmosphere or coolers.

■ **Flash Point:** The flash point of hydraulic oil is defined as the temperature at which flashes will be generated when the oil is brought into contact with any heated matter.

■ **Prevent Rust Formation:** Moisture and oxygen cause rusting of iron parts in the system that can lead to abrasive wear of system components and also act as a catalyst to increase the rate of oxidation of the fluid. Fluids with rust inhibitors minimize rust formation in the system.

■ **Demulsibility:** Demulsibility is the ability of a fluid to separate out water. Excessive water in the oil promotes the collection of contaminants, cause sticky valves, and accelerated wear.

Types of Hydraulic Fluids

Hydraulic fluids are a large group of liquids made of many kinds of chemicals. They are used in automobile automatic transmissions, brakes, and power steering; fork lift trucks; tractors; bulldozers; industrial machinery; and airplanes. Water was the first hydraulic fluid used during the early stages of the industrial revolution. The common types of fluid used in hydraulic systems are:

1. **Petroleum:** Petroleum-based oils are the most commonly used stock for hydraulic applications where there is no danger of fire, no possibility of leakage that may cause contamination of other products, no wide temperature fluctuations, and no environmental impact.

2. **Fire resistant:** In applications where fire hazards or environmental pollution are a concern, water-based or aqueous fluids offer distinct advantages. The fluids consist of water-glycols and water-in-oil fluids with emulsifiers, stabilizers, and additives.

(a) **Water-glycol:** Water-glycol fluids contain from 35 to 60 percent water to provide the fire resistance, plus a glycol antifreeze such as ethylene, diethylene, or propylene which is nontoxic and biodegradable, and a thickener such as polyglycol to provide the required viscosity. These fluids also provide all the important additives such as antiwear, foam, rust, and corrosion inhibitors.

(b) **Water-oil emulsions:** These are of two types:

    (*i*) *Oil-in-water.* These fluids consist of very small oil droplets dispersed in a continuous water phase. These fluids have low viscosities, excellent fire-resistance, and good cooling capability due to the large proportion of water. Additives must be used to improve their inherently poor lubricity and to protect against rust.

    (*ii*) *Water-in-oil.* The water content of water-in-oil fluids may be approximately 40 percent. These fluids consist of very small water droplets dispersed in a continuous oil phase. The oil phase provides good to excellent lubricity while the water content provides the desired level of fire-resistance and enhances the fluid cooling capability. Emulsifiers are added to improve stability. Additives are included to minimize rust and to improve lubricity as necessary. These fluids are compatible with most seals and metals common to hydraulic fluid applications.

(c) **Synthetic fire-resistant fluids:** These fluids are usually a blend that consists of phosphate esters, chlorinated hydrocarbons, and hydrocarbon based fluids. These fluids do not contain water or volatile materials, and they provide satisfactory operation at high temperatures without loss of essential elements (in contrast to water-based fluids). The fluids are also suitable for high-pressure applications.

HYDRAULIC RESERVOIR

Hydraulic systems need a finite amount of liquid fluid that must be stored and reused continually as the system works. This finite amount of liquid fluid is stored in a tank or reservoir. A reservoir is a storage area from where the hydraulic fluid is supplied to the system. Hydraulic reservoir performs several functions such as:

1. Heat dissipation: The hydraulic fluid absorbs the heat generated by the hydraulic systems. Hydraulic reservoir dissipates and transfers heat from the fluid by radiation and convection.

2. Settling basin: Hydraulic reservoir acts as a settling basin as the contaminants in the fluid settle down.

3. Reserve for operation: A reservoir should be able to hold the entire system fluid together with the reserve fluid for the operation. The reservoir capacity should be two to three times the pump delivery.

4. Accessory mounting: The reservoir surface can be used as a mounting platform to support the pump, motor, and other system components. This saves floor space.

Hydraulic reservoirs are usually rectangular, cylindrical, T-shaped, or L-shaped and made of steel, stainless steel, aluminum, or plastic. Hydraulic reservoirs vary in terms of capacity, but need to be large enough to accommodate the thermal expansion of fluids and changes in fluid level due to normal system operation. Large hydraulic reservoirs provide cooling and reduce recirculation. Some reservoirs include stationary or removable baffles to direct fluid flow and ensure proper circulation.

HYDRAULIC FILTER

Filters are components used in hydraulic systems for contamination control. Filter consists of porous material, which traps contaminants above a particular size in the system fluid. Hydraulic filters are available in a multitude of shapes, sizes, micron ratings, and construction materials. Hydraulic filters provide inbuilt protection and minimize hydraulic system breakdowns that are quite often caused by contamination. Efficient filtration helps prevent system failure and makes a significant contribution to low cost of ownership. These are fitted for both low- and high-pressure hydraulic applications. The filter used in a hydraulic system should be subjected to periodic and routine cleaning and maintenance. The life of a filter in a hydraulic system depends primarily on the system pressures, level of contamination, and nature of contaminants.

Construction and Working of Hydraulic Filter

Construction of a typical hydraulic filter is shown in Figure 5.10 along with its important parts. It works the same as a pneumatic filter. Hydraulic fluid entering from an inlet port is entered into the filter body and bound to pass out only from the filter element. After passing out from the filter element, all the impurities which were present in the hydraulic fluid gets trapped in the filter element and clean hydraulic fluid is obtained at outlet of the hydraulic filter. Hydraulic filters are designed to tolerate higher pressures then in case of pneumatics, *i.e.*, why they are sometimes called as heavy duty filters.

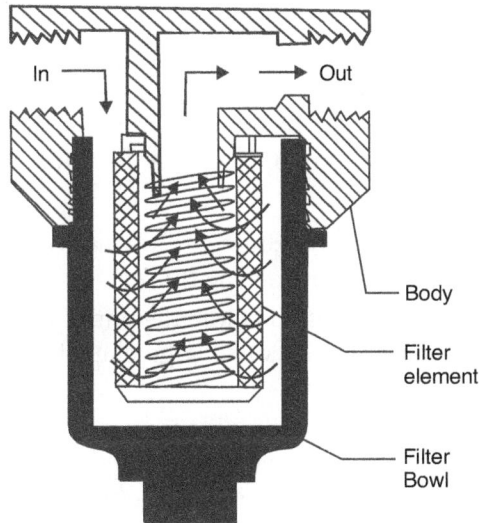

FIGURE 5.10 Hydraulic Filter.

Types of Filters

In a hydraulic system, there are two main types of filters that are frequently used. These are:

1. **Surface filters:** These are simple screens used to clean oil passing through the pores. The dirty unwanted particles are collected at the top surface of the screens when the oil is passed.

2. **Depth filters:** These are thick walled filter elements through which the oil is made to pass retaining the undesirable foreign particles. The capacity of depth filters is much higher than surface filters as much finer materials have a chance of being arrested by these filters.

Filters can also be classified as:

1. **Full-flow filter:** In a hydraulic system it is necessary that all the flow must be through the filter element. So, the oil must enter the filter element at its inlet side and get sent out through the outlet after crossing the filter element fully. In case of full flow filtration, the filter is sized to accommodate the entire oil flow at that part of the circuit.

2. **By-pass filter:** At times, the entire volume of oil need not be filtered and thus, only a portion of the oil is passed through the filter element. The main portion is directly passed without filtration through a restricted passage.

Filter Location

The location of a filter in the hydraulic system is of much significance. Various locations are used to arrange a filter either in the return line or in the pressure line. A filter may be used either in the intake side of the pump or in the outlet side of the pump. Generally, for a filter, the preferred locations are:

- **Return line filter:** It may reduce the amount of dirt ingested through the cylinder and seals from reaching the tank.
- **Intake filter:** These are fitted before the pump so that they can prevent random entry of contaminants like large chips into the pump and thus prevent damage to it.
- **Pressure filter:** A pressure filter is used sometimes at the pump outlet to prevent entry of contaminants generated in the pump, into other components like valves, etc. and thus help in avoiding the spread of such undesirable elements into the whole system. This will thus protect valves, cylinders, etc.
- **Final control filter:** To keep the large debris out of a component that can cause the component to fail through in-built arrangement or by additional protective design.

Filtering Material and Elements

The general classes of filter materials are mechanical, absorbent inactive, and absorbent active.

1. *Mechanical filters* contain closely woven metal screens or discs. They generally remove only fairly coarse particles.

2. *Absorbent inactive filters*, such as cotton, wood pulp, yarn, cloth, or resin, remove much smaller particles; some remove water and water-soluble contaminants. The elements often are treated to make them sticky to attract the contaminants found in hydraulic oil.

3. *Absorbent active materials*, such as charcoal and fuller's earth (a clay-like material of very fine particles used in the purification of mineral or vegetable-base oils), are not recommended for hydraulic systems.

PRESSURE GAUGES AND VOLUME METERS

Pressure gauges are used in liquid-powered systems to measure pressure to maintain efficient and safe operating levels. Pressure is measured in psi. Flow measurement may be expressed in terms of total quantity—gallons or cubic feet.

Pressure Gauges

Figure 5.11 shows a simple pressure gauge. Gauge readings indicate the fluid pressure set up by opposition of forces within a system. Atmospheric pressure is negligible because its action at one place is balanced by its equal action at another place in a system.

Flow Meters

Measuring flow depends on the quantities, flow rates, and the types of liquid involved. All liquid meters (flow meters) are made to measure specific liquids and must be used only for the purpose for which they were made. Each meter is tested and calibrated.

HYDRAULIC ACCUMULATOR

An accumulator is essentially a pressure storage reservoir in which a noncompressible hydraulic fluid is retained under pressure from an external source. Like an electrical storage battery, a hydraulic accumulator stores potential power, in this case liquid under pressure, for future conversion into useful work. That external source can be a spring, a raised weight, or a compressed gas. Because of their ability to store excess energy and release it when needed, accumulators are useful tools in developing efficient hydraulic systems. In certain circuit designs, the accumulator will permit a pump motor to be completely shut down for an extended period of time while the accumulator supplies the necessary fluid to the circuit. Accumulators are

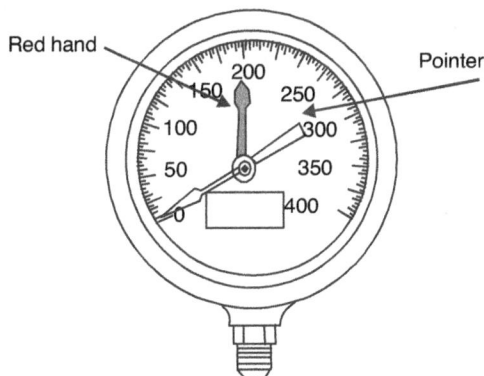

FIGURE 5.11 Pressure Gauge.

used in conjunction with hydraulic systems on large hydraulic presses, farm machinery, diesel engine starters, power breaks on airplanes, lift trucks, etc.

Applications/Functions of an Accumulator

Hydraulic accumulators are used in hydraulic systems as:

- **Compensation for large flow variation:** During operation of hydraulic installation, if requirements of supply of large volumetric flows appear, then these requirements can be supplied by an accumulator.

- **To smooth out pressure surges:** Pressure surges are caused in a hydraulic system when fluid flow is suddenly changed. The high inertia of the moving fluid sets up shock waves in the pipes causing hammer and vibration. Accumulators will absorb the energy and dampen out these surges thus reducing vibrations and shock.

- **To provide emergency power source:** The fluid energy stored in an accumulator may be sufficient to give an emergency supply in the case of major electrical failure causing the pumps to stop.

Types of Accumulators

There are three main types of accumulators used in hydraulic systems:

1. Weight-loaded Accumulators
2. Spring-loaded Accumulators
3. Gas charged Accumulators

1. Weight-Loaded Accumulator

The weight-loaded accumulator consists of a vertical cylinder containing fluid connected to the hydraulic system (Refer to Figure 5.12). The cylinder is closed by a piston on which a series of weights are placed that exerts a downward force on the piston and thereby energizes the fluid in the cylinder. The weight may be of some heavy material such as iron, concrete block, pig or scrap iron, etc. The pressure on the fluid depends on the weight on the piston and the size of the piston, and can be changed by adding or removing weights from the piston.

2. Spring-Loaded Accumulator

A spring-loaded accumulator works on the same principle as the weight-loaded accumulator. A spring-loaded accumulator consists of a cylinder

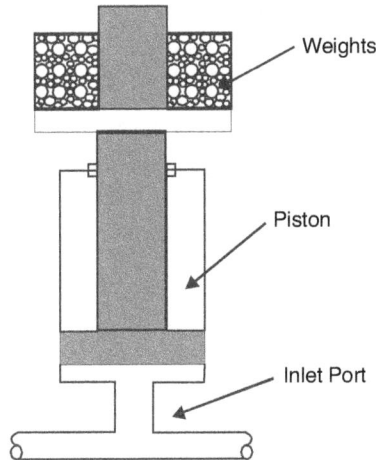

FIGURE 5.12 Weight Loaded Accumulator.

body, a movable piston, and a spring. The spring applies force to the piston. As fluid is pumped to it, the pressure in the accumulator is determined by the compression rate of the spring. The spring force increases as it is compressed by more fluid entering the chamber.

The advantages of these accumulators are that these are usually less expensive than the dead weight type and mounting is easy. They are built directly into the power unit. The disadvantage is that the spring force and the resulting pressure range cannot be easily adjusted. This accumulator is also not suitable for applications, which require large volume of fluids.

3. Gas Charged Accumulators

Gas charged or compressed gas accumulators consist of a cylinder with two chambers that are separated by an elastic diaphragm or by a floating piston. One chamber contains hydraulic fluid and is connected to the hydraulic line. The other side contains an inert gas under pressure that provides the compressive force on the hydraulic fluid. Inert gas is used because oxygen and oil can form an explosive mixture when combined under high pressure. As the volume of the compressed gas changes the pressure of the gas, and the pressure on the fluid, changes inversely. Gas charged accumulators are of two types:

1. Piston type
2. Bladder type

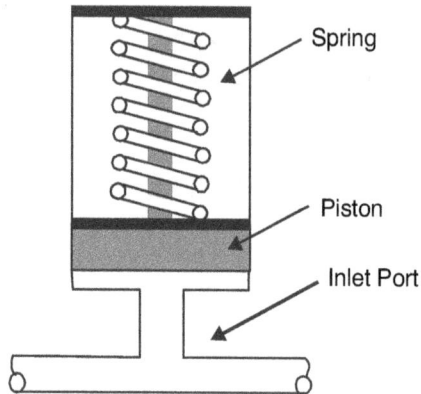

FIGURE 5.13 Spring Type Accumulator.

1. **Piston type:** These have an outer cylinder tube, end caps, a piston element, and sealing system. The cylinder holds fluid pressure and guides the piston, which forms the separating element between gas and fluid. Charging the gas side forces the piston against the end cover at the fluid end. As system pressure exceeds the minimum operating level for the accumulator, the piston moves and compresses gas in the cylinder.

2. **Bladder type:** It consists of a pressure vessel and an internal elastomeric bladder that contains the gas. The bladder is charged through a gas valve at the top of the accumulator, while a poppet valve at the bottom prevents the bladder from being ejected with the outflowing fluid. The poppet valve is sized so that maximum volumetric flow cannot be exceeded. To operate, the bladder is charged with nitrogen to a pressure specified by the manufacturer according to the operating conditions. When system pressure exceeds gas precharge pressure of the accumulator, the poppet valve opens and hydraulic fluid enters the accumulator. The change in gas volume in the bladder between minimum and maximum operating pressure determines the useful fluid capacity.

INTENSIFIER

Intensifiers, also known as boosters, use a large quantity of low-pressure fluid to produce a smaller quantity of higher-pressure fluid. The unit has a piston, which reciprocates automatically. The large piston is exposed to the force delivered by the low-pressure pump and this piston moves the large piston back and forth as the valve stem shifts. The small area piston on each end of the large piston forces oil into the system under high pressure. There are three classes of intensifiers: air-to-oil, oil-to-oil, and air-to-air. Hydraulic boosters can develop and maintain high pressure for long periods of time without using power or generating heat in the circuit. They deliver fluid only when the cylinder demands it. Because all the oil from the booster is directed to the cylinder, there are no relief-valve losses. Generally, boosters can be used as a hydraulic power source only in single cylinder applications. A booster can drive more than one cylinder if the cylinders work in unison. However, to sequence two or more cylinders, additional boosters are required. Three major steps are involved in selecting the right booster for an application.

1. **Booster size:** If a cylinder requires high-pressure delivery throughout the stroke, a single-pressure booster circuit is needed. A double-headed booster can be used if the number of cycles per minute is low, and automatic bleeding and filling are not important. A triple-headed type must be used where rapid cycling is required.

2. **Tank size:** The air-oil tanks in booster circuits perform three general functions:

   1. Make up for leakage.
   2. Act as pressure sources to traverse or return cylinder.
   3. Provide outlets for entrained air.

When the tank functions as a reservoir, its size depends on how much the system leaks. However, tanks are also outlets for entrained air. Here, the tank is not pressurized and acts primarily as a reservoir.

When functioning as a pressure source, tanks must have a volume slightly greater than the displaced volume of the cylinder. Volume of the tank should be enough to preclude oil level reaching the upper tank baffle at the high-level point. When at the low-level point, the lower tank baffle should not be exposed to air pressure.

Pressurized tanks must also serve as an outlet for entrained air.

3. **Cylinder speed:** If rapid cylinder action is required, the hydraulic cylinder should be sized so that the reaction force (force required to do work) is 50 to 60% of available cylinder force at calculated pressure. Consider both high-pressure and low-pressure work reactions.

LINES

Pipes and fittings, with their necessary seals, make up a circulatory system of liquid powered equipment. Properly selecting and installing these components are very important. If improperly selected or installed, the result would be serious power loss or harmful liquid contamination. The following is a list of some of the basic requirements of a circulatory system:

- Lines must be strong enough to contain liquid at a desired working pressure and the surges in pressure that may develop in the system.
- Lines must be strong enough to support the components that are mounted on them.
- Terminal fittings must be at all junctions where parts must be removed for repair or replacement.
- Line supports must be capable of damping the shock caused by pressure surges.
- Lines should have smooth interiors to reduce turbulent flow.
- Lines must have the correct size for the required liquid flow.
- Lines must be kept clean by regular flushing or purging.
- Sources of contaminants must be eliminated.

The control and application of fluid power would be impossible without suitable means of transferring the fluid between the reservoir, the power source, and the points of application. Fluid lines are used to transfer the fluid to the points of application.

The *three common types of lines in liquid-powered systems are pipes, tubing, and flexible hose,* which are also referred to as rigid, semi rigid, and flexible line.

1. **Piping:** The piping can be used that can be threaded with screwed fittings with diameters up to 1 1/4 inches and pressures of up to 1,000 psi. Where pressures will exceed 1,000 psi and required diameters are over 1 1/4 inches, piping with welded, flanged connections and socket-welded size are specified by nominal inside diameter (ID) dimensions. The thread remains the same for any given pipe size regardless of wall thickness. Piping is used economically in larger-sized hydraulic systems where large flow is carried. It is particularly suited for long, permanent straight lines.

2. **Tubing:** The two types of tubing used for hydraulic lines are seamless and electric welded. Both are suitable for hydraulic systems. Seamless

tubing is made in larger sizes than tubing that is electric welded. Seamless tubing is flared and fitted with threaded compression fittings. Tubing bends easily, so fewer pieces and fittings are required. Unlike pipe, tubing can be cut and flared and fitted in the field. Generally, tubing makes a neater, less costly, lower-maintenance system with fewer flow restrictions and less chances of leakage.

3. **Flexible Hosing:** When flexibility is necessary in liquid-powered systems, hose is used. Examples would be connections to units that move while in operation to units that are attached to a hinged portion of the equipment or are in locations that are subjected to severe vibration. Flexible hose is usually used to connect a pump to a system. Flexible hose should not be twisted while installing it. Doing so reduces its lift and may cause its fittings to loosen. Flexible hose should be installed so that it will be subjected to a minimum of flexing during operation. Hose should not be stretched tightly between two fittings.

The lines should be kept as short and free of bends as possible. All the lines should be installed so that it can be removed without dismantling a circuit's components or without bending or springing them to a bad angle. Supports should be added to the lines at frequent intervals to minimize vibration or movement.

FITTINGS AND CONNECTORS

Fittings are used to connect the units of a fluid-powered system, including the individual sections of a circulatory system. Many different types of connectors are available for fluid-powered systems. The type used depends on the type of circulatory system (pipe, tubing, or flexible hose), the fluid medium, and the maximum operating pressure of a system. Some of the most common types of connectors are described below:

1. **Threaded Connectors:** Threaded connectors are used in some low-pressure liquid-powered systems. They are usually made of steel, copper, or brass, in a variety of designs (Refer to Figure 5.14). The connectors are made with standard female threading cut on the inside surface. The end of the pipe is threaded with outside (male) threads for connecting. Standard pipe threads are tapered slightly to ensure tight connections.

(Socket, Tee, Elbow, Reducer, Hexagonal nipple, Reducing bush, Reducing threaded nipple, 90° bend)

FIGURE 5.14 Threaded Connectors.

2. Flared Connectors: The common connectors used in circulatory systems consist of tube lines. These connectors provide safe, strong, dependable connections without having to thread, weld, or solder the tubing. A connector consists of a fitting, a sleeve, and a nut. (Refer to Figure 5.15.)

3. Hose Couplings: A hydraulic hose consists of an inner tube through which the fluid is conveyed and comes into direct contact with the

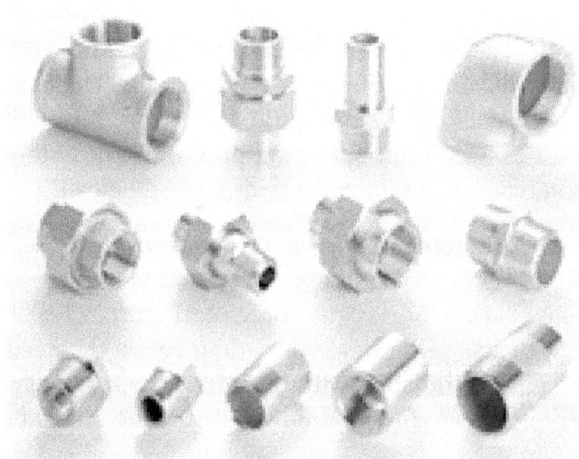

FIGURE 5.15 Flared Connectors.

hydraulic fluid. The rubber or other synthetic material used for this needs to be compatible and able to withstand the range of working temperature without losing its chemical and physical stability. The hydraulic hoses can be made using a variety of materials, depending on the factors of oil compatibility, abrasion resistance, etc. These include:

(a) *Plastic:* The plastic materials comprise of: Nylon, Braided nylon hose, Polyvinyl chloride, Textile braided hoses, Thermoplasts, Teflon, Chloro sulfonated polyethylene (Hypalon), Ethylene propylene diene (EPDM), Chlorinated polyethylene, etc.

(b) *Homogeneous Synthetic Rubber:* This category includes hose materials like: Buna N, Neoprene, Natural rubber, Butyl, etc.

The hydraulic hose is an important element in the system that needs to be checked for its reliability and durability during the manufacture process. In many hydraulic systems, it is necessary to move the drive unit along with the oil lines. Here, synthetic rubber is suited as the material for the hydraulic hoses. Plastic tunings are also used in hydraulic systems to convey the fluid.

HYDRAULIC SEALS

Seals are packing materials used to prevent leaks in liquid-powered systems. A seal is any gasket, packing, seal ring, or other part designed specifically for sealing. Sealing keeps the hydraulic oil flowing in passages to hold pressure and keep foreign materials from getting into the hydraulic passages. To prevent leakage, a positive sealing method should be used, which involves using actual sealing parts or materials. The strength of an oil film that the parts slide against provides an effective seal.

Hydraulic seals are exposed to hydraulic fluids such as hydrocarbon and phosphate ester and are designed for high-pressure dynamic applications such as hydraulic cylinders. Hydraulic seals usually need to be higher friction seals than pneumatic seals but often operate under lower operating speeds. Rod seals, piston seals, u-cups, vee-cups, and flange packing are just some of the sealing designs that can be used for a hydraulic seal. Sometimes a composite seal is used as a hydraulic seal. A composite seal is a product, which has two or three materials manufactured into one seal. Often there will be an elastomer ring and a PTFE ring giving the seal the advantages of both materials.

The life of hydraulic seals is dependent on many factors including the maximum operating speed, maximum operating temperature, maximum operating pressure, and the vacuum rating. When ordering hydraulic seals one must know the shaft outer diameter or seal inner diameter, housing bore diameter or seal outer diameter, the axial cross section, and the radial cross section.

FIGURE 5.16 Seals.

EXERCISES

1. Why is a receiver used in a pneumatic system?

2. Discuss the various kinds of air receiver.

3. List the various benefits of inter cooling.

4. Classify the various intercoolers and give at least two advantages of each.

5. Explain the construction and working of pneumatic filter along with a neat sketch.

6. What is the function of an accumulator? How is it different from a reservoir?

7. What are the properties of hydraulic fluids?

8. Write a short note on seals used in hydraulics and pneumatics.

9. Explain the construction and working of hydraulic filter along with a neat sketch.

10. What is a FRL unit in pneumatic systems?

11. What are deliquescent and desiccant air dryers? Why are they used in industries?

12. With the help of a neat sketch, explain the construction and working of air lubricator.

13. Differentiate between the hydraulic and pneumatic filter.

14. State at least two differences between pressure and flow regulator.

15. Differentiate between dead weight and spring-loaded accumulators.

16. Why are accumulators used in pneumatic and hydraulic systems?

17. Differentiate between strainers and filters.

18. Differentiate between full flow and by pass filters.

19. What are the properties of fluid used in hydraulics?

20. Distinguish between safety and relief valves.

21. What is the function of the seal? Briefly explain the various types of seals used in the fluid power system.

22. What is the function of intercooler in pneumatics?

23. List the reasons for using filters in hydraulics and pneumatics.

CHAPTER 6

CYLINDERS AND MOTORS

INTRODUCTION

An actuator is an essential component in all fluid power systems. **Actuator** is last in a fluid power system. The purpose of all downstream equipment is to convert the fluid power into mechanical power by means of an actuator. The output of the fluid power system is through an actuator. The output can be in the form of linear or rotary motion. Actuators that produce linear output are called *cylinders* and actuators that produce a rotary output are called *motors*. Figure 6.1 shows a system having two actuators.

Cylinders can be classified as hydraulic and pneumatic based on the working fluid used, similarly cylinders can be single acting or

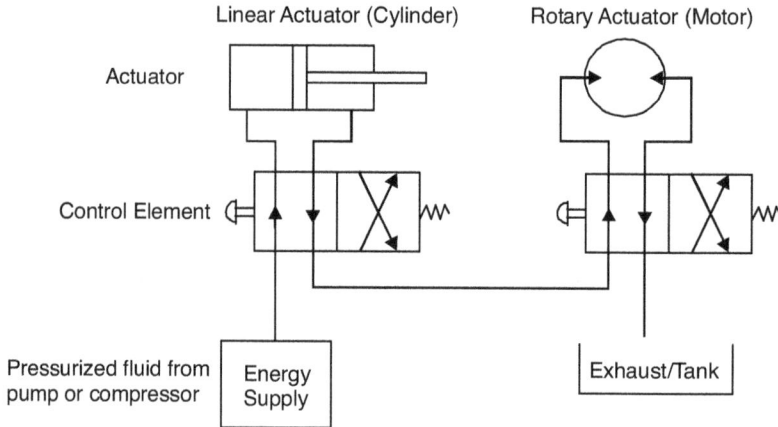

FIGURE 6.1 System Showing Two Actuators.

double acting depending upon the constructional features. These are generally specified by bore and stroke; they can also have options like cushions installed. Cushions slow down the cylinder at the end of the stroke to prevent slamming.

A fluid motor is a device that converts energy of fluid to rotary motion. Fluid motors are used for power transmission in various pneumatic and hydraulic devices in industry. Fluid motors may be classified as *unidirectional and bi-directional motors*. The motors that can rotate the output shaft in one direction only are called unidirectional motors and those can rotate the output shaft in either direction are called bi-directional motors. Further the fluid motors can be classified as fixed and variable displacement motors. Fixed displacement motors deliver constant torque and variable speeds. Variable speeds can only be achieved by controlling the amount of flow input. On the other hand in variable displacement motors the relative position of internal and external parts can be changed to obtain different torque speed combinations.

CYLINDERS

A cylinder is one of the simplest components used in hydraulic systems. Cylinders are linear actuators, which convert fluid power into mechanical power. The linear motion and high force produced by cylinders are big reasons why designers specify hydraulic and pneumatic systems in the first place. One of the most basic of fluid power components, cylinders have evolved into an almost endless array of configurations, sizes, and special designs. This versatility not only makes more innovative designs possible, but also makes many applications a reality that would not be practical or possible without cylinders. A cylinder consists of a cylinder body and a rod and piston assembly that moves inside it as shown in Figure 6.2. The fluid force acting on the piston causes the movement of the piston assembly.

A cylinder is one of the devices that create movement. When pressure is applied to a port it causes that side of the cylinder to fill with fluid. If the fluid pressure and area of the cylinder are greater than the load that is attached, then the load will move. If the pressure remains constant a larger diameter cylinder will provide more force because it has more surface area for the pressure to act on. Cylinders are sometimes called linear actuators or linear motors.

CLASSIFICATION OF CYLINDERS

There are many ways of classifying cylinders but, broadly, cylinders or linear actuators can be classified on the basis of:

▪ Construction
▪ Working fluid

CLASSIFICATION OF CYLINDERS ON THE BASIS OF CONSTRUCTION

Depending on the construction, cylinders can be classified as:

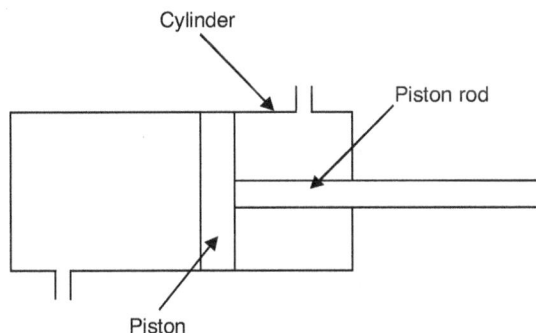

FIGURE 6.2 A Cylinder with Rod and Piston Assembly.

1. Single-Acting Cylinders
2. Double-Acting Cylinders

The basic difference between the two single cylinders is that single act-ing cylinders can be pressurized from one end, while the double acting cylinders can be pressurized from both ends.

SINGLE-ACTING CYLINDER

A single-acting cylinder is shown in Figure 6.3 (*a*). The cylinder is only powered in one direction and needs another force to return it such as an external load or a spring. This cylinder only has a head-end port and is operated hydraulically/pneumatically in one direction. When pressurized fluid is send into a port, it pushes the plunger/piston rod

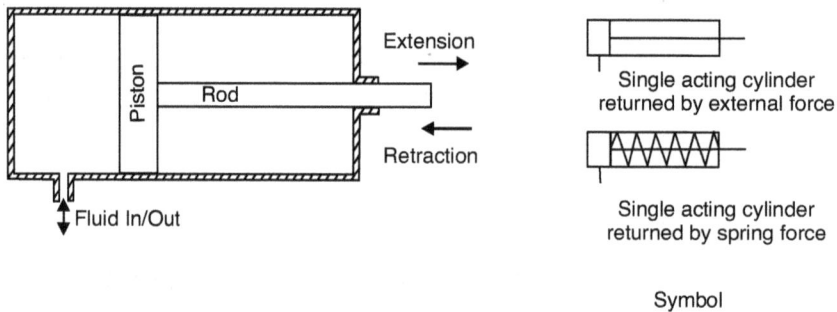

FIGURE 6.3 (a) Single-Acting Cylinder.

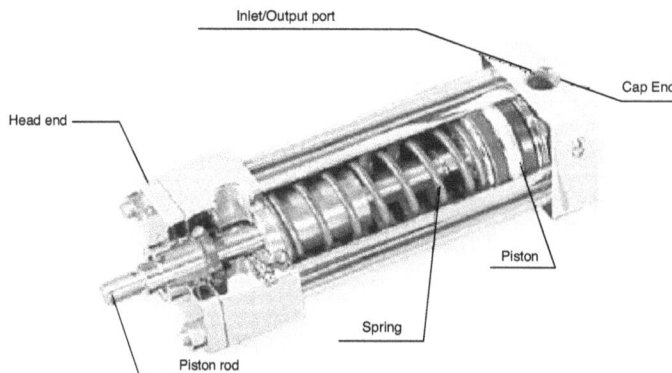

FIGURE 6.3 (b) Cut section view of a Pneumatic Single-Acting Cylinder (Spring Returned).

thus extending it. To return or retract a cylinder, fluid must be exhausted out. A plunger returns either because of the weight of a load or from some mechanical force such as a spring. In mobile equipment, flow to and from a single-acting cylinder is controlled by a reversing directional valve of a single-acting type. This type of cylinder is commonly used for jacks and lifts. Figure 6.3 (*b*) shows the cut section view of a spring returned pneumatic single-acting cylinder.

DOUBLE-ACTING CYLINDER

Double-acting cylinders are the most commonly used cylinders in hydraulic and pneumatic applications. In this design, pressure can be applied to either side of the cylinder. The symbol for double-acting cylinder shows two ports where fluid pressure can be applied as shown in Figure 6.4. The extension and retraction of cylinder is due to hydraulic/pneumatic pressure.

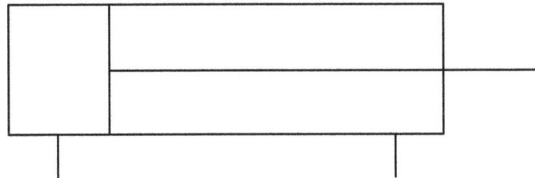

FIGURE 6.4 Symbol of Double-Acting Cylinder.

Construction of a Double-Acting Cylinder

Construction of a double-acting cylinder is shown in Figure 6.5.

A double-acting cylinder is made up of following components:

1. Cylinder tube
2. Piston
3. Piston rod
4. Cylinder end
5. Cylinder head
6. Piston seal
7. Piston rod seal

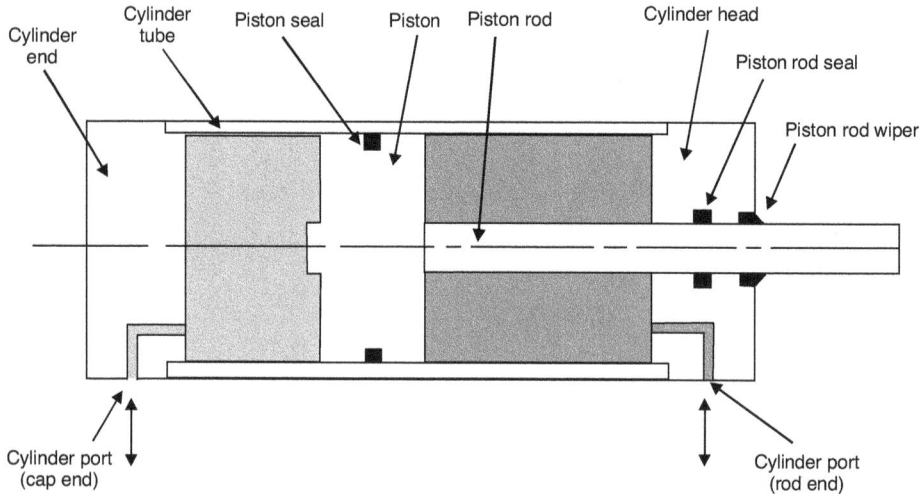

FIGURE 6.5 Double-Acting Cylinder.

8. Piston rod wiper
9. Cylinder port (cap end)
10. Cylinder port (rod end).

A cylinder is constructed of a barrel or tube, a piston and rod (or ram), two end caps, and suitable oil seals. A barrel is usually seamless steel tubing, or cast, and the interior is finished very true and smoothly. The moving part inside the cylinder body is highly polished; chrome plated steel piston rod and the solid cast iron piston assembly. The body and the heads are held together by steel tie rods and nuts. The heads on either end contain ports for fluid flow. It is supported in the end cap by a polished surface. Seals and wipers are installed in the rod's end cap to keep the rod clean and to prevent external leakage around the rod. Other points where seals are used are at the end cap and joints and between the piston and barrel. Mounting provisions often are made in the end caps, including flanges for stationary mounting or clevises for swinging mounts. Internal leakage should not occur past a piston. It wastes energy and can stop a load by a hydrostatic lock (oil trapped behind a piston).

Figure 6.6 shows the cut section view of a double-acting hydraulic cylinder.

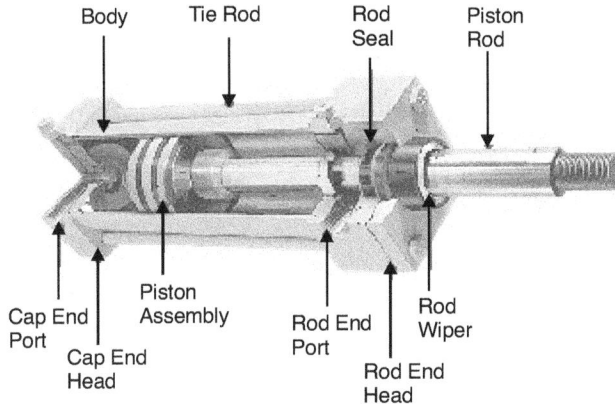

FIGURE 6.6 Cut Section View of Double-Acting Hydraulic Cylinder.

TYPES OF SINGLE-ACTING CYLINDERS

There are three types of single-acting cylinders, which are commonly available:

1. Ram type cylinder
2. Telescopic cylinder
3. Spring return cylinder

Ram Type Cylinder

The ram type of cylinder is the simplest type of cylinder used in hydraulic systems. It consists of a fluid chamber and fluid inlet. The ram is mounted vertically and is pushed up by hydraulic pressure as shown in Figure 6.7. When the pressure is removed, it retracts back due to force of gravity. Ram type cylinders are used in applications where a single-acting force is required usually in vertical direction. These can carry a heavy load and have a large stability on heavy load. These are mostly used in elevators, jacks, and automobile hoists.

Telescoping Cylinder

Telescoping cylinders are used in applications, which require a lengthy operation. These consist of a series of nested cylinders called sleeves (Refer to Figure 6.8). These sleeves are progressively smaller in diameter and each one can fit into the next larger sleeve. All the sleeves can collapse into the

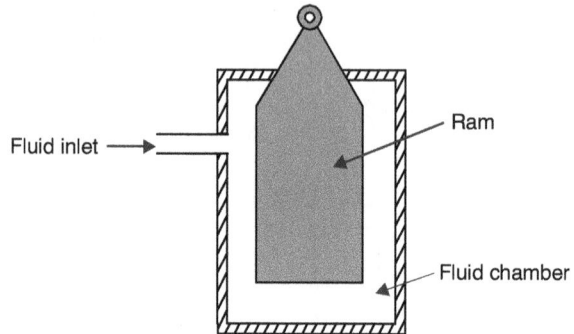

FIGURE 6.7 Ram Type Cylinder.

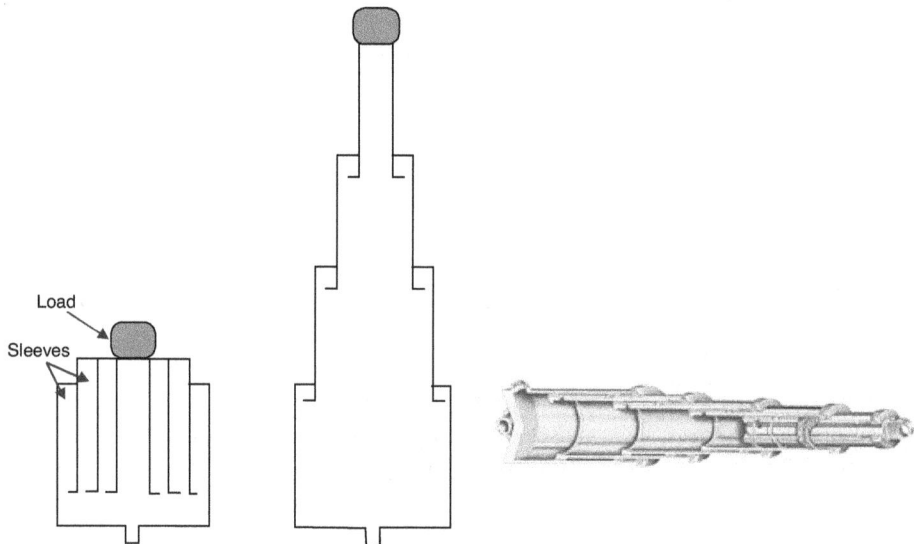

FIGURE 6.8 Telescoping Cylinder.

sleeve of larger diameter, which gives it a very compact size. The cylinder can achieve a long stroke when all the sleeves are extended. In the extended position, the force on the load depends on the smallest sleeve in use. Force output varies with rod extension: highest at the beginning, when full piston area is used; lowest at the end of the stroke, when only the area of the final stage can be used to transmit force. Telescoping cylinders are widely used for vehicle applications such as forklift trucks and dump trucks.

Spring Return Cylinder

Spring return cylinder is another variation of single-acting cylinder, in which pressure is applied on one end. It consists of a piston inside a fluid chamber, which is kept in one extreme position by spring (Refer to Figure 6.9). When pressure is applied to the piston, it extends by overcoming the spring force. The piston retracts back due to spring force when the pressure applied on the piston is removed.

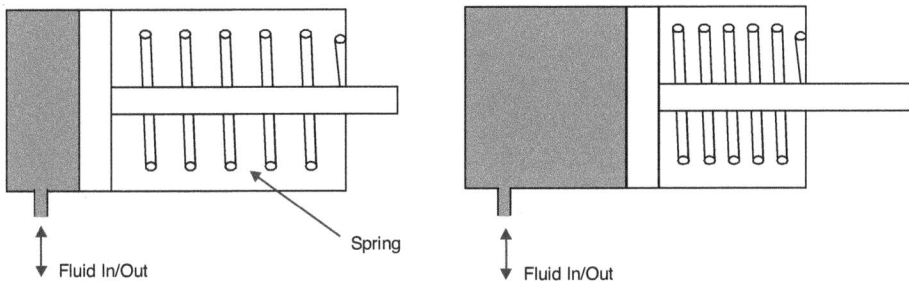

Spring

Fluid In/Out

Fluid In/Out

FIGURE 6.9 Spring Return Cylinder.

TYPES OF DOUBLE-ACTING CYLINDERS

The commonly available types of double-acting cylinders are:

1. Double rod cylinder
2. Tandem cylinder

Double Rod/Through Rod Cylinder

It consists of rods on both end of the piston. Due to this, the area exposed to the fluid on both sides of the piston is identical. Therefore, these cylinders produce equal force and equal speed in either direction, while performing two operations with one stroke. It contains two pressure ports as shown in Figure 6.10. Double rod cylinders are available with a hollow rod, so that fluid or another machine element can be passed through the cylinder. These cylinders are used in applications where two loads have to be moved through equal distance.

Tandem Cylinder

A tandem cylinder consists of two cylinders, which have a common rod as shown in Figure 6.11. This means that two cylinders are being used in

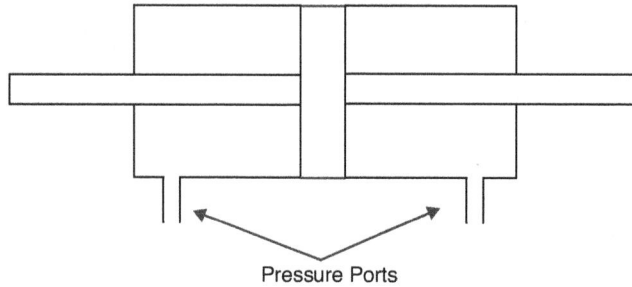

FIGURE 6.10 Double Rod Cylinder.

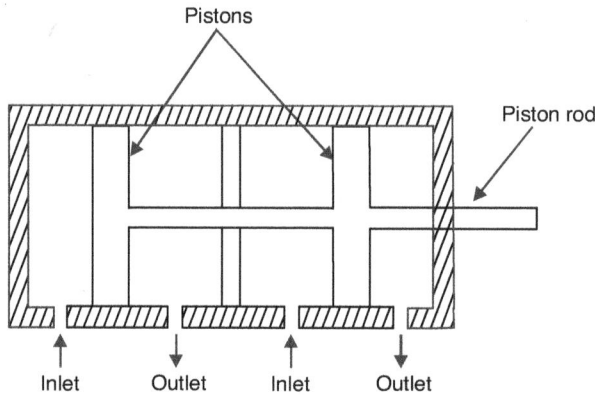

FIGURE 6.11 Tandem Cylinder.

tandem with a common rod. With these types of cylinders, higher force can be developed with the same bore size and fluid pressure. This is because the fluid pressure acts on double the area at the same time that are joined with one another to work as single unit.

OTHER TYPES OF CYLINDERS

Tie-Rod Cylinder

Tie-rod cylinders are the oldest and most commonly used cylinders in industrial jobs. The cylinder body is held together by four or more tie rods that extend the full length of the body and pass through the end caps or mounting plate. In operation, they may perform any of the common cylinder functions except telescoping. (Refer to Figure 6.12.)

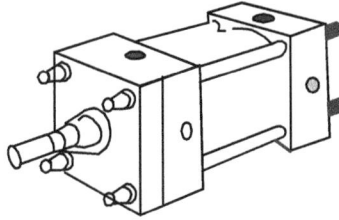

FIGURE 6.12 Tie-Rod Cylinders.

One-Piece Cylinder

These are most often used on mobile equipment and farm machinery. The body is either cast integrally, or head and body may be welded together. This is the least expensive type of cylinder. It is compact and simple. But it cannot be repaired when damaged or worn. (Refer to Figure 6.13.)

Threaded Head Cylinder

These types of cylinders offer a compromise between tie-rod and one-piece units. Threaded units are relatively compact and streamlined, yet can be disassembled for repair by unthreading either or both ends from the cylinder body.

Diaphragm Cylinder

These are used in either hydraulic or pneumatic service for applications that require low friction, no leakage across the piston, or extremely sensitive response to small pressure variations. In diaphragm type of cylinders, the piston is attached to a diaphragm, which is attached to the end cap as shown in Figure 6.14. Because there is no friction between the cylinder piston and the cylinder barrel as compressed air enters the cylinder, the

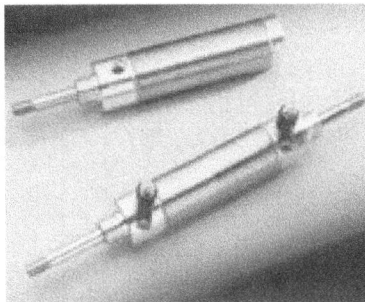

FIGURE 6.13 Single-Acting and Double-Acting One-Piece Cylinders.

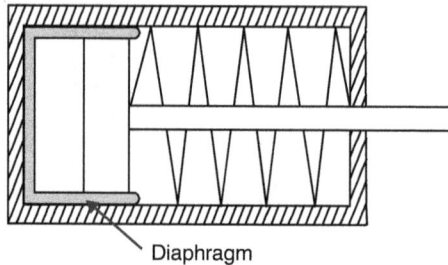

FIGURE 6.14 Diaphragm Cylinder.

piston and the rod start to move almost instantly. Movement is free, almost frictionless. As the piston travels the length of the cylinder barrel, the diaphragm unrolls to allow the piston to travel. Upon the return stroke, the diaphragm re-rolls to its former shape. This type of cylinder requires no external lubrication source. They are frequently used as pneumatic actuators in food and drug industries because they require no lubrication and do not emit a contaminating oil mist.

Rotating Cylinder

The rotating cylinder shown in Figure 6.15 imparts linear motion to a rotating device. They are often used to actuate rotating chucks on turret lathes. Fluid is ported to the rotating cylinder through a stationary distributor. Rotating cylinders are available both with solid and hollow pistons.

Non-Rotating Cylinder

Non-rotating cylinders are used in applications that demand both accurate linear position and precise angular orientation. Special guides can be added to standard cylinders to prevent rod rotation, but this is often expensive and unwieldy. More often, twin-rod or rectangular cylinders are

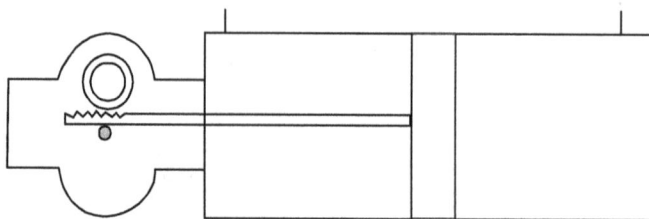

FIGURE 6.15 Rotating Cylinder.

used. Sometimes it is also preferable to use a unique square piston rod with rounded corners that prevents rotation better than other rod configurations.

Duplex Cylinder

Duplex cylinders also have two or more cylinders connected in line, but the pistons of a duplex cylinder are not physically connected (Refer to Figure 6.16). The rod of one cylinder protrudes into the non-rod end of the second, and so forth. A duplex cylinder may consist of more than two in-line cylinders and the stroke lengths of the individual cylinders may vary. This makes them useful for achieving a number of different fixed stroke lengths, depending on which individual pistons are actuated.

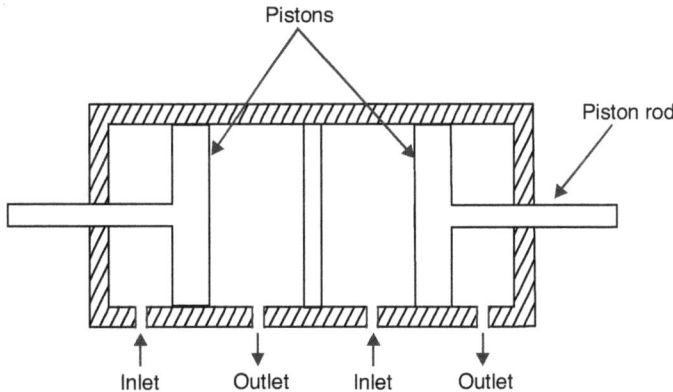

FIGURE 6.16 Duplex Cylinder.

CLASSIFICATION OF CYLINDERS ON THE BASIS OF WORKING MEDIUM

Depending on working medium cylinders can be classified as:

1. Hydraulic Cylinders
2. Pneumatic Cylinders.

HYDRAULIC CYLINDERS

Hydraulic cylinders are linear hydraulic motors that turn the hydrostatic power of a fluid into mechanical power. The hydraulic fluid utilized in the cylinder is oil, most commonly used in industrial applications. Hydraulic cylinders consist of a plunger or piston inside a cylindrical housing. The bore of the cylinder must be perfectly cylindrical to prevent any obstruction of the piston. Piston seals prevent internal leakage from the high-pressure side of the hydraulic cylinder to the low side. Rod seals are used to prevent external leakage. Piston rods, which move in and out of the cylinder, are usually hard chrome plated to provide protection from corrosion and wear resistance. Most hydraulic cylinders are made out of steel, stainless steel, or aluminum.

Hydraulic cylinders can be single acting or double acting. A single-acting hydraulic cylinder is configured for motion in only one direction, either pulling in or pushing out. An internal spring is used to provide plunger return when the hydraulic pressure is removed. In single-action hydraulic cylinders, less valving and plumbing is needed than in double-acting cylinders. The more complex double-acting cylinder is pressurized to move in both directions along the vertical or horizontal plane or any other plane as needed. These cylinders provide higher speed operations and tighter control than single-acting cylinders and are less sensitive to system back-pressures that can occur due to long tube lengths. A typical hydraulic double-acting cylinder along with all the parts is shown in Figure 6.17 and the part list is given in Table 6.1.

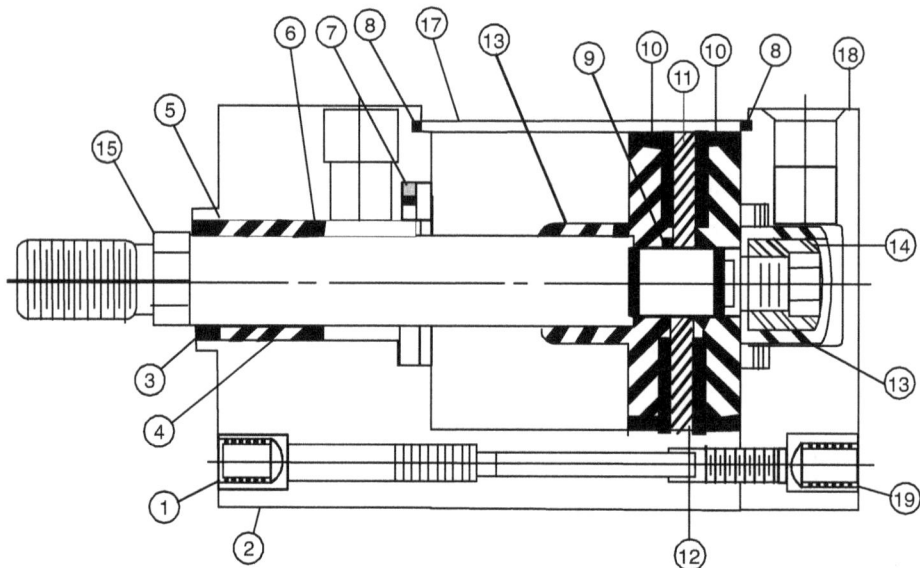

FIGURE 6.17 Parts of Double-Acting Hydraulic Cylinder.

TABLE 6.1 Bill of Material.

S. No.	Name of component	Quantity
1	Head screw	4
2	Head	1
3	Retaining ring	1
4	Rod bearing	1
5	Rod wiper	1
6	Rod seal	1
7	Cushion kit with needle valve	2
8	Tube seal O-ring	2
9	Piston rod seal	1
10	Piston seal	2
11	Piston wear strip	1
12	Piston magnet	1
13	Piston follower	2
14	Retaining nut for piston & rod	1
15	Piston rod	2
16	Main stud for retaining nut	1
17	Tube, cylinder body	1
18	Cap	1
19	Cap screw	4

PNEUMATIC CYLINDERS

Pneumatic cylinders are the final component in a pneumatic or compressed air control or power system. Air or pneumatic cylinders are devices that convert compressed air power into mechanical energy. This mechanical energy produces linear or rotary motion. In this way, the air cylinder functions as the actuator in the pneumatic system, so it is also known as a pneumatic linear actuator. Main components of a pneumatic cylinder consist of steel or stainless steel piston, a piston rod, a cylinder barrel, and end covers. Pneumatic cylinders can be used for pressure ranges between 5 bars and 20 bars. Consequently they are constructed from lighter materials such as aluminum and brass. Because gas is a compressible substance, the motion of a pneumatic cylinder is hard to control precisely. The basic theory of hydraulic and pneumatic cylinders is otherwise the same. Pneumatic cylinders are a proven way to provide quick, clean, reliable, and inexpensive linear motion, and a multitude of available designs, styles, and options can suit most any conceivable application. In Figure 6.18, a typical pneumatic double-acting cylinder is shown along with its all the parts. The assembly parts list is given in Table 6.2.

APPLICATIONS OF CYLINDERS

Cylinders found their applications in food processing equipment, chemical and pharmaceutical machinery, power plants, waste-to-steam generating plants, valve actuators, marine use in dams and pollution control, printing machinery, medical and surgical tables, packaging machinery, damper controls, earth moving machinery, mining industry, construction machinery, plant engineering, defense technology, automotive engineering, mechanical engineering, textile industries, railways, power plants, agricultural machinery, etc.

CYLINDER CUSHIONING

Cylinders (pneumatic/hydraulic) are widely used for mechanical handling systems. Pneumatic cylinders operate at much higher speed than hydraulic cylinders. Due to this, there is a tendency of the piston to ram against the end covers as the piston approaches the ends at high velocity. Often, the impact occurs at the both end points of cylinder and generates the destructive shock within the structural operating members of the

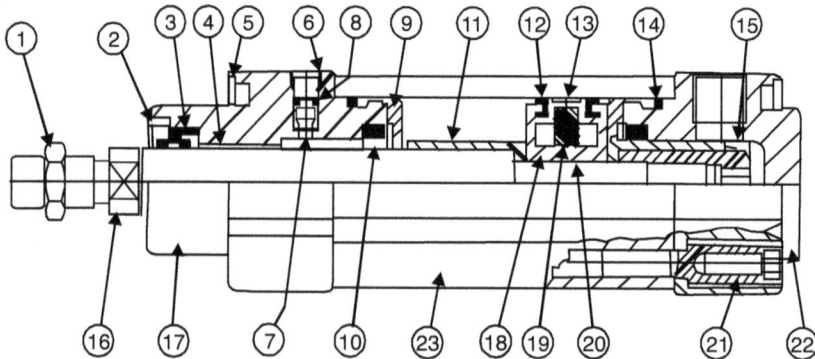

FIGURE 6.18 ISO Standard Double-Acting Pneumatic Cylinder.

TABLE 6.2 Bill of Material.

S. No.	Name of component	Quantity
1	Locknut	1
2	Retaining ring	1
3	Rod seal	1
4	Rod bearing	1
5	Dust seal	1
6	Retainer cushion screw	1
7	Cushion screw	1
8	O-ring	1
9	Shock absorber	1
10	Cushion seal	1
11	Cushion spear	1
12	Piston seal	2
13	Piston wear strip	1
14	Barrel seal	2
15	Piston retaining nut	2
16	Piston rod	1
17	Rod end cover	1
18	Piston	1
19	Magnet	1
20	O-ring	1
21	Cover bolt	4
22	Head end cover	1
23	Barrel	1

machine or equipment. Some form of cushioning is normally required to reduce the rate of travel of a cylinder before the piston strikes the end cover. Reducing the piston velocity at the end of its travel lowers the stresses on the cylinder while reducing vibration in the structure of which it is a part. To slow an action and prevent shock at the end of a piston stroke, some actuating cylinders are constructed with a cushioning device at either or both ends of a cylinder. This cushion is a flow control valve that does not operate until the cylinder piston reaches a certain point in the cylinder. Then, the cushion restricts airflow to slow the cylinder movement. This allows it to move to the end of its travel at a slower speed. This adjustment is normally on the end of the cylinder head. Cushioning of cylinders at one or both ends of piston stroke is used to reduce the shock and vibration. All the double-acting cylinders except for small sizes are provided with end position cushioning arrangement.

Working of Cushions

The working of cushions is explained with the help of a double-acting cylinder with cushions at both ends (Refer to Figure 6.19).

The piston moving in the cylinder has cushion noses installed on both the sides. These cushion noses are designed such that they mate with the cavity made in cap end, which is called as cushion chamber. Cushion assembly consists of a needle valve mechanism. Needle valve adjusts the cushioning effect, *i.e.*, by rotating needle of the needle valve; the intensity of shock can be decreased or increased. When the piston is moving backwards, *i.e.*, it is proceeding in the direction of end cap shown in figure, the cushion nose enters inside the cushion chamber. The locked up fluid just before the cap end is allowed to pass through a carefully designed flow resistance path. This produces the back up pressure hence opposing movement of piston and piston rod. This force decelerates the motion and result in a smooth operation.

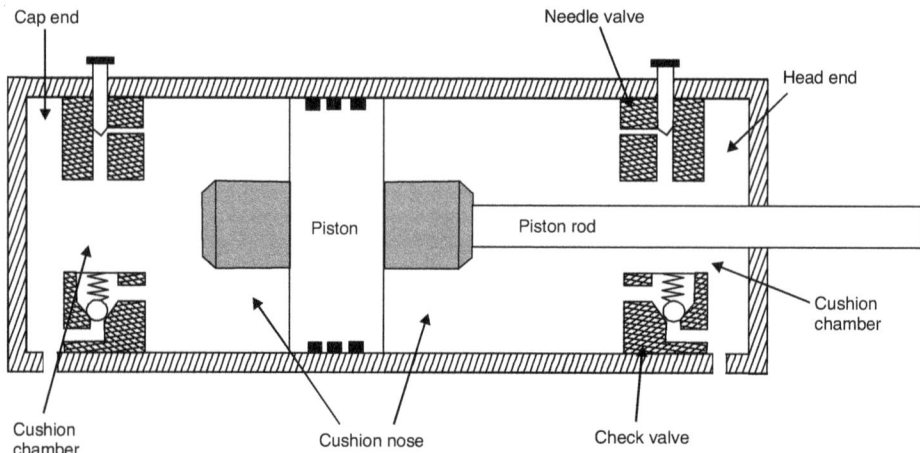

FIGURE 6.19 Double-Acting Cushioned Cylinder with Cushions at Both Ends.

Now when the process is over and cushion nose fully enters into the cushion chamber then the forward stroke will take time to start because the area of nose is great, whereas the entrance from where fluid is coming in is intentionally kept small. In order to overcome this problem sometimes a check valve gallery is provided below the cushion chamber to initiate the process without consuming much time.

Ideal Cushioning means that there is no end of stroke bounce, *i.e.*, the direction of travel of the piston is the same throughout the entire cushioning sequence, and that its velocity is exactly zero when it reaches the end of its travel. The sound of end cover contact is negligible and the total cycle time is minimized. Cushioning is an important aspect of cylinder life. Cushioning not only ensures smooth operation but also extends the life of cylinders by preventing shocks. In Figure 6.20, a pneumatic cylinder is shown which has both conventional pneumatic and elastic cushions.

CYLINDER MOUNTINGS

Cylinders are manufactured in innumerable combinations of bore, rod, and stroke sizes with various standard-mounting styles. There is a direct relationship between the cylinder mounting style and the effects of the operating pressure upon the unit. Standard mounting styles for fluid power cylinders are divided into two general classifications based upon the combined effects of force direction and mounting conditions. Those classifications are as follows:

1. **Cylinders which produce a straight-line transfer of force:** This class consists of models having fixed mounting styles that are secured in a rigid condition. Cylinder mountings of this type are divided into

FIGURE 6.20 A Pneumatic Cylinder.

two groups: those models, which absorb force on the cylinder centerline and those styles, which do not absorb force on the centerline of the unit.

2. Pivoted cylinders which transfer force along a variable path: This group consists of mounting styles, which pivot around a fixed pin and are able to compensate for alignment changes in one plane during operation.

The way in which a cylinder is mounted is critical to its performance and service life. Improper mounting can result in damage to the cylinder as well as to the equipment on which it is being used. Inaccurate installation, or the use of an inappropriate mounting style may result in misalignment, which causes harmful side loading and bending. Some of the common ways in which cylinders can be mounted are:

- Extended tie-rods mounted cylinders
- Flange-mounted cylinders
- Side or lug mounted cylinders
- Pivot mounted cylinders
- Trunnion mounted cylinders

Extended Tie-Rods Mounting

Extended tie-rod cylinders are suitable for straight-line force transfer and at the places where the space for installation is limited. The arrangement is shown in Figure 6.21. For pull applications (where tensile forces set up in piston rod) head end tie-rod mounts are suitable solution, on the other hand for push applications (where compressive forces set up in piston rod) cap end tie-rod mounts are preferred. This mounting style provides

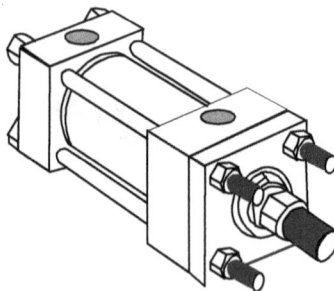

FIGURE 6.21 Extended Tie-Rod Mounting.

a stable means of handling thrust loads and may be used whenever available clearances permit the installation of this model. Consideration should be given to providing added support to the cylinder body on horizontally mounted units having long stroke lengths.

Mounting styles recommended by NFPA (National Fluid Power Association) for tie-rod mounted cylinders are shown in Figure 6.22.

Flange-Mounted Cylinders

Flange-mounted cylinders are also suitable for use on straight-line force transfer applications. The flange mount is one of the strongest, most rigid methods of mounting (Refer to Figure 6.23). With this type of mount there is little allowance for misalignment, though when long strokes are required, the free end of the mounting should be supported to prevent sagging and possible binding of the cylinder. The best use of a cap end flange is in a thrust load application (rod in compression). Head end flange mounts are best used in tension applications. Proper selection of a flange mounting style depends upon whether the primary work direction results in tension or compression loads being carried by the cylinder rod. Front flange-mounted units are preferred for "pull type" applications where the rod is held in tension. "Push type" applications, which subject the cylinder rod to compression loads, are best suited for rear flange style mountings.

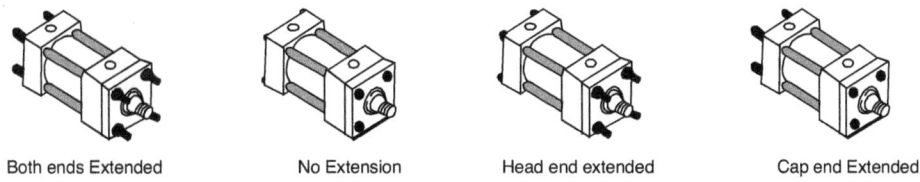

| Both ends Extended | No Extension | Head end extended | Cap end Extended |

FIGURE 6.22 Preferred Mounting Styles Using Tie Rods.

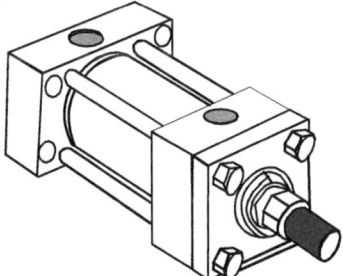

FIGURE 6.23 Flange Mounting.

Mounting styles recommended by NFPA (National Fluid Power Association) for flange-mounted cylinders are shown in Figure 6.24.

Side or Lug Mounted Cylinders

Side or Lug lug mounted cylinders do not absorb force on their centerlines; due to this the movement of the cylinder applies a turning moment, which tends to rotate the cylinder about its bolts. The side or lug mounted cylinder provides a fairly rigid mount. These types of cylinders can tolerate a slight amount of misalignment when the cylinder is at full stroke, but as the piston moves toward the cap end, the tolerance for misalignment decreases. It is important to note that if the cylinder is used properly (without misalignment), the mounting bolts are either in simple shear or tension without any compound stresses. (Refer to Figure 6.25.)

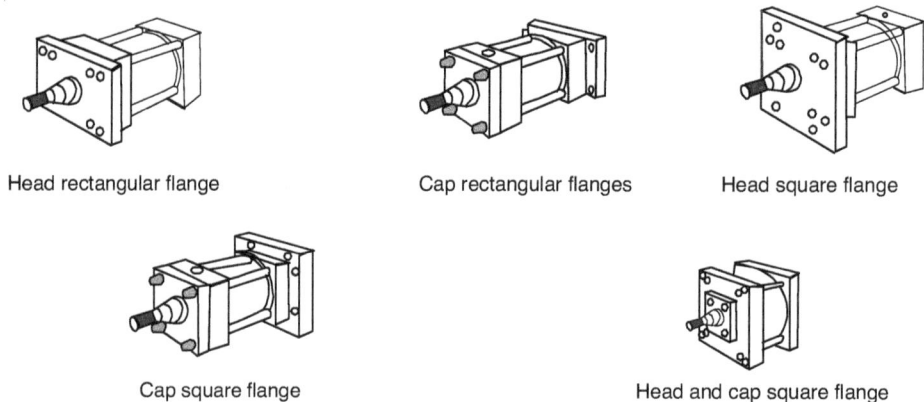

Head rectangular flange   Cap rectangular flanges   Head square flange

Cap square flange   Head and cap square flange

FIGURE 6.24 Preferred Mounting Styles Using Flanges.

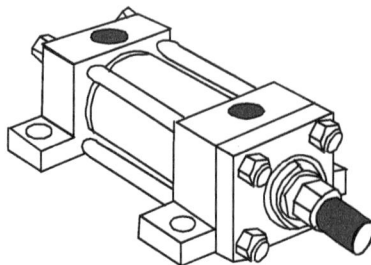

FIGURE 6.25 Side Lug Mounting.

Mounting styles recommended by NFPA (National Fluid Power Association) for cylinders using lugs are shown in Figure 6.26.

Pivot-Mounted Cylinders

All pivoted cylinders need a provision on both ends for pivoting. These types of cylinders are designed to carry shear loads. Long-stroke, pivot-mounted cylinders will unavoidably have high-side loads because of the rod weight. In these applications, a stop tube or dual piston becomes essential. Both spread the distance between the rod bearing and piston, reducing the effective load at these two points.

Mounting styles recommended by NFPA (National Fluid Power Association) for pivot-mounted cylinders are shown in Figure 6.28.

Trunnion-Mounted Cylinders

These cylinders are designed to absorb force on their centerlines (Refer to Figure 6.29). They are suitable for both push (compression) and pull (tension) applications. They may also be used where the machine member to

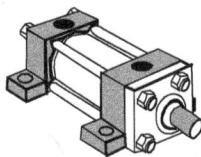

Side lug Mounting   Foot lug Mounting   Centerline lug Mounting

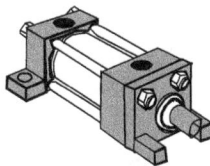

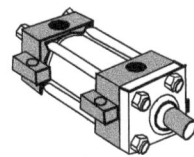

FIGURE 6.26 Preferred Mounting Styles Using Lugs.

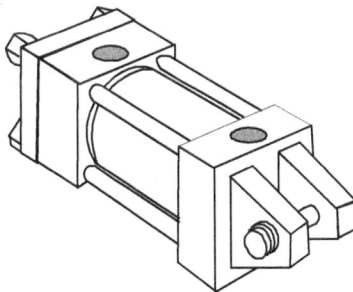

FIGURE 6.27 Pivot Mounting.

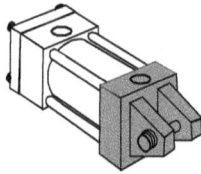

Clevis Mounting   Fixed Eye mounting

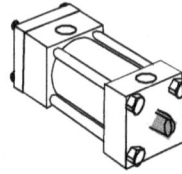

FIGURE 6.28 Preferred Mounting Styles Using Pins.

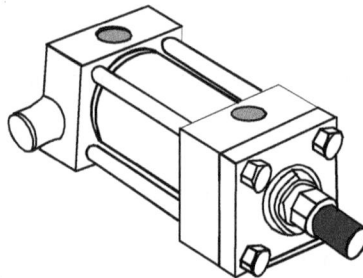

FIGURE 6.29 Trunnion Mounting.

be moved follows a curved path. The trunnion pivot pins should be covered by bearings that are rigidly held and closely fitted for the entire length of the pin. On trunnion-mounted cylinders, it is necessary to fit pillow blocks or mated bearings as close to the cylinder head as possible, to minimize bending stresses in the heads. Spherical-bearing pillow blocks on trunnion mounts should be avoided because they introduce bending stresses.

Mounting styles recommended by NFPA (National Fluid Power Association) for trunnion-mounted cylinders are shown in Figure 6.30.

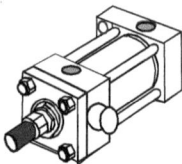

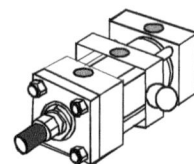

Head Trunnion   Cap Trunnion   Intermediate Trunnion

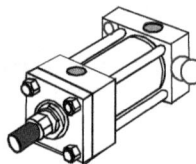

FIGURE 6.30 Preferred Mounting Styles Using Trunnions.

CYLINDER SIZING

To determine the cylinder size that is needed for a particular system, certain parameters must be known. First of all, a total evaluation of the load must be made. This total load is not only the basic load that must be moved, but also includes any friction and the force needed to accelerate the load. Also included must be the force needed to exhaust the air from the other end of the cylinder through the attached lines, control valves, etc. Any other force that must be overcome must also be considered as part of the total load. Once the load and required force characteristics are determined, a working pressure should be assumed. This working pressure that is selected must be the pressure seen at the cylinder's piston when motion is taking place. It is obvious that cylinder's working pressure is less than the actual system pressure due to the flow losses in lines and valves. With the total load (including friction) and working pressure determined, the cylinder size may be calculated using Pascal's Law. Force is equal to pressure being applied to a particular area. The formula describing this action is:

$$\text{Force = Pressure n Area}$$

Area of a Cylinder

There are two type of areas in one cylinder, *i.e.*, area on rod side and area on non-rod side of the cylinder. Both the force and speed of a cylinder are dependent on knowing the area of the cylinder. It is important to add here that the area on the rod end of a cylinder is different than that of the non-rod end. Area on non-rod side of piston is equal to the area of bore it can be given mathematically (in square inches) as:

$$\text{Area of cylinder} \quad \frac{\pi}{4}D^2$$

The same formula can be used to find out the area of rod and thus the area on the rod side of piston can be given as:

$$\text{Area (rod end) = Area of Piston − Area of the rod itself}$$

It can be added here that when pressure is applied to the rod end of cylinder it will move faster and have less force as compared to rod less side of cylinder.

Forces of a Cylinder

A cylinder usually has two forces. When fluid at the same pressure is entered in both input and output ports, the force acting on the side "without the rod" of the piston is more because the area is more in that case. P_1 and P_2 are the pressures acting on two sides of the piston with the condition that $P_1 = P_2$. Let us assume that A_1 and F_1 are the area and force acting on the piston on "without the rod" side respectively and similarly A_2 and F_2 are the area and force acting on the rod side of piston. Mathematically:

$$P_1 = P_2$$

$$\frac{F_1}{A_1} = \frac{F_2}{A_2}$$

$$F_1 = \frac{F_2 A_1}{A_2}$$

$$F_2 = \frac{F_1 A_2}{A_1}$$

Therefore, from the above expression, it is clear that force F_1 on the side of the piston "without the rod" is more than the force F_2 on the "rod" side of piston.

CYLINDER SPECIFICATION

Cylinders can be specified on the basis of following:

- Cylinder bore
- Piston rod diameter
- Stroke length
- Pressure range
- Force output at max pressure
- Mounting styles
- Rod ends
- Cushions (at one end or both ends)
- Standard operating temperature

INTRODUCTION TO MOTORS

A fluid power motor is a device that converts fluid power energy to rotary motion and force. Pressure is converted into torque and flow rate is converted into speed. The function of hydraulic motor is opposite to that of a pump. However, the design and operation of fluid power motors are very similar to pumps. The difference being, instead of pushing the fluid as the pump does in a hydraulic motor, the rotating elements, *i.e.*, vanes, gears, pistons, etc. are pushed by the oil pressure to enable the motor shaft to rotate and thus develop the necessary turning torque and continuous rotating motion. Figure 6.31 shows the working principle of fluid motor.

Fluid motors may be either fixed or variable displacement. Fixed displacement motors provide constant torque and variable speed. The speed is varied by controlling the amount of input flow. Variable displacement motors are constructed so that the working relationship of the internal parts can be varied to change displacement. With the input flow and operating pressure remaining constant, varying the displacement can vary the ratio between torque and speed to suit the load requirements. The majority of motors used in fluid power systems are fixed displacement type; however, variable displacement piston motors are in use mainly in hydrostatic drives.

Although most fluid power motors are capable of providing rotary motion in either direction, some applications require rotation in only one direction. In these applications, one port of the motor is connected to the system pressure line and the other port to the return line or exhausted to the atmosphere. The flow of fluid to the motor is controlled by a flow control valve, a two-way directional control valve, or by starting and stopping

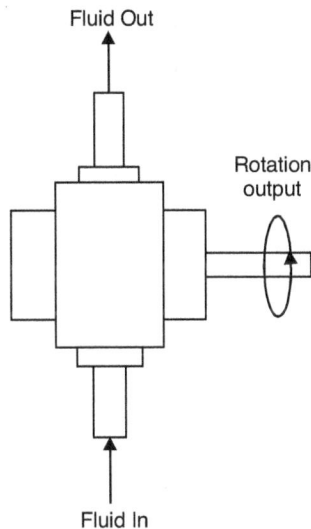

FIGURE 6.31 Working Principle of Fluid Motor.

the power supply. The speed of the motor may be controlled by varying the rate of fluid flow to it. In most fluid power systems, the motor is required to provide actuation power in either direction. In these applications, the ports are referred to as working ports, alternating as inlet and outlet ports. The flow to the motor is usually controlled by either a four-way directional control valve or a variable-displacement pump.

MOTOR RATINGS

Hydraulic motors are rated according to the following parameters:

- **Torque:** Torque is the turning force developed at the motor shaft due to its rotation. The torque increases with an increase in operating pressure and decreases with a decrease in operating pressure.
- **Pressure:** The pressure required by a motor depends on the torque requirement and its displacement. A motor with a higher displacement requires lower pressure to generate a specific torque as compared to a motor of lower displacement.
- **Displacement:** Displacement is defined as the amount of fluid required to turn the motor shaft one revolution. It is expressed in cubic inches per revolution, the same as pump displacement. Hydraulic motors can be a fixed or variable displacement type. A fixed displacement motor gives a constant speed and torque with a fixed flow rate and operating pressure. A variable displacement motor gives a variable speed and torque even with constant flow rate and operating pressure.
- **Speed:** Speed of a motor is the function of its displacement and the input flow rate. The maximum speed of a motor is the speed it can sustain for a specified time without damage. The minimum speed of a motor is the slowest speed at which a motor rotates smoothly and continuously. The torque output of motor is independent of its speed.

HYDRAULIC AND PNEUMATIC MOTORS

Hydraulic motors convert hydraulic energy into mechanical energy. In industrial hydraulic circuits, pumps and motors are normally combined with proper valves and piping to form a hydraulic powered transmission. A pump, which is mechanically linked to a prime mover, draws fluid from a reservoir and forces it to a motor. Usually, a pump is connected via a carrier line to a motor, which then draws fluid from a reservoir and forces it into the motor. The fluid forces the movable components of the motor into motion, which in turn rotates the attached shaft. The shaft, which is mechanically linked to the workload, provides rotary mechanical motion. Finally, the fluid is discharged at low pressure and transferred back to the pump. Hydraulic motors are part of hydrostatic power transmission systems. Depending upon specific applications, these motors prove more efficient, suitable, and economical over their electrical or pneumatic counterparts.

Figure 6.32 shows the basic operations of a hydraulic motor.

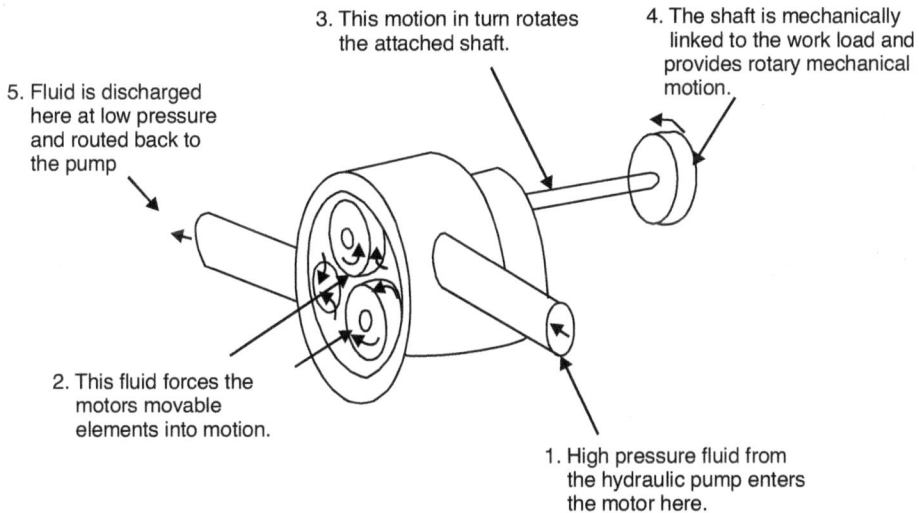

FIGURE 6.32 Hydraulic Motor.

Advantages of Hydraulic Motors

- The torque of the hydraulic motor is independent of the speed of the motor. Thus, a constant torque can be obtained from a hydraulic motor even with changing speeds.

- A hydraulic motor is protected against any damage due to overloading. All hydraulic systems are equipped with relief valves, which open when the system pressure rises too high due to overloading.

Pneumatic/Air Motors are used to produce continuous rotary power from a compressed air system. These motors are similar in construction and function to hydraulic motors. Pneumatic motors are used in applications where low to moderate torque is required. Advantages of air motors over electric motors:

- They do not require electrical power; air motors can be used in volatile atmospheres.
- They generally have a higher power density, so a smaller air motor can deliver the same power as its electric counterpart.
- Unlike electric motors, many air motors can operate without the need for auxiliary speed reducers.
- Air motor speed can be regulated through simple flow-control valves instead of expensive and complicated electronic speed controls.
- Air motor torque can be varied simply by regulating pressure.
- Air motors generate much less heat than electric motors.

SYMBOL OF MOTORS

Figure 6.33 shows the symbols for hydraulic and pneumatic motors.

CLASSIFICATION OF FLUID MOTORS

There are three basic designs of rotary pumps or compressors; the gear pump, the vane pump, and various designs of piston pump or compressor, which have already been explained in Chapter 4. These can also be used as the basis of rotary actuators. The principles of hydraulic and pneumatic motors are very similar but hydraulic motors give large available torques and powers despite lower rotational speeds because of high pressures. Fluid motors are usually classified according to the type of internal element, which is directly actuated by the flow. The most common types of

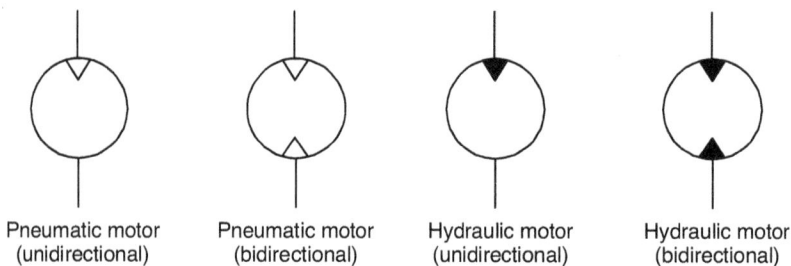

| Pneumatic motor | Pneumatic motor | Hydraulic motor | Hydraulic motor |
| (unidirectional) | (bidirectional) | (unidirectional) | (bidirectional) |

FIGURE 6.33 Symbols for Hydraulic and Pneumatic Motors.

elements are the gear, the vane, and the piston. So the fluid motors are classified as:

- Gear motors
- Vane motors
- Piston motors

Each of these types can be produced by motor manufacturers as either unidirectional or reversible, although most motors used in mobile equipment are the reversible.

GEAR MOTORS

Gear motors are the most common hydraulic units. The operation of a gear motor is shown in Figure 6.34. These consist of a pair of matched spur or helical gears enclosed in a case. Both gears are driven gears, but only one is connected to the output shaft. Operation of gear motor is essentially the reverse of that of a gear pump. Flow from the pump enters the inlet port and flows in either direction around the inside surface of the casing, forcing the gears to rotate as indicated. This rotary motion is then available for work at the output shaft. Like the gear pump, the gears in a gear motor are

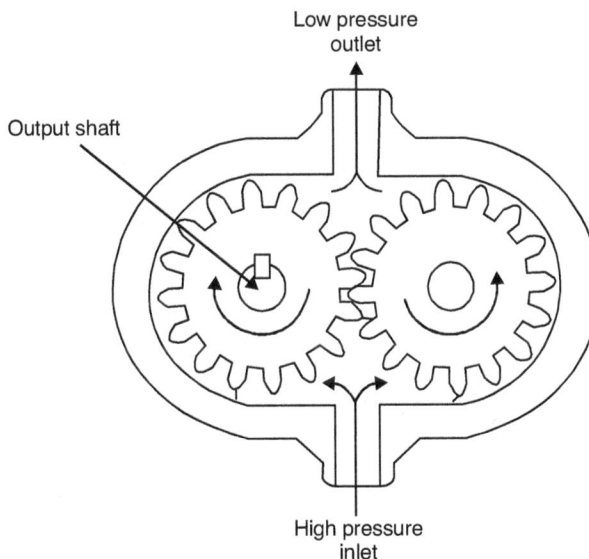

FIGURE 6.34 Gear-Type Motor.

closely fitted in the housing end, and for this reason, flow of fluid through the motor from the inlet to the outlet can occur only when the gears rotate. Gear motors are of fixed displacement type which means that the output shaft speed varies only when the flow rate through the motor changes. These motors are generally two directional, the motor being reversed by direction of fluid through the motor in the opposite direction.

The motor shown in Figure 6.34 is operating in one direction; however, the gear-type motor is capable of providing rotary motion in either direction. To reverse the direction of rotation, the ports may be alternated as inlet and outlet. When fluid is directed through the outlet port, the gears rotate in the opposite direction. Gear motors are the least expensive but the noisiest of the hydraulic motors. These have the ability to operate at high speeds; however, they are inefficient at low speeds.

VANE MOTORS

A typical vane-type motor is shown in Figure 6.35. This particular motor provides rotation in only one direction. When pressure oil enters the inlet port, the unequal area of the vanes results in the development of a torque in the motor shaft. The larger the exposed area of the vane, or higher the pressure, the more torque is developed, which makes the shaft rotate. Because no centrifugal force exists until the motor begins to rotate,

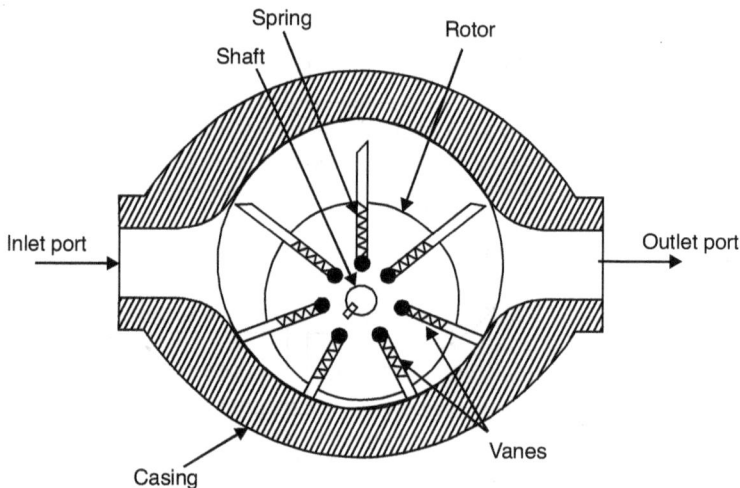

FIGURE 6.35 Vane Motor.

some method must be provided to initially hold the vanes against the casing contour. Springs are often used for this purpose. However, springs usually are not necessary in vane-type pumps because a drive shaft initially supplies centrifugal force to ensure vane to casing contact. Vane motors rotate in either direction but they do so only when the flow rate through the motor is reversed. Hydraulic vane motors are the most popular general purpose motor, but they are limited by their tolerance to high pressure systems and the higher percentage of slippage or internal leakage relative to the lower total fluid flow at low speeds.

PISTON MOTORS

Piston motors are the most commonly used in hydraulic systems. They are basically the same as hydraulic pumps except they are used to convert hydraulic energy into mechanical (rotary) energy. The most commonly used hydraulic motor is the fixed displacement piston type. These motors are positive displacement motors that can develop an output torque at the shaft by allowing oil pressure to act on the pistons. Some equipment uses a variable displacement piston motor where very wide speed ranges are desired. Hydraulic piston motors can be either axial or radial and are generally the most expensive of the hydraulic motors. They have advantages over the other motors, however, in that piston motors are far more adaptable to high torque, low speed operation and higher system pressure applications.

There are three types of piston motors:

- Swash plate in-line-axis piston-type motors
- Swash plate bent axis piston type motors
- Radial-piston motors

Swash Plate In-Line-Axis Piston-Type Motors

The axial piston motor is of the swash plate type and has a bank of cylinders arranged in a circle (360 degrees) parallel to each other (Refer to Figure 6.36). Each cylinder has a piston, which reciprocates with one end of the piston pushing against an eccentric swash-plate located at one end of the bank of cylinders. There is a mechanical arrangement through which the eccentric plate is connected to an output shaft that is axially aligned with the cylinders. During motor operation, the cylinders are filled with high-pressure hydraulic fluid in a particular sequence making the pistons move outwards to push sequentially against the swash-plate causing it to rotate. On the return stroke of the piston, the fluid is swept back at low pressure to return to a reservoir. The operation imparts rotational movement to the output shaft, of which one end is connected to the swash-plate and the other to the workload. This is a design that caters to a very compact cylindrical hydraulic motor.

Axial-piston motors have high volumetric efficiency, combined with excellent operation at both high and low speeds.

Swash Plate Bent Axis Piston Type Motors

These motors are almost identical to the pumps. They are available in fixed- and variable-displacement models in several sizes (Refer to Figure 6.37). Variable-displacement motors can be controlled mechanically or by pressure compensation. These motors operate similarly to in-line motors except that piston thrust is against a drive-shaft flange. A parallel component of thrust causes a flange to turn. Torque is maximum at maximum displacement; speed is at a minimum. This design of piston motor is very heavy and bulky, particularly the variable-displacement motor. Using these motors on mobile equipment is limited. Although some piston type motors are controlled by directional-control valves, they are often used in combination with variable-displacement pumps. This pump-motor combination (hydraulic transmission) is used to provide a transfer of power between a driving element, such as an electric motor, and a driven element. Hydraulic transmissions may be used for applications such as a speed reducer, variable speed drive, constant speed or constant torque drive, and torque converter.

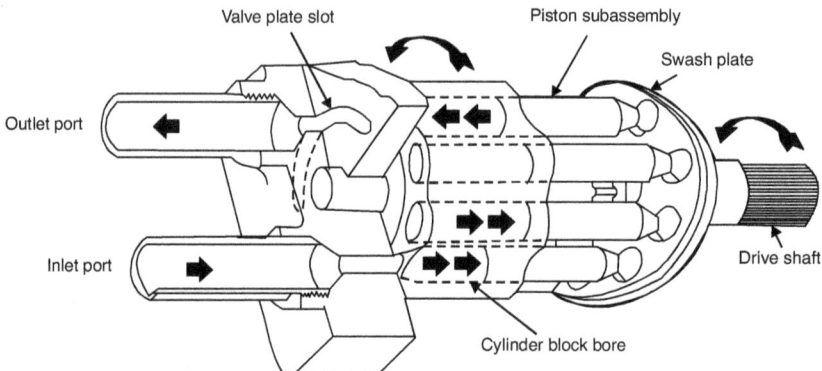

FIGURE 6.36 Swash Plate In-Line-Axis Motors.

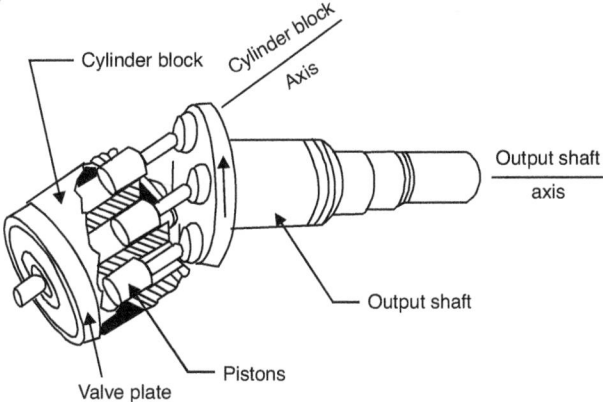

FIGURE 6.37 Swash Plate Bent Axis Motors.

Radial-Piston Motors

Radial-piston motors produce very high torque at low speed. These motors have a cylinder barrel attached to a driven shaft. The barrel contains a number of pistons that reciprocate in radial bores. The radial-piston motor operates in reverse of the radial-piston pump. In the radial-piston pump, as the cylinder block rotates, the pistons press against the rotor and are forced in and out of the cylinders, thereby receiving fluid and pushing it out into the system. In the radial motor, fluid is forced into the cylinders and drives the pistons outward. The pistons pushing against the rotor cause the cylinder block to rotate. Figure 6.38 shows a design with radial cylinders

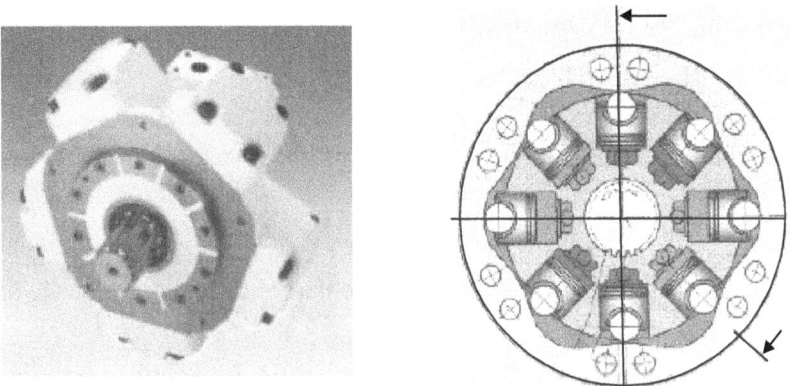

FIGURE 6.38 Radial-Piston Motors.

each in a separate block. The pistons are connected to the shaft by a crank or some other mechanism. Suitable valve designs allow the oil into cylinders and force the pistons to reciprocate and turn the shaft. These produce high power and torque.

APPLICATION OF MOTORS

Hydraulic motors provide solutions in applications involving infinite speed control, stalling under full torque, high power-to-weight ratio, and small size. Their characteristics make them useful in a wide variety of industries. Some of the applications of motors are:

- In the aerospace industry.
- In the food processing industry.
- In construction and agricultural equipment.
- In heavy earth moving equipment like excavators, skids, forklifts, heavy dump trucks, bulldozers, etc.

Cost-wise, piston type hydraulic motors are the costliest whereas gear type motors are the least expensive. However, each has its own advantages depending on how it is used.

EXERCISES

1. What is an actuator?

2. Classify various types of actuators.

3. Classify various types of cylinders.

4. Explain the construction and working principle of a single-acting hydraulic cylinder with a neat sketch.

5. Explain with a neat sketch the construction and working of double-acting cylinder.

6. Draw the symbol for double-acting cylinder.

7. With the aid of sketches, explain the constructional features of a hydraulic cylinder.

8. Differentiate between double-acting hydraulic cylinder and double-acting pneumatic cylinder. Explain the working of the double-acting pneumatic cylinder.

9. Discuss various configurations in single-acting cylinders.

10. List various differences between single-acting and double-acting cylinders.

11. How is the hydraulic cylinder different from the pneumatic cylinder?

12. With the help of a neat sketch, show the various parts of hydraulic and pneumatic double-acting cylinders.

13. Explain the construction, working, design, and mounting of hydraulic and pneumatic cylinders with the help of neat sketches.

14. What are the functions of the ram type and telescoping cylinders?

15. What is a tandem cylinder?

16. List the various standard mounting styles being used in industry.

17. What are the applications of the tandem and duplex cylinders?

18. What is a cylinder cushion? State the purpose of cushioning in cylinders.

19. List various materials being used for the manufacturing of a cylinder.

20. Sketch a double-acting cushioned hydraulic cylinder and label the components.

21. Briefly discuss the types of motors.

22. List the parameters for motor ratings.

23. With the help of a neat sketch, explain the construction and working of gear and vane motors.

24. Sketch the standard symbol for a pneumatic motor.

CONTROL VALVES

INTRODUCTION

In a pneumatic or hydraulic system, the objective is generally to supply power from a compressor or pump to the fluid actuator (*i.e.*, motor or cylinder). But the problem is not always the same, example, sometimes the stroke of the cylinder moves very fast whereas sometimes it is extends very slowly. The solution to these fluid problems can be obtained by using a combination of valves.

In almost every fluid system, valves are required to control the direction, flow, pressure, or quantity of fluid. Based on the application, valves can be divided into four categories: *direction control valves, flow control valves, pressure control valves,* and *special valves*. Valves can be classified according to their construction, *i.e., poppet valves* and *spool valves*.

We can also classify the valves on the basis of mode of actuation, *i.e.*, they can be actuated manually, electrically, electromagnetically, or with the help of fluids. This chapter concentrates on explaining identification, function, construction, and location of valves in relation to hydraulic and pneumatic circuits.

Hydraulic systems are high-pressure systems and pneumatic systems are low-pressure systems. Hydraulic valves are made of strong materials (e.g., steel) and are precision manufactured. Pneumatic valves are made from cheaper materials (e.g., aluminum and polymer) and are cheaper to manufacture.

CLASSIFICATION OF VALVES

Valves are classified as:

- Direction control valve
- Flow control valve
- Pressure control valve
- Check valve

Direction control valves are most important among all other valves because of their frequent use in a fluid power system. As the name implies, they are used to direct the flow of fluid in a desired direction.

Flow control valves are those in which the flow of fluid is varied according to requirements. They are of fixed and adjustable flow control type.

Pressure control valves are used for regulating the pressure of fluids in a fluid system. Pressure relief valve, regulating valve, etc. comes under this category.

All other valves in which the free flow of fluid is allowed only in one direction come under the category of *check valves*, e.g., non return valve, quick exhaust valve, twin pressure valve, non return flow control valve, etc.

Some of the important valves are discusses below.

DIRECTION CONTROL VALVES

Direction control valves are designed to direct the flow of fluid, at the desired time, to the point in a fluid power system where it will do work. Direction control valves are intermediators, which supply pressurized fluid from component or pump to an actuator (cylinder or motor) as per its requirements. It starts, stops, and regulates the direction of fluid and directs the fluid in the line in which it is required. Direction control valves are specified by the number of ports and the number of its possible positions.

SYMBOL AND DESIGNATION OF DIRECTION CONTROL (DC) VALVE

1. A DC valve is specified by the number of fluid ports and the number of positions a valve can take. The number before the slash identifies the number of ports, and the second number refers to number of positions a valve can take, e.g., a 3/2 DC valve is a 3 port and 2 position direction control valve, 4/3 DC valve is a 4 port and 3 position direction control valve, etc.

2. The symbol for a DC valve is built around a series of boxes or rectangles, one box for each usable position of the valve. The number of adjacent boxes indicates whether a valve is a two or three position valve. (Refer to Figure 7.1.)

3. Ports are ways or openings in the DC valve from where fluid under pressure enters and exits. Usually a DC valve may have 2, 3, 4, or 5 ports. Each valve port should be shown on the symbol. The ports are shown on only one of the boxes, the box that represents the flow path that exists at the start of the cycle. Some examples are shown in Figure 7.2.

4. A DC valve may take a maximum of 3 positions. They are ON, OFF, or Neutral (1, 2, or 0) and each position of the valve are shown by a separate box. (Refer to Figure 7.3.)

A 2-Position valve is shown by two boxes.

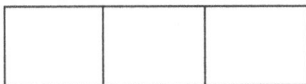

A 3-Position valve is shown by three boxes.

FIGURE 7.1 Symbols for 2-Position and 3-Position Valves.

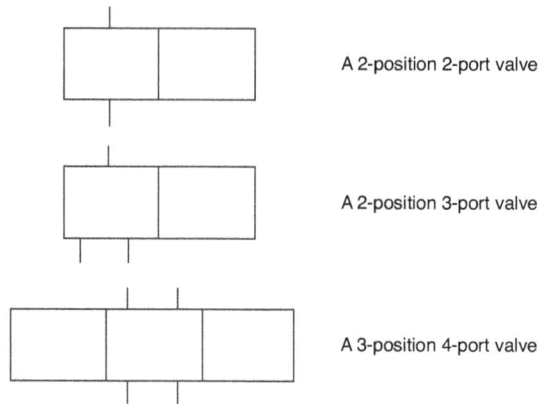

A 2-position 2-port valve

A 2-position 3-port valve

A 3-position 4-port valve

FIGURE 7.2 Symbols for 3 Types of Valves.

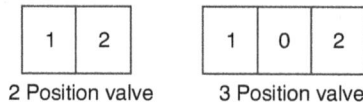

1	2

2 Position valve

1	0	2

3 Position valve

FIGURE 7.3 Symbols for Valves with Positions Shown.

5. The naming of ports is standardized, *i.e.*, it is followed all over and is given below:

P	Pressure port/supply port.
R, T	Exhaust ports, R for pneumatics and T for hydraulics.
A, B, or C	Working ports, which supply pressure to actuators (cylinder or motors).
L	Leakage port. (Only in hydraulics)
X, Y, Z	Pilot supplies. (Shown by dotted lines)
⟶	Arrows used for hydraulics are filled ones.

6. Lines and arrows are used to represent the flow and direction of fluid across the valves. Each box contains a group of lines that represent the flow paths the valve provides when it is in that position. If a port is blocked, it is shown by the symbol as shown in Figure 7.4. If two ports are connected and fluid can flow, this is shown by a line drawn between the two ports. The direction in which fluid flows during a normal operating cycle is shown by putting arrowheads at the ends of the flow paths next to the ports from where the fluid will come out. The above can be explained with the help of a 2/2 DC valve, which is shown in Figure 7.4.

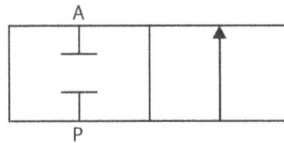

FIGURE 7.4 2/2 DC Valve.

In the above figure, the left box shows the conditions that exist at the start of the cycle. Port P is blocked, and port A is blocked. When the valve is shifted, the flow condition shown in the right hand box exists, *i.e.*, port P is open to port A.

7. The normal position of a valve is that position at which the naming is done. In the valve symbol shown in Figure 7.5, right-side position is the normal position of the valve. X and Y are the pilot supplies.

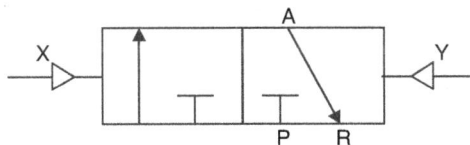

FIGURE 7.5 3/2 Pilot Operated Normally Closed DC Valve.

CLASSIFICATION OF DC VALVES

DC valves are classified:

▨ On the basis of method of valve actuation.

▨ On the basis of construction.

CLASSIFICATION OF DC VALVES ON THE BASIS OF METHODS OF VALVE ACTUATION

The flow of fluids is controlled by changing the position of DC valves. Knowledge of these control methods is very important because actuation of valves is always subject to the application of valves. The actuators are devices or methods that cause the valve to shift from one position to another. Valve actuators are selected based upon a number of factors such as the torque necessary to operate the valve and the need for automatic actuation. There are different methods of actuation, which can be used to actuate DC valves. The different types of actuators available are:

1. *Manual Actuation:* It means when the designed system is not fully automatic and no relationship can be defined between operation of valves and time lag between two consecutive operations. Manual actuators are capable of placing the valve in any position but do not permit automatic actuation. There are four types of manual actuations:

   (a) *Push Button:* A button is pushed until the pressure is required to flow from that valve is reached and then the button is released; the valve again attains its previous position under the action of a spring (because a spring is used in combination on the other side of the valve).

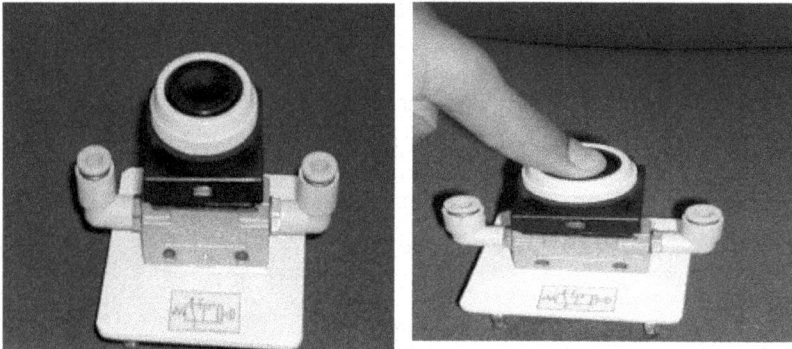

FIGURE 7.6 DC Valve with Push Button.

   (b) *Push/Pull Button:* In this case, spring return is not provided on the other side of the valve hence the operator has to pull the valve again when the position has to be changed.

   (c) *Pedal:* Valves can be actuated by pressing a foot pedal.

   (d) *Lever:* Valves can be actuated by pulling the lever backward and pushing it forward.

2. *Mechanical Actuation:* This category includes all the mechanical components, which are operated by an external force. Types of mechanical actuations are:

   (a) *Roller:* For changing the position of the valve, a spring return lever is used. The force of push is applied on the lever and the position of the valve is changed. In order to make friction free contact of the lever and force applying member, a roller is mounted on the lever. The external force-applying member may be the piston rod of the cylinder when the piston is extending or it may be a cam, which continuously provides motion to the lever via the roller. (Refer to Figure 7.7.)

(b) *Plunger*

(c) *Spring:* A spring is used in most directional control valves to hold the flow-directing element in neutral position. Spring holds the non actuated valve in one position until an actuating force great enough to compress the spring shifts the valve. When the actuating force is removed, the spring returns the valve to its original position.

3. *Pilot Operation/Actuation:* When the pressurized fluid is used to operate the valves in addition to extending and retracting of the cylinder, it is called pilot actuation or pilot operation and the additional valves used for this purpose are called pilot valves. Pilot-operated valves (Refer to Figure 7.8) can be mounted at any position to which pressurized fluid can be piped. Depending upon the type of fluid used, pilot actuation can be of two types: hydraulic, pneumatic.

Hydraulic actuators use oil under pressure for its operation. Because the pressurized fluid needs to be controlled by a DC valve, such valves cannot be used alone. Pneumatic actuators use compressed air for its operation.

4. *Detent:* Detent are locks that hold a valve in its last position after the actuating force is removed until a stronger force is applied to shift the valve to another position. The detent may then hold this new position after the actuating force is again removed. (Refer to Figure 7.9.)

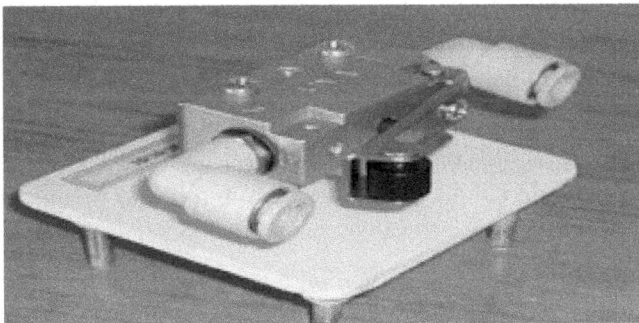

FIGURE 7.7 Roller Operated DC Valve.

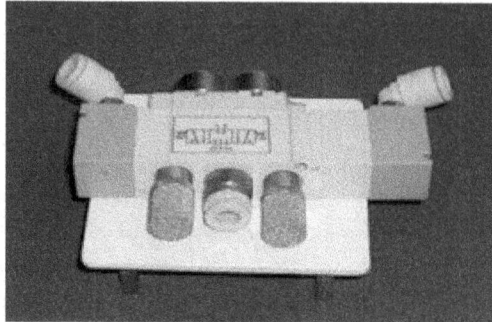

FIGURE 7.8 Pilot-Operated DC Valve.

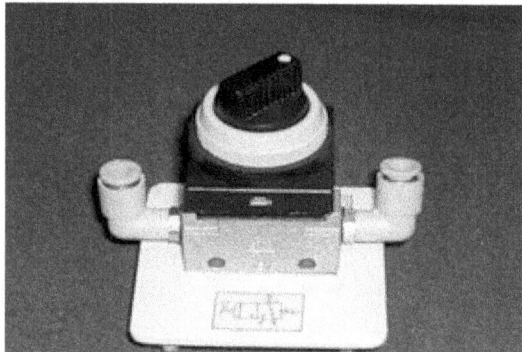

FIGURE 7.9 DC Valve with Detent.

5. *Electromagnets:* Electromagnets are commonly known as solenoids. Solenoid actuated valves are popular for industrial machines because of ready availability of electrical power in industries. A solenoid consists of a coil and an armature. When the coil is energized by an electric current, a magnetic field is created around it, which attracts the armature. When the armature moves, it pushes the spool. (Refer to Figure 7.10.)

Electromagnets are nothing but temporary magnets. They behave like magnets when a current is passed through them. In the same way a plunger is made to move up and down in a frame by energizing it, this motion is used to actuate valves.

6. *Pilot Solenoid Operation:* When pilot valves and solenoid valves are used in combination, it is called pilot solenoid operation. For example, a 3/2 pilot operated DC valve (pneumatic or hydraulic), which is used to operate the main 5/2 DC valve, is itself operated by a solenoid valve.

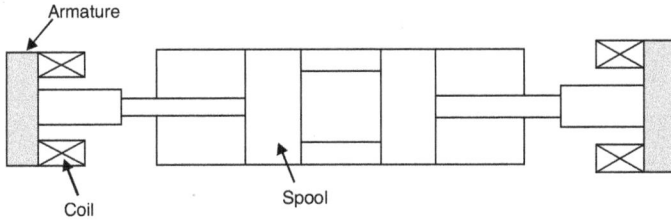

FIGURE 7.10 Solenoid.

SYMBOLS FOR VALVE ACTUATORS

The symbols for valve actuators are drawn next to the end of the valve boxes. There are standard symbols for valve actuators. These symbols may be drawn on either end of the valve without altering their meaning. The symbols for valve actuators are shown in Table 7.1.

TABLE 7.1 Symbols for Valve Actuators.

Manual	*Mechanical*	*Pilot Operation*	*Electrical*	*Pilot Solenoid Operation*	*Detent*
Push Button	Roller	Pneumatic Actuation	Solenoid	Electro Pheumatic	
Pedal	Plunger			Electro Hydraulic	
Lever	Spring	Hydraulic Actuation			
Push/Pull Button					

EXAMPLES OF DC VALVES WITH ACTUATORS

Figure 7.11 shows a 4/2 solenoid-operated spring-returned DC valve. The rule is that each valve actuator is drawn next to the box that exists when that actuator is in command.

In the above figure, when the spring has control of the valve, the flow paths in the left-hand box exist. When the solenoid (the right-hand actuator) is in command, the flow paths in the right-hand box exist.

A typical DC valve is made up of three parts. (Refer to Figure 7.12.)

Some of the common DC valves with actuators are shown in Table 7.2.

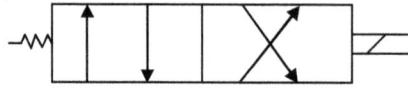

FIGURE 7.11 4/2 Solenoid-Operated Spring-Return DC Valve.

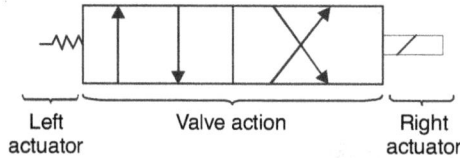

Left actuator   Valve action   Right actuator

FIGURE 7.12 A Sample DC Valve.

TABLE 7.2 DC Valves with Actuators.

Name of DC Valve	Symbol
2/2 push button operated spring returned DC valve	
3/2 push button operated spring returned DC valve	
3/2 roller operated spring returned DC valve	
3/2 lever operated spring returned DC valve	
3/2 air pilot operated spring returned DC valve	
4/2 pilot operated DC valve	
4/2 push button operated spring returned DC valve	
4/2 solenoid air operated spring returned DC valve	
4/3 solenoid operated spring returned DC valve	

CLASSIFICATION OF DC VALVES ON THE BASIS OF CONSTRUCTION

Valves may be classified on the basis of construction into three types:

1. Rotary valves
2. Poppet/seat valves
3. Spool/sliding valves

1. Rotary Valves: A rotary valve consists of a rotary spool that is placed inside a valve body. A rotary spool has passages for fluid to flow. The spool can be rotated manually or mechanically. When the spool rotates, it either blocks the valve port or connects them to the passages. The working of rotary valves is explained in Figure 7.13 by taking all the positions of valves.

Initially pressure port P is connected to working port A and port B is connected to the tank as shown in Figure 7.13 *(a)*. All the ports are blocked in the intermediate position, when the valve rotates as shown in Figure 7.13 *(b)*. In the extreme position, when the valve spool is rotated, the flow from pressure port P goes to port B and flow from port A goes to tank as shown in Figure 7.13 *(c)*. Rotary valves are compact, simple, and mainly used for hand operation in pneumatic systems.

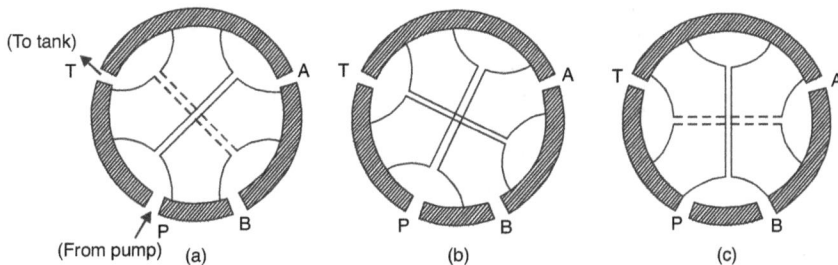

FIGURE 7.13 Working of Rotary Valve.

2. Poppet Valves: Poppet or seat valves have a ball or a piston cone as a seating element, which is normally kept closed by a light spring. In these valves, the poppet is forced off its seat allowing flow in only one direction. No flow is allowed in the opposite direction. Figure 7.14 shows a simple poppet valve. When the push button is pressed, as shown in Figure 7.14 *(b)*, the ball lifts off its seat and allows fluid to flow. When the push button is released, the spring forces the ball up again closing the valve. (Refer to Figure 7.14 *(a)*.) Seat valves have the advantage of perfect sealing over spool valves.

3. *Spool Valves:* A spool type of directional valve is the most common type of valve used in a hydraulic system. A spool valve consists of a spool and valve housing. Different ports are cut in the valve housing and the spool is constrained to move axially inside the housing. Fluid is routed to or from the work ports as the spool slides between passages to open and close flow paths, depending on the spool position. Spool valves are cheap and easy to manufacture. They require less finishing as compared to seat valves. Spool valves are compatible with all the methods of valve actuation, whereas this is not the case of seat valves. Perfect sealing in spool valves cannot be achieved because of their specific construction. These valves are widely used because they can be shifted to 2, 3, or more positions for moving fluid between different combinations of inlet and outlet ports.

Figure 7.15 shows the typical directional control valve, which consists of a valve body with internal flow passages within the valve body and a sliding spool. When the spool is in position as shown in Figure 7.15 (*a*), the flow path is open, *i.e.*, pressure port P is connected to working port A. When the spool is shifted on either side, the flow path is closed as shown in Figure 7.15 (*b*).

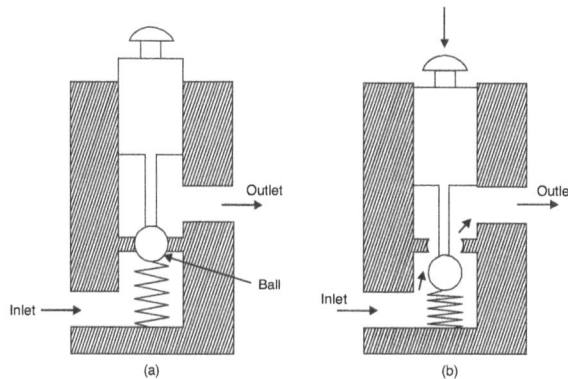

FIGURE 7.14 Working of Poppet Valve.

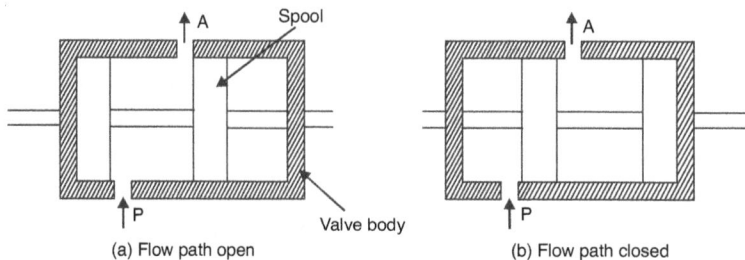

FIGURE 7.15 Working of Spool Valve.

CONSTRUCTION OF 2/2, 3/2, AND 4/2 DIRECTION CONTROL VALVES

2/2 DC VALVES

A 2/2 DC valve is a 2 port and 2 position direction control valve. A 2/2 DC valve consists of two ports connected to each other with passages, which are connected or disconnected. A 2/2 DC valve can be of two types:

- 2/2 spool type DC valve
- 2/2 seat type DC Valve

2/2 Spool Type DC Valve

The 2/2 spool valve is explained with the help of the diagram given below. Initially in Figure 7.16 (*a*), the 2/2 DC valve is at its normally closed position, *i.e.*, pressure port P is not connected with working port A. As the spool moves back and forth, it either allows or prevents fluid flow through the valve. In Figure 7.16 (*b*), when the manual push button is pressed, the spool moves inside and thus connects pressure port P with working port A. Positions are shown with symbolic representation.

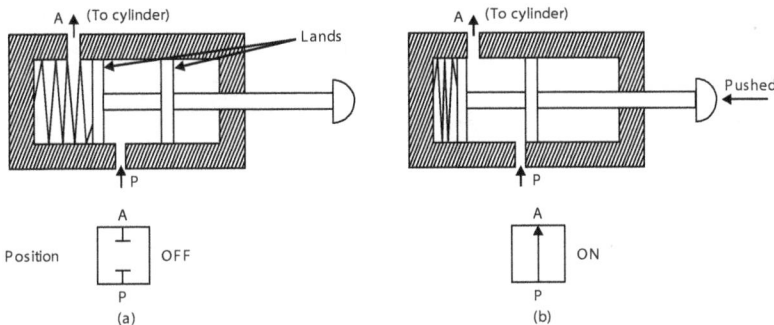

FIGURE 7.16 2/2 Spool Type DC Valve.

Symbol

Figure 7.17 shows the symbol of a 2/2 DC valve.

2/2 Seat Type DC Valve

To understand the 2/2 seat type valve refer to Figure 7.18 (*a*) and (*b*). In this type of construction, a poppet valve with its valve seat is resting on

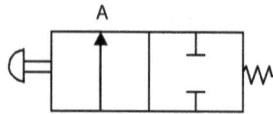

FIGURE 7.17 2/2 Push Button Operated Spring Returned DC Valve.

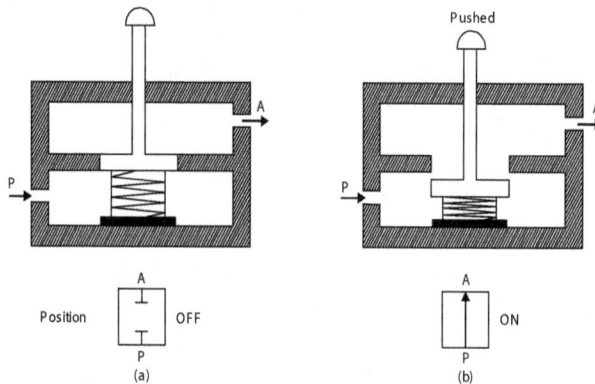

FIGURE 7.18 2/2 Seat Type DC Valve.

a spring. A manual push button is provided to change the position of the valve. In Figure 7.18 (*a*), the valve is in its normally closed position as port P and port A are not connected, so fluid cannot pass through this valve. This represents the 'OFF' position of the valve and when the manual push button is pressed as shown in Figure 7.18 (*b*), the valve moves downward hence making way for fluid to pass, and in this way the 'ON' position of the valve can be attained.

3/2 DC VALVES

A 3/2 DC valve is a 3 port and 2 position direction control valve. A 3/2 DC valve consists of three ports connected to each other with passages, which are connected or disconnected. 3/2 DC valve can be of two types:

- 3/2 spool type DC valve
- 3/2 seat type DC valve

3/2 Spool Type DC Valve

Construction of a 3/2 DC spool valve is similar to a 2/2 DC spool valve with a small difference that outlet port R or T (R for pneumatics and T for hydraulics) is added. Again the position of the spool is shifted for the extending and retracting of a cylinder. (Refer to Figure 7.19.)

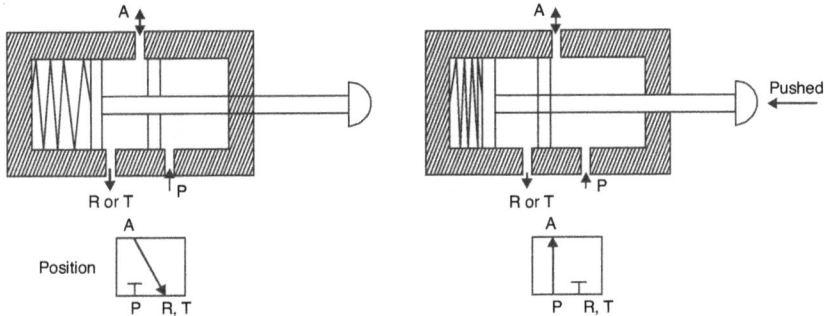

FIGURE 7.19 3/2 Spool Type DC Valve.

Symbol

Figure 7.20 shows the symbol of a 3/2 DC valve.

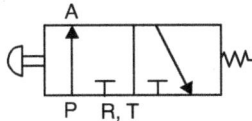

FIGURE 7.20 3/2 Push Button Operated Spring Returned DC Valve.

3/2 Seat Type DC Valve

The general arrangement of a 3/2 seat type DC valve is shown in Figure 7.21. A lever is used as a means of actuation. When the lever is pulled, the left-side pressure port P gets connected with working port A and fluid from the tank or source is supplied to the actuator and similarly when this lever is pushed, the right-side port P is closed and working port A gets connected with exhaust port R or T and fluid from the cylinder is exhausted.

Symbol

Figure 7.22 shows the symbol of a 3/2 DC valve.

4/2 DC VALVES

A 4/2 DC valve is a 4 port and 2 position direction control valve. A 4/2 directional valve consists of four ports connected to each other with passages, which are connected or disconnected. Four-way valves are used to

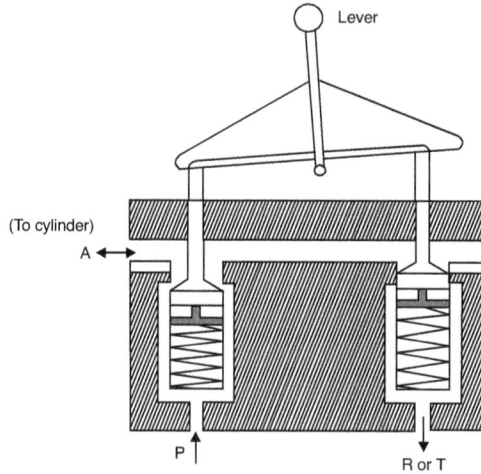

FIGURE 7.21 3/2 Seat Type DC Valve.

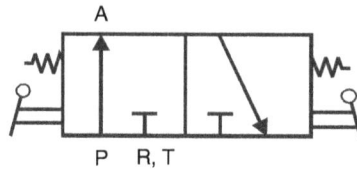

FIGURE 7.22 3/2 Lever Operated Spring Returned DC Valve.

control the direction of fluid flow in a hydraulic/pneumatic circuit. A 4/2 DC valve can be of two types:

- 4/2 spool type DC valve
- 4/2 seat type DC valve

4/2 Spool Type DC Valve

A 4/2 spool type DC valve is very common and its main application is in the control of a double acting cylinder. This valve consists of four ports and two positions as shown in Figure 7.23. The working of a 4/2 DC valve is explained by taking all positions of valve. In the figure shown below, the spool is actuated by means of pilot supplies from both sides.

As shown in Figure 7.23 (a), when the spool is in a position where the supply pressure P is connected to port A and port B is connected to the exhaust port, the cylinder will extend. As the spool moves to the other extreme position, i.e., left, the pressure port P is connected to port B and

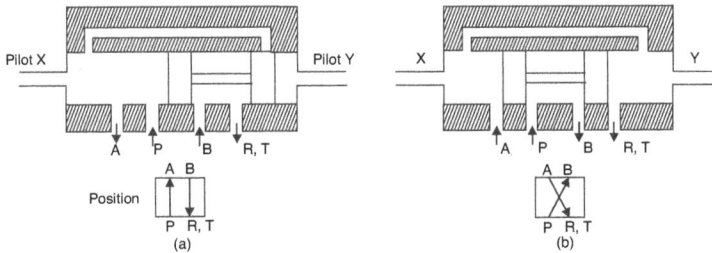

FIGURE 7.23 4/2 Spool Type DC Valve.

port A is connected to the exhaust port (Refer to Figure 7.23 (*b*)), now the cylinder retracts. With a directional control valve in a circuit, the cylinder's piston rod can be extended or retracted and work is performed.

Symbol

Figure 7.24 shows the symbol of a 4/2 DC valve.

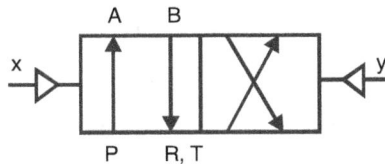

FIGURE 7.24 4/2 Pilot Operated DC Valve.

4/2 Seat Type DC Valve

A 4/2 seat valve is shown in Figure 7.25. It has four ports (namely P, R, A, and B) and two positions. A push button with a spring return is used to control the vertical movement of the seat valves inside the guides. The ports are alternately connected and departed by moving the seat valves up and down. Initially when the button is not pressed, the compressed air can pass from pressure port P to the working port B from the passage between the valve seats and compressed air may exhaust out from working port A to exhaust port R on other side. When the push button is pressed, pressure port P is joined with working port A and working port B gets linked with exhaust port R by means of internal galleries.

Symbol

Figure 7.26 shows the symbol of a 4/2 DC valve.

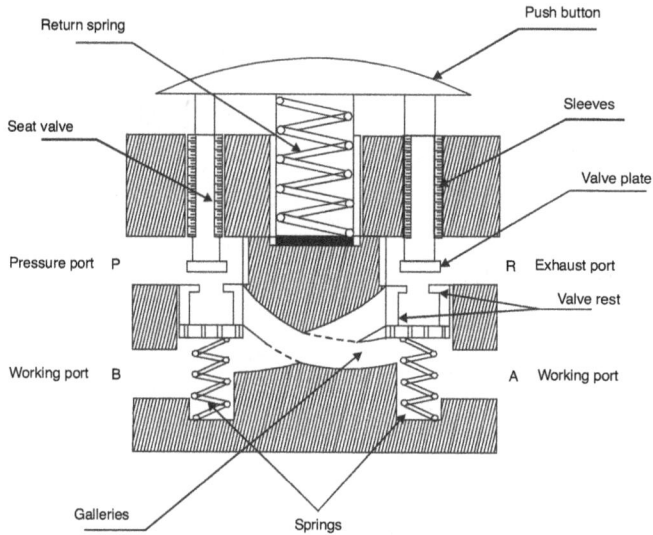

FIGURE 7.25 4/2 Seat Type DC Valve.

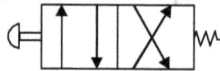

FIGURE 7.26 4/2 Push Button Operated Spring Returned DC Valve.

CENTER CONDITIONS IN 4 WAY DC VALVES

The default position of a spool in a 4 way (4 ports) DC valve is called its center position. The center position of the spool determines the operation of the direction control valve. The different center conditions in DC valves are:

- Open center
- Closed center
- Tandem center
- Float center

Open Center Condition: In the open center condition, all ports (namely A, B, P, and T) are interconnected to each other as shown in Figure 7.27. In this condition, the actuator is not held under pressure and the pump flow goes to the tank at low pressure. The symbol for the open center condition shows all the four ports connected to each other. This type of

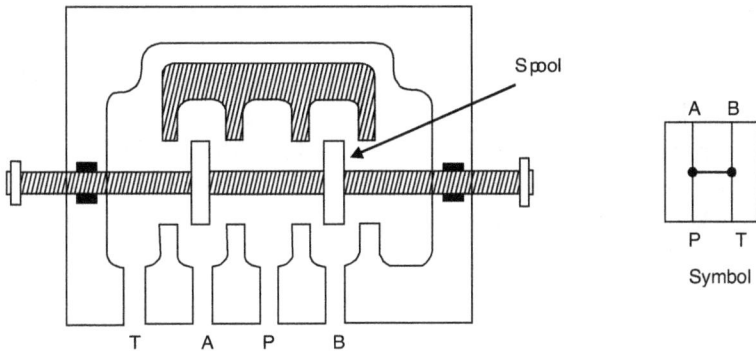

FIGURE 7.27 Open Center Condition in a 4 Way DC Valve.

center condition is used when the pump flow is supposed to perform a single operation only. This condition is suitable for applications where the actuator is not held under pressure. This spool helps to minimize shocks while being moved from one position to other.

Closed Center Condition: In the closed center condition, all ports are blocked as shown in Figure 7.28. In this condition, the fluid

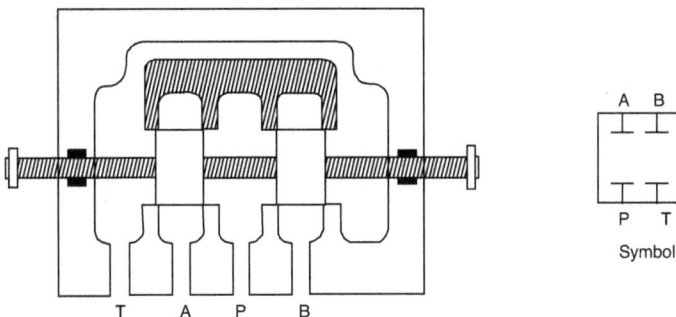

FIGURE 7.28 Closed Center Condition in a 4 Way DC Valve.

cannot pass through the valve and can be used for any other operation. The valves with closed center condition are used when a single pump performs more than one operation and where there must be no pressure loss. The symbol for closed center condition shows all the four ports blocked. The disadvantage of closed center condition is that the pump flow cannot go to the tank under this condition and there is possibility of leakage.

Tandem Center Condition: In the tandem center condition, the pressure port is connected to the tank port and both the actuator ports are blocked as shown in Figure 7.29. In this condition, the pump flow goes to tank at low pressure. This center condition is used in applications where two or more actuators are in series with the same power source. The symbol of a tandem center condition shows actuator ports A and B blocked and pressure port P connected to the tank port T.

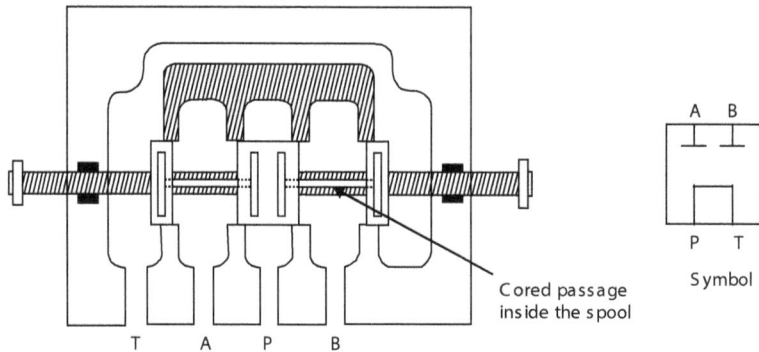

Cored passage inside the spool

Symbol

T   A   P   B

FIGURE 7.29 Tandem Center condition in a 4 Way DC Valve.

Float Center Condition: In the float center condition, the pressure port is blocked and the actuator ports are connected to the tank port as shown in Figure 7.30. In this condition, pressure is maintained in the pressure port and removed from the actuator port to the tank. This center condition is used in applications using pilot operated directional valves. The symbol of a float center condition shows the actuator ports A and B connected to tank port T and pressure port P is blocked.

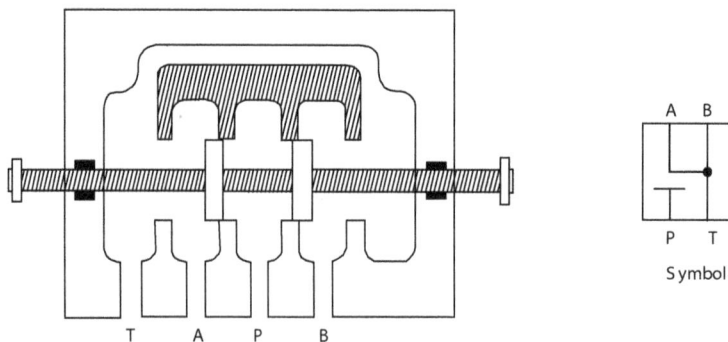

Symbol

T   A   P   B

FIGURE 7.30 Closed Center Condition in a 4 Way DC Valve.

CHECK VALVE

Check valve is a one-way directional valve. Its function is to allow the flow in one direction only. It does not permit the flow in the opposite direction. (Refer to Figure 7.31.) For this reason, they are also known as non-return valves. In order to avoid any leakage, these valves are always of poppet type construction. They consist of a ball or a poppet, which is kept in its normally closed position by a spring. When the fluid pressure overcomes the spring force, the poppet is forced off its seat, allowing flow in that direction. Flow is not possible in the opposite direction. This valve is used frequently in hydraulic and pneumatic circuits because of its number of applications.

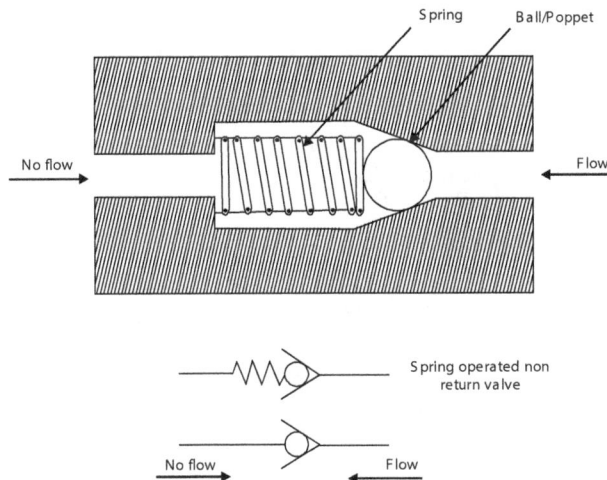

FIGURE 7.31 Check Valve.

Applications of Check Valve

1. *To prevent return flow in hydraulics:* Check valves are used to prevent return flow of fluid from the hydraulic system. Its application is shown in the basic hydraulic circuit diagram for control of single acting cylinder (Refer to Chapter 8). Like the pressure relief valve, a check valve is required in every hydraulic system.

2. *Speed control:* These valves can also be used to control the speed of a pneumatic or hydraulic cylinder. This application is shown in speed control in circuits (Refer to Chapter 8).

Speed in pneumatic and hydraulic circuits can also be controlled by using a check valve rectifier. Working of check valve rectifiers can be understood by studying the example given in Figure 7.32. This circuit is drawn to obtain the slow forward and backward motion to the ram. When pressurized fluid (either hydraulics or pneumatics) is fed to pressure port P, it escapes out of the DC valve via working port A and go

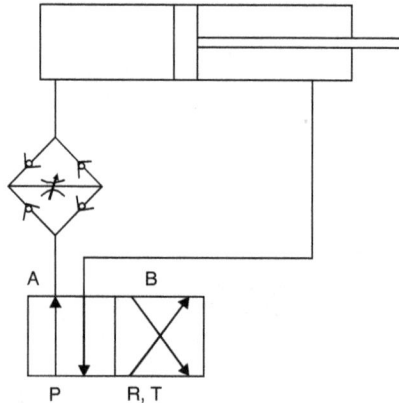

FIGURE 7.32 Check Valve Rectifier.

the rectifier. In this, check valves are so arranged that fluid have to pass through variable speed control when flowing from the pump/compressor to the cylinder. As a result the speed is reduced, *i.e.*, the piston moves slow during forward stroke.

When the position of the DC valve is changed, the pressurized fluid starts flowing through the rod side of the cylinder and is discharged from the other side of the cylinder. The pressurized fluid coming out of the cylinder has to pass through check valves and variable restrictors. This results in the lowering of speed of the piston during the backward stroke. Hence, speed of the cylinder can be reduced by the use of flow control valves.

3. *To bypass the blocked lines in hydraulics:* Sometimes check valves are also used to bypass the flow from blocked lines as shown in Figure 7.33.

PILOT OPERATED CHECK VALVE

This type of check valve allows flow in one direction but prevents flow in the opposite direction unless pilot pressure is applied. A pilot operated

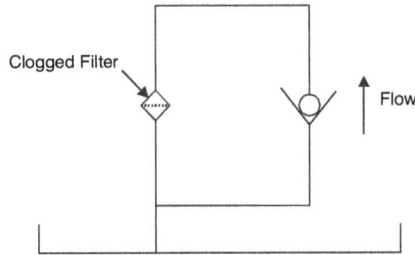

FIGURE 7.33 A Check Valve Can be used to Bypass a Clogged Filter.

check valve comprises of a movable poppet and a piston, which acts on the poppet based upon the pilot pressure. It consists of pilot port X, one inlet port, and one outlet port. (Refer to Figure 7.34.)

As shown in Figure 7.34 (*a*), when fluid pressure is applied through the inlet port, it causes the poppet to shift, allowing flow in that direction. No flow is possible in the opposite direction. The reverse flow is possible only when pilot pressure is applied. As shown in Figure 7.34 (*b*), as pilot pressure is applied through pilot port X, the fluid pressure forces the piston to shift the poppet allowing flow in reverse direction.

Symbol

The symbol of a pilot operated check valve is shown in Figure 7.35. A pilot pressure line is shown acting in a direction, which opens the valve.

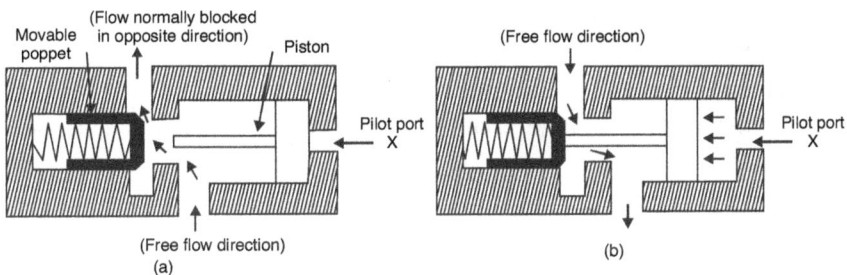

FIGURE 7.34 Pilot Operated Check Valve.

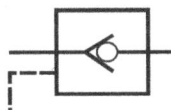

FIGURE 7.35 Symbol for Pilot Operated Check Valve.

PRESSURE CONTROL VALVES

Pressure control valves are found in every hydraulic system, where they act as safety devices. These valves assist in a variety of functions, from keeping system pressures safely below a desired upper limit to maintaining a set pressure in the part of a circuit. Most pressure control valves are classified as normally closed. This means that flow to a valve's inlet port is blocked from an outlet port until there is enough pressure to cause an unbalanced operation. Valves that come under this category include relief, reducing, sequence, counterbalance, and unloading valves. Pressure control valves can be designed either as a poppet/seat type or spool type valve. Spool type valves have the advantage of short stroke and therefore rapid response. Seat type valves have the advantage of being totally leak proof. On the other hand, spool type pressure control valves provide precision control.

PRESSURE RELIEF VALVE

Pressure relief valves are the most common types of pressure control valves used in hydraulics. It is always required that pressure in the hydraulic line should not increase the design limit of components in a hydraulic system. A pressure relief valve is required in a hydraulic line to limit the pressure in line and to ensure safety of the components of a hydraulic system from the danger of overload and bursting. It is generally connected between the pump line and the tank line so as not to obstruct the fluid passage and comes under action when required. This valve is also called a maximum pressure valve or safety valve due to its function.

The pressure relief valve shown in Figure 7.36 consists of a body section containing a piston/plunger that is retained by a spring and a cover section

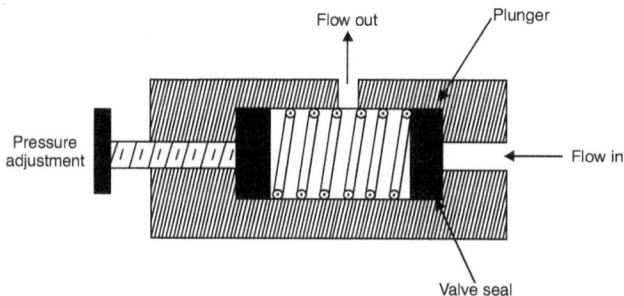

FIGURE 7.36 Pressure Relief Valve.

that controls the piston body movement. It is set at a particular pressure by adjusting a spring compression provided in the valve. Initially, valve is shown in its rest position. When pressure in the line increases, the fluid pressure overcomes the set pressure and the piston/plunger (spring loaded) is forced to move inside by making way for fluid to enter the valve. The fluid is exhausted out via another port directly to the tank, thereby saving the system components from the danger of bursting. When the pressure in the hydraulic system normalizes again, *i.e.*, pressure outside the valve is less than pressure inside the valve, the plunger again comes back to its initial position and passage is blocked.

Symbol

Symbol of a pressure relief valve is shown in Figure 7.37. A box represents the symbol of a relief valve with a spring on one side. This means that the valve is spring operated. The arrow on the spring means that it is adjustable.

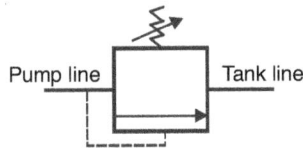

FIGURE 7.37 Symbol for Pressure Relief Valve.

Cracking Pressure and Pressure Override

- *Cracking pressure:* It is defined as the pressure at which the relief valve opens first and allows fluid to pass through it.
- *Full flow pressure:* It is defined as the pressure at which maximum fluid is passing through the valve.
- *Pressure Override:* The difference between full flow pressure and cracking pressure is known as pressure override.

PRESSURE REDUCING VALVE

Pressure reducing valves are used to attain low pressures in a hydraulic system where required. Here the emphasis is on the precise control of fluid. For example, if in a hydraulic system, the branch circuit pressure is limited to 250psi, but the main circuit is operating at 750psi, a pressure relief valve is adjusted in a main circuit to a setting above 750 psi to meet main circuit's requirements. However, it would surpass a branch circuit pressure of 250psi. Therefore, besides a pressure relief valve in the main circuit, a pressure-reducing valve must be installed in a branch circuit and set at 250psi.

A pressure reducing valve is shown in Figure 7.38. The branch circuit pressure is set by adjusting the spring's compression. The fluid from the main circuit enters through the inlet port and enters the branch circuit through the outlet port. (Refer to Figure 7.38 (a).) The pressure of fluid passing from the valve is almost equivalent to pressure exerted by the spring from upwards. Hence, the spring movement is negligible. In Figure 7.38 (b), pressure of fluid passing from the valve is very high as compared to system pressure, i.e., pressure set by the spring. Hence, fluid pressure will be exerted on the bottom of the spool. As a result, the spool will move upward, thereby reducing the passage of high-pressure fluid at the inlet port. This will lower down the pressure at the outlet port.

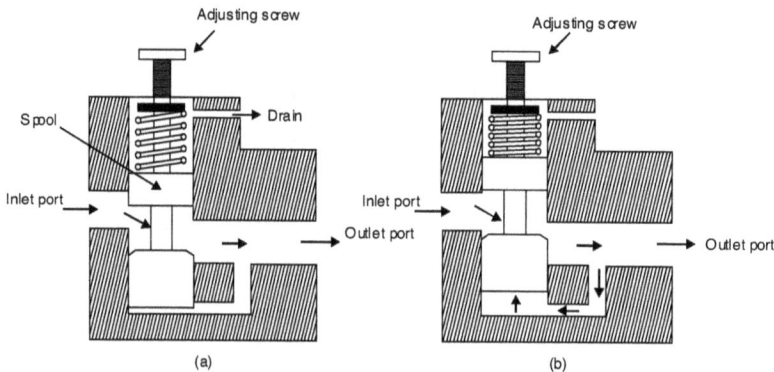

FIGURE 7.38 Pressure Reducing Valve.

SEQUENCE VALVE

Sequence valves used in hydraulic systems are also pressure control valves. A sequence valve is used to control various operations in a sequence/order. For example, a sequence valve is used for sequencing of two cylinders to perform a drill operation. One cylinder clamps the workpiece and the other cylinder performs the drilling operation. A sequence valve in principle and in operation is very similar to a pressure relief valve. In a pressure relief valve, once the set pressure is reached, the relief valve drains the fluid back to the tank, whereas in a sequence valve, once the set pressure is reached, the pressurized fluid is made available for the next operation. Sequence valves are normally closed 2-way valves. They regulate the sequence in which various functions occur in a circuit.

The operation of a typical pressure-controlled sequence valve is illustrated in Figure 7.39. The opening pressure is obtained by adjusting the tension of the spring that normally holds the piston in the closed position. The top part of the piston has a larger diameter than the lower part. Fluid enters the valve through the inlet port, flows freely around the lower part of the piston, and exits through the outlet port to perform the required operation in the primary unit as shown in Figure 7.39 (*a*). This fluid pressure also acts against the lower surface of the piston.

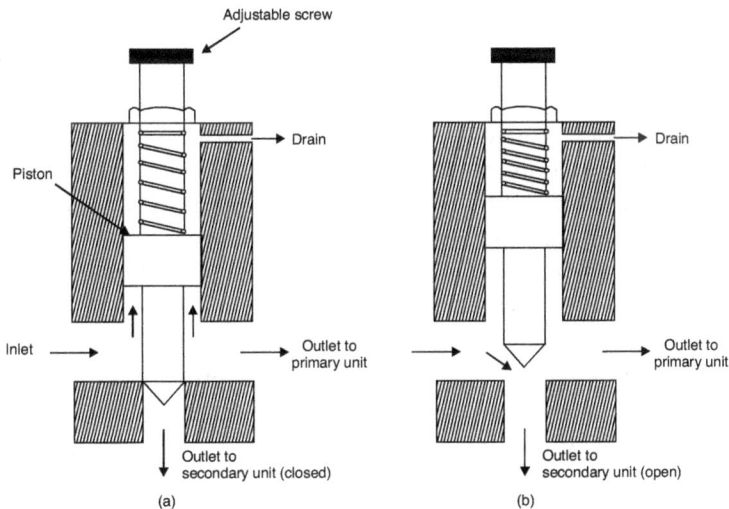

FIGURE 7.39 Sequence Valve.

When the primary actuating unit completes its operation, pressure in the line increases sufficiently to overcome the force of the spring, and the piston rises. The valve is then in the open position, as shown in Figure 7.39 (*b*). The fluid entering the valve takes the path of least resistance and flows to the secondary unit. A drain passage is provided to allow any fluid leaking past the piston to flow from the top.

Symbol

Symbol of a sequence valve is shown in Figure 7.40.

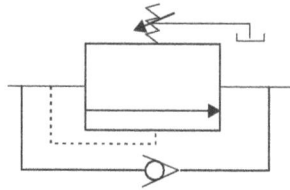

FIGURE 7.40 Symbol for Sequence Valve.

COUNTERBALANCE VALVE

A counterbalance valve is a pressure control valve that maintains control of a vertical cylinder to prevent it from falling freely because of gravity or due to a load attached to it. A counterbalance valve allows free flow of fluid in one direction and maintains a resistance to flow in another direction until a certain pressure is reached. This valve is normally located in a line between a directional-control valve and an outlet of a vertically mounted actuating cylinder, which supports weight or must be held in position for a period of time. A counterbalance valve is set at a higher pressure than the system load pressure. If the cylinder has to extend, the pressure has to rise above the set pressure and open the counterbalance valve. A counterbalance valve serves as a hydraulic resistance to an actuating cylinder. The main difference between the pilot operated check valve and counterbalance valve is that the opening pressure of the pilot operated check valve depends on the pressure behind the valve and the opening pressure of counterbalance valve depends on the spring pressure behind the valve. Figure 7.41 shows a counterbalance valve.

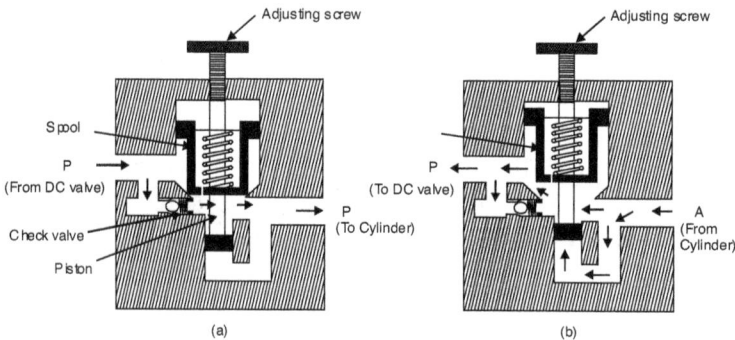

FIGURE 7.41 Counterbalance Valve.

In the initial position, the counterbalance valve is in a closed position as shown in Figure 7.41(*a*). The pressurized fluid coming from the DC valve enters through port P. In the closed position, the spool valve blocks the discharge port A and the fluid cannot flow through the valve. The fluid will go to the discharge port through the check valve and lift the cylinder. It is the tendency of the cylinder to fall because of the load attached to it or by gravity. Because of the higher-pressure setting of the counterbalance valve, the cylinder remains in the upward position.

During reverse flow, the fluid from the cylinder enters the counterbalance valve as shown in Figure 7.41(*b*). The pressurized fluid cannot pass to the DC valve because of the closed position of the valve. When there is an increase in pressure due to resistance to flow in the line, the flow is diverted through the pilot line. The force exerted by the pressurized fluid acts at the bottom of the piston. The fluid pressure overcomes the pressure setting of the valve and forces the spool to move in an upward direction. It opens and allows free flow around the shaft of the spool valve and fluid escapes out from inlet port P.

A counterbalance valve is used in some hydraulically operated forklifts. It offers a resistance to the flow from an actuating cylinder when a fork is lowered. It also helps support a fork in the up position.

Symbol

Figure 7.42 shows the symbol of a counterbalance valve.

FLOW CONTROL VALVES

As the major application of flow control valves in hydraulics and pneumatics is to control the speed of the cylinder and fluid motor, so these are also known as speed control valves. Flow control valves control the speed of

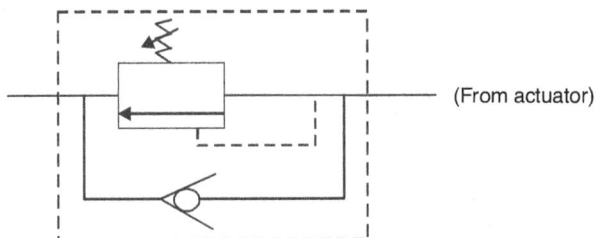

(From actuator)

FIGURE 7.42 Symbol for Counter Balance Valve.

an actuator by regulating the flow rate. There are two types of flow control valves:

- Fixed throttle
- Adjustable throttle

Constant flow rate is required if constant speed is required. *Fixed throttle valves* are used where some reduction in flow is required. As shown in Figure 7.43, fluid has to pass through a reservoir, which is in the form of a long cylinder. This type of design is appreciable in case of pneumatics because of increased viscosity influence. This would result in losses due to internal friction and generation of heat.

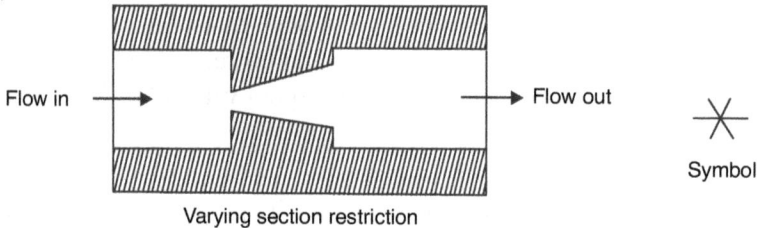

Varying section restriction

FIGURE 7.43 Varying Section Resistance Illustrated.

Arrangement shown in Figure 7.44 is a modification over the former. With this, influence of viscosity is reduced and results in lesser generation of heat.

Adjustable throttle valves are used where there is a need of infinite speeds of cylinders and motors or where a cylinder or motor is to run at constant speed under changing loads. In case of adjustable throttle valves,

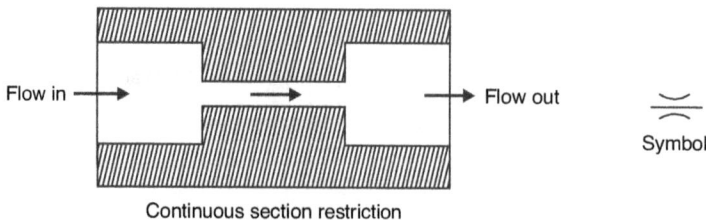

Continuous section restriction

FIGURE 7.44 Continuous Section Restriction Illustrated.

usually an adjustable needle is used to vary flow configuration in comparison to requirements. General arrangement is shown with the help of Figure 7.45.

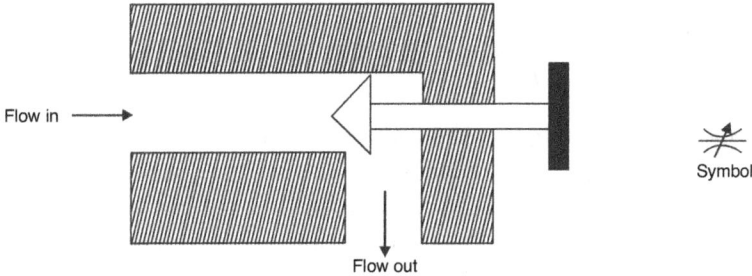

FIGURE 7.45 Adjustable Throttle Flow Control Valve.

NON RETURN FLOW CONTROL VALVE

This is a combination of variable flow control and a check valve. This arrangement is used to control the flow in one direction only, that direction depends on the position and direction of the check valve. As illustrated in Figure 7.46, there will be no restriction on flow control when fluid has to flow from a lower port to a higher port but fluid flow is controlled when it has to flow from upper port to lower port.

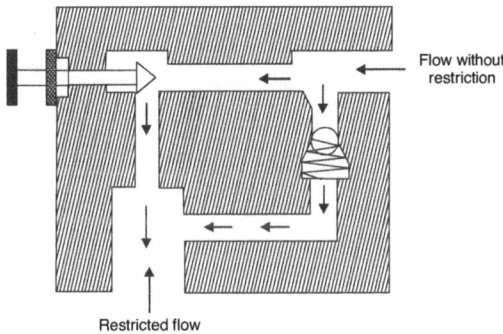

FIGURE 7.46 Non Return Flow Control Valve.

Symbol

Figure 7.47 shows the symbol of a non return flow control valve.

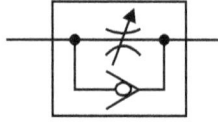

FIGURE 7.47 Symbol for Non Return Flow Control Valve.

Application of Non Return Flow Control Valve for Speed Control in a Cylinder

Speed of a cylinder can be reduced by either entering the pressurized fluid at slow speed or by exhausting it at a slow speed. This task can be accomplished by employing a non return flow control valve at the inlet or exit of the cylinder, respectively. This topic is covered in detail in Chapter 8 under the heading "Throttle In and Throttle Out Circuit." As the speed of the cylinder can be slowed down by providing a flow control valve, the speed of the cylinder can also be increased by the use of a quick exhaust valve in case of pneumatics.

QUICK EXHAUST VALVE

Quick exhaust valve is used to exhaust pressurized fluid from the cylinder at a faster rate in pneumatics. This is done by preventing the exhaust air from the cylinder to pass through the DC valve. A quick exhaust valve is shown in Figure 7.48.

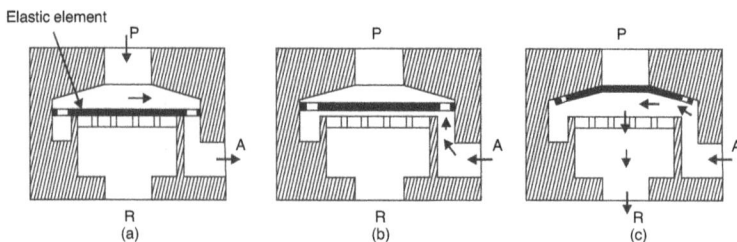

FIGURE 7.48 Quick Exhaust Valve.

It consists of pressure port P from where the fluid is coming from the main line, the working port A from where the fluid goes to the cylinder, and exhaust port R. When the pressurized fluid (air) is entered into the cylinder, *i.e.*, when pressurized fluid goes from port P to port A (from main line to cylinder), it simply allows the fluid to pass as it comes as shown in Figure 7.48 *(a)*. When the fluid from the cylinder is to be returned, then the elastic element is displaced by the fluid (air) pressure itself as shown in Figure 7.48 (*b*). The fluid gets exhausted through port R without passing through port P as shown in Figure 7.48 (*c*). Hence, the speed of the cylinder can be increased by the use of the quick exhaust valve. This valve is limited for its use in compressed air applications because one need not recollect the air in the system.

Symbol

Figure 7.49 shows the symbol of a quick exhaust valve.

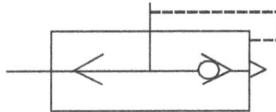

FIGURE 7.49 Quick Exhaust Valve.

TIME DELAY VALVE/AIR TIMER

This valve is popularly known as air timer because the time of delaying a signal can be adjusted by 5–15 sec with this valve. Time delay valves are pilot-operated valves in which the pilot air is supplied through a variable restrictor so that it takes time for the operating pressure to build up. The time delay is adjusted by adjusting the variable restrictor. A time delay valve consists of a 3/2 pilot operated spring return DC valve, an inbuilt air reservoir, and a non return flow control valve as shown in Figure 7.50.

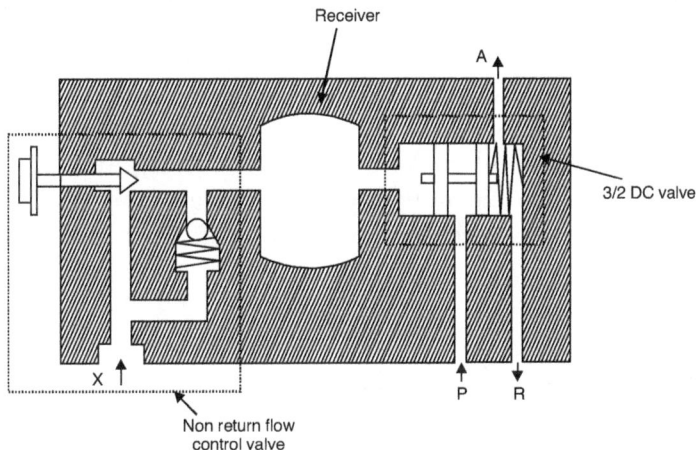

FIGURE 7.50 Time Delay Valve.

To send a signal (in the form of pressurized fluid) to some other actuator in the system, the valve spool has to be shifted. The shifting depends on pressure in pilot line X. The pilot line is connected with a non return flow control valve and an air receiver. Air starts accumulating in the receiver. When the pressure required to shift the spool is built up in the receiver, the spool of the DC valve shifts and the pressure port P gets connected to working port A. The size of the receiver and the setting of the non return flow control valve decides the time required to shift the spool of the DC valve. This time is the amount of delay time offered by the time delay valve.

Symbol

Figure 7.51 shows the symbol of a time delay valve.

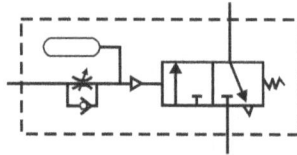

FIGURE 7.51 Symbol for Time Delay Valve.

PNEUMATIC LOGIC VALVES

Pneumatics is always preferred over hydraulics where precise control is required. Pneumatic valves are capable of performing logic functions as required in various applications of industry. A machine can be fully automated by using electronic logic gates, timers, switches, etc. and finally programming their PLCs. Then that machine is bound to do the same work again and again. Similarly, pneumatic logic elements can be used as an artificial brain to start, stop, check, warn at emergencies. All the gates (NOT, YES, AND, OR, NOR, NAND, MEMORY) are derived from four basic logic functions, *i.e.*, NOT, AND, OR, and MEMORY. Whereas from pneumatic valves, twin pressure valves and shuttle valves are important from the subject point of view. They are described below.

TWIN PRESSURE VALVE

It consists of a valve body having 3 ports and a movable spool as shown in Figure 7.52. This valve is also called an AND gate because it requires two equal inputs to deliver a signal. Output at the C port is attained if input from ports A and B is fed and is equal in strength. Then there will only be

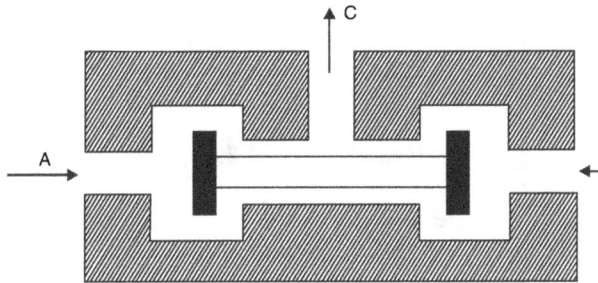

FIGURE 7.52 Twin Pressure Valve.

flow at C. If there is input from A port only then the spool will shift towards the right and block the passage from A to C and similarly, if there is an input from B port only then the spool will shift to the left side and close flow passage from B to C.

Symbol

Figure 7.53 shows the symbol of a twin pressure valve.

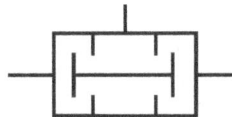

FIGURE 7.53 Symbol for Twin Pressure Valve.

SHUTTLE VALVE

It consists of a valve body having 3 ports (A, B, and C) as shown in Figure 7.54. This valve is also called an OR gate because it requires a single input or two equal inputs to deliver a signal. A and B are input ports and C is an output port and a control element in the form of a ball. Here, at least, one input is required to attain an output at the C port. If air is blown through port A, a ball closes port B and air escapes to port C, and if air is blown through port B, again the spherical element moves and closes port A and air will escape through port C. And if the air is blown from both A and B ports at same time, even then there will be flow to C port, *i.e.*, either from A or from B or from both.

Symbol

Figure 7.55 shows the symbol of a shuttle valve.

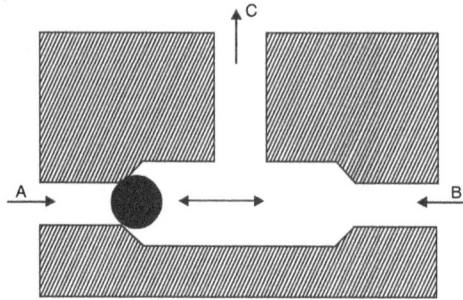

FIGURE 7.54 Shuttle Valve.

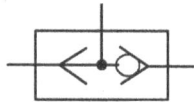

FIGURE 7.55 Shuttle Valve.

SERVO VALVES

Servo valves are generally used for providing a pressure response to an electrical or electronic control signal. They can be infinitely positioned to control the amount, pressure, and direction of fluid flow. Servo control by definition is *the process of controlling greater energies by using smaller energies.*

In general, servo valves have a pilot valve and a slave valve. This pilot valve use electrical torque motors or force motors as a mode of its actuation. A pilot valve sends the amplified signal to the slave valve (also called the main valve) in the form of pressurized fluid. Feedback is also provided by some means to see whether the spool of the main valve has acquired its desired position or not. This kind of valve, which performs the function of controlling fluid in a certain direction, is called a two-stage servo valve. Servo valves are available in *one, two, or three stage designs.* A single stage is a directly operated valve. Two valve stages are comprised of a pilot stage and final/main stage. Three stage valves are similar, except that the pilot itself is a two-stage servo valve. Three-stage servo valves are used in situations where very high flow is anticipated.

The term servo valve was traditionally understood as mechanical feedback valves, where a torque motor is connected to the main valve spool by a spring element (feedback wire). Spool displacement causes the wire to impart a torque onto the pilot valve motor. The spool will hold position when torque from the feedback wire's deflection equals the torque from an electromagnetic field induced by the current through the motor coil. These two-stage valves contain a pilot stage controlled by torque motor, and a main or second stage controlled by fluid discharging from the first stage. Use of servo valves in industry is increasing day by day because of their capacity to control large amount of fluids.

TORQUE MOTOR AND ARMATURE ASSEMBLY

Torque motor and armature assembly is shown in Figure 7.56. The torque motor consists of an armature mounted at the center in the air gap of magnetic field produced by a set of permanent magnets as shown in Figure 7.56 (a). By passing a current from the armature coils, the armature ends become polarized and are attracted to one magnet pole piece and repelled by the other. This sets up the torque in the armature assembly, which rotates about the fixture sleeve as shown in Figure 7.56 (b). The upward movement of the armature assembly is transmitted to the main valve spool in the form of linear motion with the help of a linkage whether it is a jet pipe or a flapper.

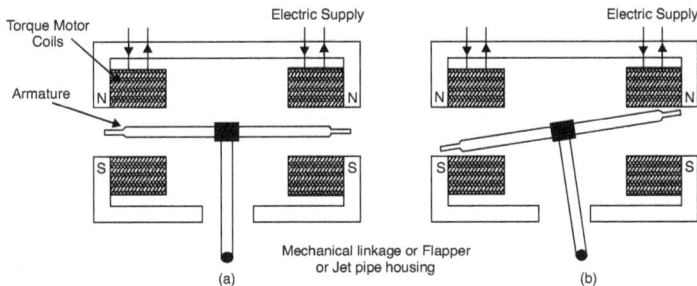

FIGURE 7.56 Torque Motor and Armature Assembly.

CLASSIFICATION OF SERVO VALVES

Servo valves can be broadly classified as:

- Single-stage servo valve.
- Multistage servo valve.

SINGLE-STAGE SERVO VALVE

A valve is said to be a single-stage servo valve when the electrical signal is directly applied to the control main valve position and there is no further intermediator required. Single-stage servo valves consist of a main valve (which is to be controlled) and a torque motor. The armature or torque motor is connected with the spool of the main valve via a mechanical linkage. The armature is pivoted along its center as shown in Figure 7.57 and is surrounded by the coils of the torque motor. Depending upon the electrical signal, whether the signal is low, medium, or high, the armature is deflected inwards or outwards. The same is true for the main spool valve.

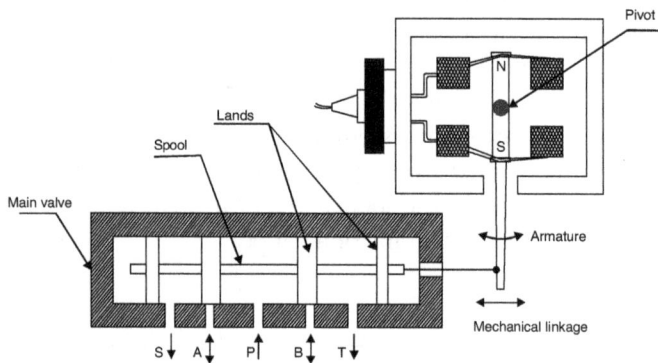

FIGURE 7.57 Single Stage Servo Valve.

(S-Discharge port, A-Working port, P-Pressure port, B-working port, T-Discharge port)

With a low electrical signal or a practically zero electrical signal, pressure port P will be connected to working port A and working port B will be connected to discharge port T. When the electrical signal is 50%, the armature becomes straight and this resembles the closed center position of the main valve. At that time, all of the ports remain departed from each other as shown in Figure 7.57. In the third case, when the signal is 100%, the pressure port P connects the working port B and working port A gets connected to the discharge port S.

For heavy load operation, two-stage servo valves are generally preferred over single-stage servo valves because the forces required to move the spool are higher for higher flows of fluid from the main valve. Further these single-stage servo valves can be of two types:

1. Flapper jet servo valve
2. Pipe jet servo valve

Flapper Jet Servo Valve

The general arrangement of a flapper jet servo valve is shown in Figure 7.58. It consists of a main valve, torque motor, a flapper, the pipeline with orifice, and nozzles where the armature is placed between the torque motor coils. The coils, by means of an electric signal, deflect the flapper in the desired direction. The other end of the flapper is used to control the pilot pressure in the main valve.

As long as the flapper remains straight, the pilot pressure on both sides of the main valve remains the same and the valve remains in the position as shown in Figure 7.58 (a). But with a change in the electrical signal, the flapper changes the position thereby increasing pressure in one side and decreasing pressure on the other side. The other position of the main valve is attained by dispositioning of the flapper is shown in Figure 7.58 (b).

Pipe Jet Servo Valve

A jet pipe servo valve also uses an electrical signal to change the position of the main valve. The valve consists of a torque motor, armature, nozzles, jet pipes, and main valve with spool type construction as shown in Figure 7.59. For obtaining rest position of this valve (position that is shown in Figure 7.59, when all the ports are closed to each other), only 50% of the control signal is applied. With this, the armature remains straight and there is an equal amount of flow in both pilot lines of the main valve.

In order to change the position of the servo valve, the control signal is changed and that deflects the armature thereby directing more fluid in one pilot line and less fluid in other pilot line. The bottom free end of the armature is connected with the spool of the main valve. This is done so to ensure that, whether the spool has attained the required position or not,

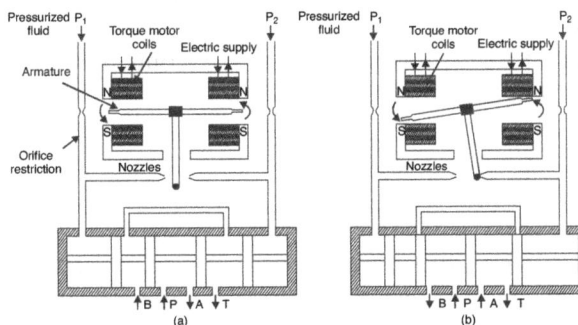

FIGURE 7.58 Flapper Jet Servo Valve.

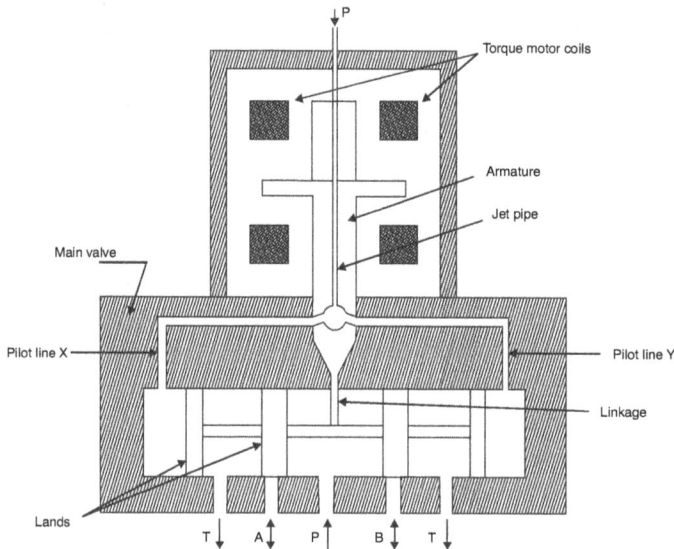

FIGURE 7.59 Pipe Jet Servo Valve.

and also after reaching the desired position, further movement of the spool and armature cancels out the electric signal automatically.

MULTISTAGE SERVO VALVE

All of the aforementioned designs can be used to design a multistage hydraulic servo valve. The idea for each design is specific to the application requirements. Usually, most designs do not exceed three stages. Mounting a nozzle flapper, jet pipe, or direct-driven valve onto a larger main stage satisfies most requirements for dynamics and flow. Sometimes, the jet-pipe valve is used in a multistage configuration where the mechanical feedback of a traditional jet pipe is replaced with electronic feedback. This servo jet style has the pilot characteristics of a typical jet pipe.

SERVO SYSTEM

The servo system as shown in Figure 7.60 consists of an amplifier, which amplifies the signal, so that it is to be used as a signal for the servo valve. There are now two inputs to the next component of the system, *i.e.*, servo valve. The first input is pressurized fluid from the pump, that is to be directed by the servo valve to the actuator and the second input is a control signal by an amplifier or pilot valve. The servo valve directs the flow of fluid to the actuator, *i.e.*, a double-acting hydraulic cylinder. The movement of the double-acting cylinder takes place, and this movement is measured electronically. Actual value is compared with the desired value and the relevant error signal is provided.

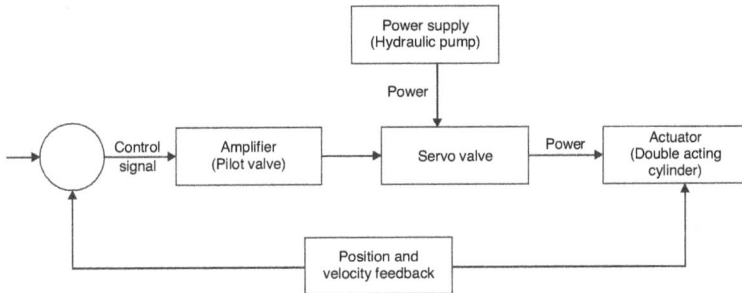

FIGURE 7.60 Block Diagram of Servo Control System.

EXAMPLE OF A SERVO CONTROL SYSTEM

Figure 7.61 shows an example of a hydraulic servo system in which an axial piston pump is being used as a source of energy, which is driven by an electric motor. Hydraulic fluid is taken from a service tank via the hydraulic filter. The pressurized fluid is fed to actuator by using a servo valve, which provides certain amplification and directs it in right direction. An accumulator can also be provided depending upon the pressure-flow characteristics of the pump. This fluid under pressure is then used to displace the worktable in some machinery.

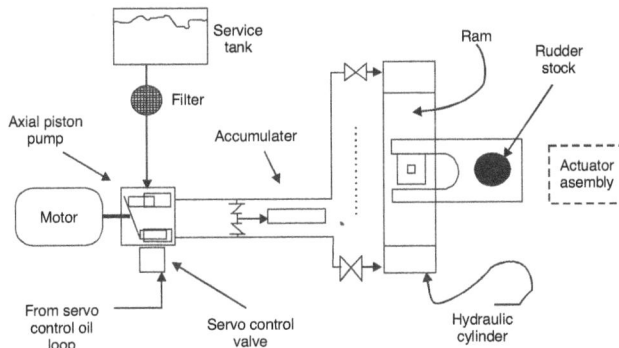

FIGURE 7.61 Shows Use of Servo System in Displacing Worktable.

HYDRAULIC AND PNEUMATIC SYMBOLS

Pumps and Compressors
Hydraulic Pumps

Unidirectional (Fixed-displacement type)

Bi-directional (Fixed-displacement type)

Unidirectional (Variable-displacement type)

Bi-directional (Variable-displacement type)

Air Compressors
Cylinders

Single acting (returned by external force)

Single acting (returned by spring force)

Double-acting cylinder

Double acting (double ended piston rod)

Single fixed cushion

Double fixed cushion

Unidirectional

Vacuum pump

Hydraulic and Pneumatic Actuators
Hydraulic motors

Unidirectional (Fixed-displacement type)

Bi-directional (Fixed-displacement type)

Unidirectional (Variable-displacement type)

Bi-directional (Variable-displacement type)

Oscillating motor

Pneumatic Motors

Unidirectional

Bi-directional

Unidirectional (Variable-displacement type)

Bi-directional (Variable-displacement type)

Oscillating motor

Check Valves

Check / non return valve

Pilot operated check valve (Pilot to close)

Pilot operated check valve (Pilot to open)

Shuttle valve

Single adjustable cushion

Double adjustable cushion

Double acting telescopic cylinder

Hydraulic and Pneumatic Valves
Direction Control Valves

2/2 direction control valve (Normally closed)

2/2 direction control valve (Normally open)

3/2 direction control valve (Normally closed)

3/2 direction control valve (Normally open)

4/2 direction control valve

4/3 direction control valve (Closed center)

5/2 direction control valve

5/3 direction control valve (Closed center)

Methods of Actuation
Manual Control

Manual control (General symbol)

Push button

Lever

Quick exhaust valve

Shut off valve

Flow Control Valves

Fixed orifice

Adjustable throttle valve

Flow control valve (With fixed output)

Flow control valve (With variable output)

Non return flow control valve

Pressure compensated

Pressure & temperature compensated

Flow dividing valve

Pressure Control Valves

Pressure relief valve (Normally closed)

Pilot operated pressure relief valve

Proportional pressure relief valve

Pressure reducing valve

Sequence valve

Combined Operation

Pilot valve actuated by solenoid

Anyone can independently actuate

Foot pedal

Mechanical Control

Plunger

Spring

Roller

Roller (One direction only)

Electrical Control

Solenoid (Single winding)

Solenoid (Double winding)

Solenoid (Double winding variable control)

Electric motor

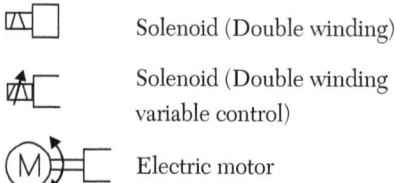

Pilot Operation

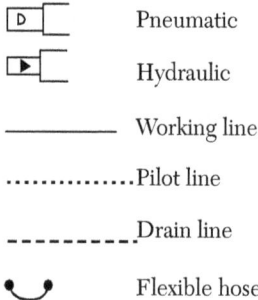

Pneumatic

Hydraulic

Working line

Pilot line

Drain line

Flexible hose

Fluid Accessories

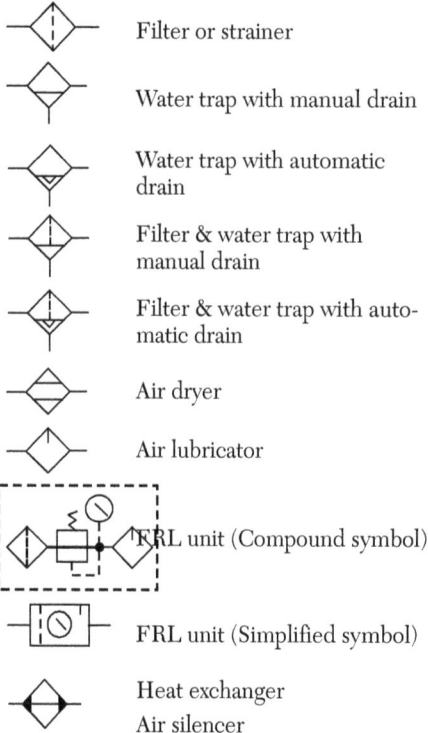

Filter or strainer

Water trap with manual drain

Water trap with automatic drain

Filter & water trap with manual drain

Filter & water trap with automatic drain

Air dryer

Air lubricator

FRL unit (Compound symbol)

FRL unit (Simplified symbol)

Heat exchanger
Air silencer

Accumulator

Energy Transmission

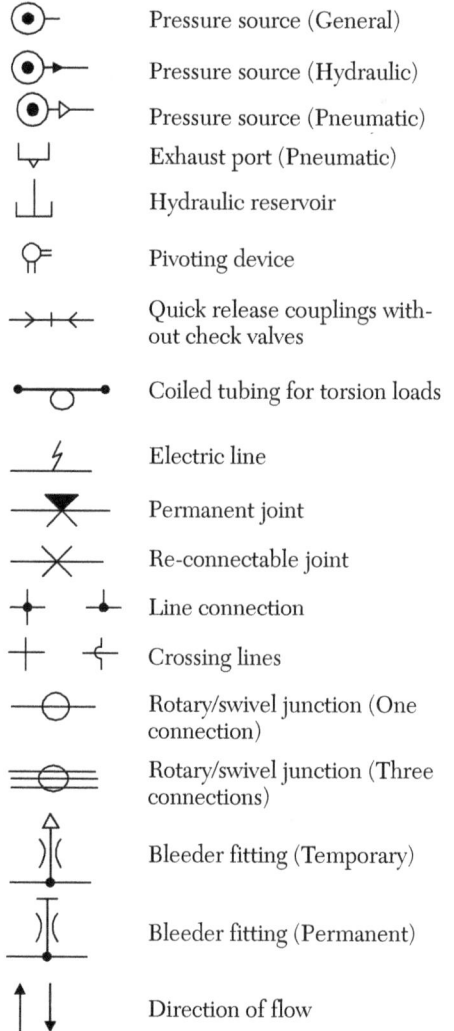

Pressure source (General)

Pressure source (Hydraulic)

Pressure source (Pneumatic)

Exhaust port (Pneumatic)

Hydraulic reservoir

Pivoting device

Quick release couplings without check valves

Coiled tubing for torsion loads

Electric line

Permanent joint

Re-connectable joint

Line connection

Crossing lines

Rotary/swivel junction (One connection)

Rotary/swivel junction (Three connections)

Bleeder fitting (Temporary)

Bleeder fitting (Permanent)

Direction of flow

Pilot Operation

Direction of flow

⟶✕	Power take off/plugged connection		Differential pressure gauge
	Shaft (Uni-directional rotation)		Thermometer
	Shaft (Bi-directional rotation)		Liquid level indicator
	Over center device to avoid stopping at center		Flow indicator
	Quick release couplings with check valves		Flow meter

Measuring Instruments

	Pressure gauge		Tachometer
			Torque meter

EXERCISES

1. What are directional control valves?

2. Identify the different types of pneumatic valves used in pneumatic circuits.

3. What is the function of a directional control valve? Identify the types of directional control valves.

4. What is a servo system?

5. What is a pressure sequence valve?

6. Describe the working principle of a pneumatic flow control valve with the help of a sketch.

7. What is a 5/2 pilot operated direction control valve?

8. Sketch any 4/2 direction control solenoid operated valve and a pneumatic pressure regulator valve.

9. What is a four way directional control valve?

10. What do you mean by a 5/2 DC valve?

11. What is a time delay valve?

12. What is a solenoid? How does it work?

13. What is a servo valve?

14. Sketch a shuttle valve.

15. Name the three types of servo valves.

16. What type of DC valve will you use for controlling a double-acting pneumatic actuator?

17. What type of DC valve will you use for controlling a single-acting spring return pneumatic actuator?

18. Explain the construction and working principle of pneumatic directional control valves with the help of a sketch.

19. What is the difference between an open center and a closed center type of directional control valve?

20. With the help of a sketch, explain how a check valve works.

21. What are the alternate names given to an AND gate and an OR gate?

22. Draw the standard symbol for a non return type flow control valve and twin pressure valve.

23. Draw the standard graphical symbol for a 4-way pilot operated spring centered directional control valve.

24. Draw the cross section of any position control spool valve and poppet valve.

25. Identify the standard graphic symbols applicable to fluid power control. (*PTU, Dec. 2002*)

26. Explain how a pressure control valve works with the help of a sketch.

27. How does a pilot check valve differ from a simple check valve?

28. Draw a sketch of pneumatic AND valve.

29. What is the function of a quick exhaust valve in pneumatics?

30. What are servo valves? Explain the construction and working of various types of servo valves with the help of sketches.

31. What are the differences between servo-controlled valves and pilot operated valves?

32. Draw the standard graphical symbol of any type of servo valve.

33. Explain how the hydraulic flow control valve works.

34. How are DC valves classified?

35. Discuss the various types of valves used for pressure control and flow control in fluid power systems.

36. Differentiate between spool and seat valves.

37. With the help of a sketch, explain how a rotary valve works.

38. What are the different center conditions in DC valves?

39. Explain how a counter balance valve works with the help of a sketch.

40. Explain the different methods of actuation of DC valves.

41. Draw a sketch of a pneumatic OR valve.

42. Differentiate between pressure and flow control valves. Explain the working of any two-pressure control valves with the help of a sketch.

43. What is the function of 5/2-way valve?

44. What is a sequence valve and its purpose?

45. What is a quick exhaust valve?

46. Explain the twin pressure valve and shuttle valve with the help of a sketch.

47. Explain the construction and working of pipe jet servo valve.

48. What is the function of a check valve in hydraulics and pneumatics? Discuss the applications of the check valve.

49. What is cracking pressure and pressure override?

50. What is a servo valve? How does it work?

51. Check valve is used to control the speed of hydraulic and pneumatic cylinder. Explain.

52. Draw the symbols of (i) check valve, (ii) sequence valve, (iii) counterbalance valve, (iv) time delay valve, (v) twin pressure valve, and (vi) shuttle valve.

53. Distinguish between pressure relief and pressure reducing valves.

54. What are the differences between servo-controlled valves and pilot operated valves?

55. With the help of a sketch, explain the working of a pneumatic cylinder of double acting type controlled by appropriate flow regulating and control valves.

CHAPTER 8

CIRCUITS

INTRODUCTION

A pneumatic or hydraulic system is designed to perform a particular job and it consists of large number of equipment like a power unit, a service unit, accessories, valves, actuators, etc. These components when combined in a logical sequence to get desired output in the form of the motion of the actuators and shown diagrammatically by using certain standard symbols of components, constitutes a fluid circuit. Fluid engineers design fluid circuits. It is very important to remember the standard symbols of every component along with the function of that component in order to draw a fluid circuit diagram. This chapter focuses on illustrating standard fluid symbols as per ISO (International Organization for Standardization) standards and explains the methodology to draw a circuit by taking suitable examples.

The fluid symbols are drawn in Chapter 7 (***Control Valves***).

There is not much difference between pneumatic and hydraulic circuit diagrams. Some of the common differences are:

- In pneumatic circuit diagrams, the return line is not necessary as in the case of hydraulic circuit diagrams because in pneumatics, the used air from the cylinder can be directly exhausted to atmosphere.
- In pneumatic circuit diagrams, R represents the exhaust port whereas in hydraulics, T represents it.

Names of the pressure and working ports remain the same in pneumatic and hydraulic circuits.

Basic circuits, which are important from a subject point of view, are given in this chapter. Pneumatics and hydraulics are explained separately for the ease of the readers.

BUILDING UP THE CIRCUIT DIAGRAM

Any circuit diagram, whether hydraulics or pneumatics, consists of certain elements, which can be divided into three categories:

- Driving elements
- Controlling elements
- Energy consuming members

Driving elements consist of pumps/compressors, filters, regulators, pressure gauges, service units, pressure relief valves, etc.

Controlling elements comprise all types of valves, *i.e.*, direction control valves, flow control valves, pressure control valves, signaling elements, etc.

Hydraulic and pneumatic actuators come under *energy consuming* members.

Some elements are monostable or bistable elements.

A *monostable* element has one stable position and automatically returns to it when the switching signal is removed. Examples of these are directional valves with a spring return, pressure switches, proximity detectors, and logic valves.

A *bistable* element has two stable positions and requires a switching signal to change it from one state to the other. Examples of these are directional control valves with no spring return such as pilot-pilot operation, solenoid-solenoid operation, valves with detents, and switches with no spring return.

Figure 8.1 represents the control chain flowchart showing how the signal flows through various elements from bottom to top in a pneumatic/hydraulic circuit. All elements required for the energy supply should be drawn at the bottom and the energy should be distributed from the bottom to the top. This flowchart means that the circuit diagram must be drawn without considering the actual physical locations of the elements and it is recommended that all the cylinders and directional control valves must be drawn horizontally.

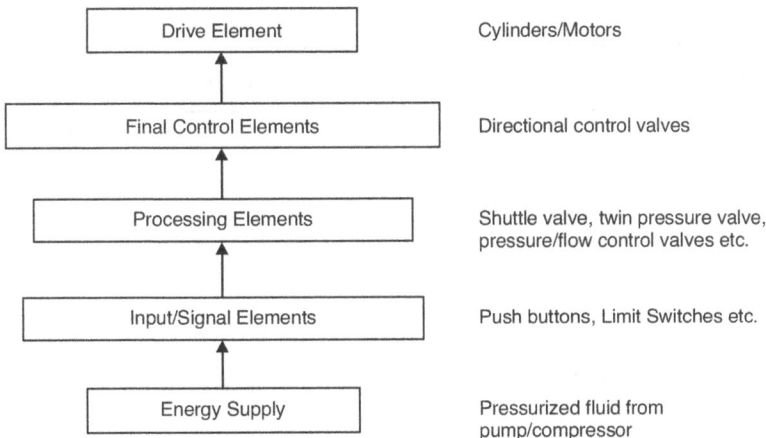

Drive Element	Cylinders/Motors
↑	
Final Control Elements	Directional control valves
↑	
Processing Elements	Shuttle valve, twin pressure valve, pressure/flow control valves etc.
↑	
Input/Signal Elements	Push buttons, Limit Switches etc.
↑	
Energy Supply	Pressurized fluid from pump/compressor

FIGURE 8.1

DESIGNATION OF COMPONENTS IN A CIRCUIT DIAGRAM

Circuit diagrams are prepared by arranging various components of a pneumatic/hydraulic system into a sequence according to their functions. Sometimes these fluid circuits are so complex that it is difficult to understand where it starts and where it ends. In order to deal with this difficulty, a system for designating components is used. There are two types of systems for designating components in a circuit diagram:

- System using digits
- System using alphabets

System Using Digits

(a) One method is to allot serial, e.g., 1, 2, 3, etc., to components of the circuit. This method avoids confusion and is best for use in complex circuits.

(b) Second method is to divide one pneumatic circuit into a number of small groups and there by allotting serial number to groups like 1, 2, 3....,, etc., and then further given a serial number within the group. The complete designation consists of group numbers and serial numbers within a group, e.g., 2.4 means Group – 2, Component number – 4.

Allocation of group numbers

Group 0	All items involved in supply of energy
Group 1,2,3,4	One group allocated to one cylinder and its sub components

Serial numbers are used by prefixing the group number before them.

.1	Control components
.01, .02	Components between moving and control components like flow control, pressure control, or check valves.
.2, .4 (even numbers)	
.3, .5 (odd numbers)	All components influencing forward stroke.
	All components influencing backward stroke.

The above said terminology is shown with the help of following example:

The circuit shown in Figure 8.2 is drawn for a special punching operation using a double-acting cylinder as ram, which starts punching operation when safety cover is secured and operator pushes the button. A cylinder extends fast and punches a hole and returns back slowly to its initial position.

The above drawn circuit is divided into 2 groups:

- Group 0 — Energy supply
- Group 1 — Double-acting cylinder as working element

In the pneumatic circuit diagram shown in Figure 8.2, compressed air is supplied from a common source such as a centralized compressor with the help of pipelines to FRL/service unit (0.1). Then this air is supplied to 3 signal elements (*3/2 DC valves*) namely 1.2, 1.3, 1.4 and a control element (*4/2 DC valve*) namely 1.1. Serial number 1.02 denotes the twin pressure valve.

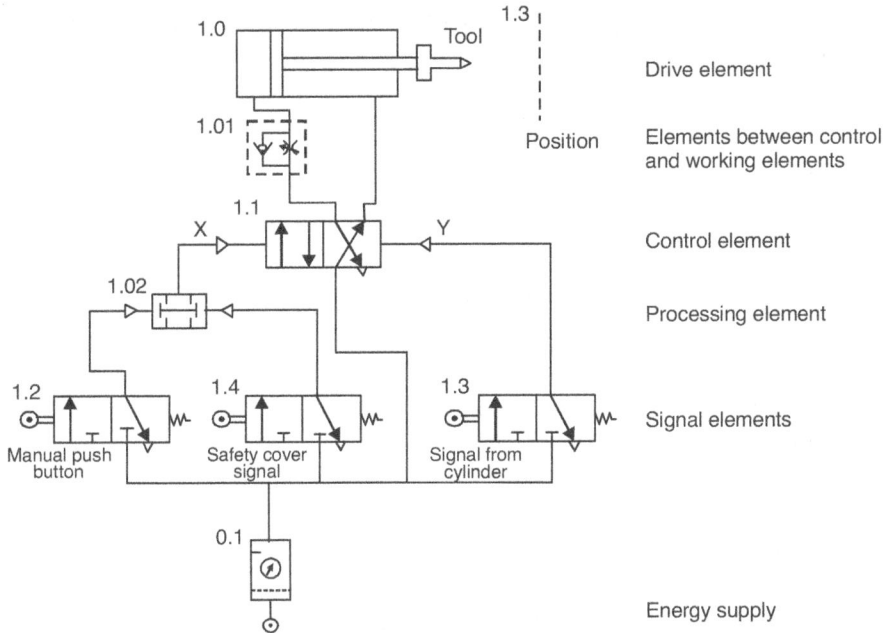

FIGURE 8.2 Circuit Showing Designation of Various Components.

Manual push button of 1.2 signal element is pressed but supply pressure cannot pass through the 1.02 twin pressure valve until the signal element 1.4 is operated. Signal element 1.4 is operated when the safety cover on a work piece is secured properly. When these two conditions are fulfilled, *i.e.*, the first manual push button is pressed and the second cover on the work piece is secured, then only pressurized fluid, *i.e.*, air passes from valve 1.02 and enters pilot line X, thereby pushing the control element and changing its position. When this piston rod starts extending outside at a normal speed because of the flow control valve in line and after reaching the extreme forward position, the piston returns back automatically and with a comparatively lower speed.

PNEUMATIC CIRCUITS

Pneumatic circuits can be defined as a graphic representation of the elements of a pneumatic system where these elements are represented by their corresponding symbols and these symbols are sequentially arranged in a specific manner depending upon the nature of output required. Some of the basic pneumatic circuits are explained below.

PNEUMATIC CIRCUIT FOR CONTROL OF SINGLE-ACTING CYLINDER

The basic pneumatic circuit for control of a single-acting cylinder is shown in Figure 8.3.

Here a single-acting cylinder (1.0) is being controlled by a 3/2 push button operated, spring return direction control (DC) valve (1.1) as the control element. The compressed air is supplied from a main compressor placed anywhere in the plant to FRL unit (serving unit), where air is filtered, lubricated, and regulated to the desired pressure and directed to the 1.1 control element. In the initial position, as shown in Figure 8.3 (a), 1.1 DC valve is in a normally closed position and the cylinder is in a fully retracted position. When the push button of the control element 1.1 is pressed, the pressure port P gets connected to he working port A and the pressurized fluid, *i.e.*, air starts moving from a 1.1 DC valve to a cylinder inlet causing it to move forward. Figure 8.3 (b) shows the fully extended position of the cylinder. As cylinder 1.0 is spring returned, the cylinder attains its previous position when the push button of 1.1 control element is released.

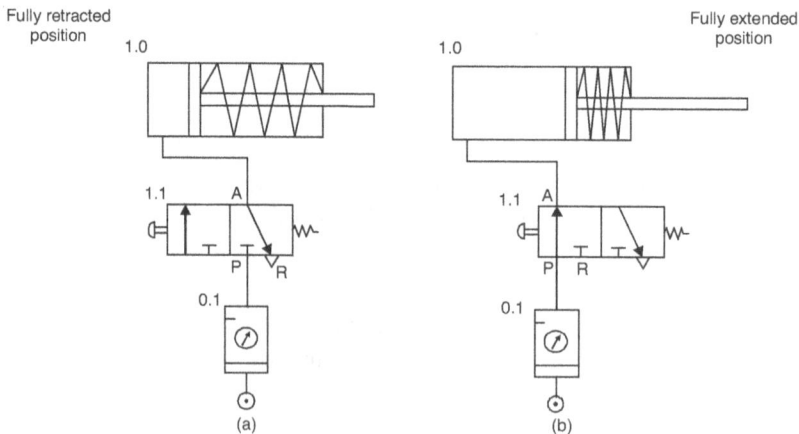

FIGURE 8.3 Pneumatic Circuit for Control of Single-Acting Cylinder.

PNEUMATIC CIRCUIT FOR CONTROL OF DOUBLE-ACTING CYLINDER

The basic pneumatic circuit for control of a double-acting cylinder is shown in Figure 8.4. Here a double-acting cylinder (1.0) is being controlled by a 4/2 push button operated, spring return direction control (DC) valve (1.1). The circuit shows the manual control over the forward and backward motion of the cylinder. In the circuit shown below, there is no return spring in the double-acting cylinder for the retraction of the piston rod, so both forward and backward motions of the cylinder are controlled by pressurized fluid, *i.e.*, air and DC valve.

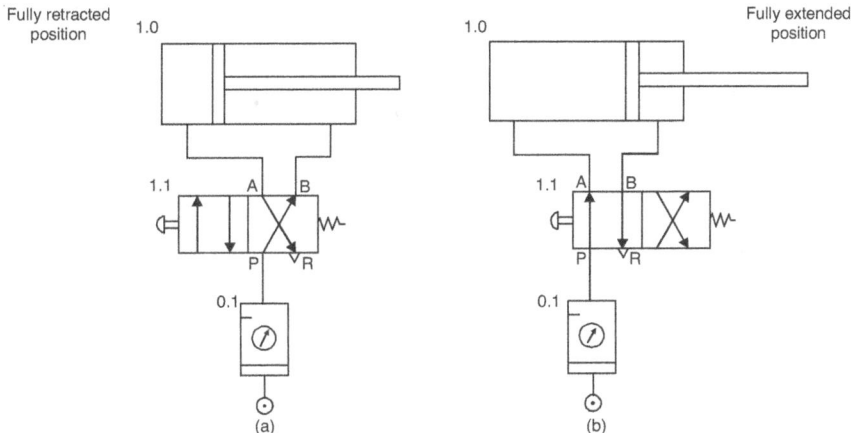

FIGURE 8.4 Pneumatic Circuit for Control of Double-Acting Cylinder.

Initially the cylinder 1.0 is in a fully retracted position as shown in Figure 8.4 (*a*). Pressure port P is connected to working port B and port A is connected to exhaust port R. When the push button of control element 1.1 is pressed, the position of the DC valve changes. The pressure port P gets connected to the working port A and the pressurized fluid, *i.e.*, air starts moving from 1.1 DC valve to the cylinder causing it to move in a forward direction. Figure 8.4 (*b*) shows the fully extended position of the cylinder. To cause the backward motion of cylinder 1.0, the push button of the 1.1 DC valve is released. Because of the action of the spring, 1.1 DC valve attains its previous position as shown in Figure 8.4 (*a*).

APPLICATION OF 2/2 AND 3/2 DC VALVES

2/2 DC valve can be used only to run those devices in which there is no return flow of fluid, example, in case of pneumatic motors, air escapes out in the atmosphere after rotating the rotor. Figure 8.5 shows the application of the 2/2 and 3/2 DC valves.

Figure 8.5 shows the application of 2/2 and 3/2 DC valves. Here, both the DC valves, 1.1 and 1.2, are in the normally closed position as shown in Figure 8.5 (*a*). When the push button of 1.2 DC valve is pressed, the pressure port P gets connected to the working port A and compressed air starts flowing from the pressure port P to the working port A of 1.2 DC valve.

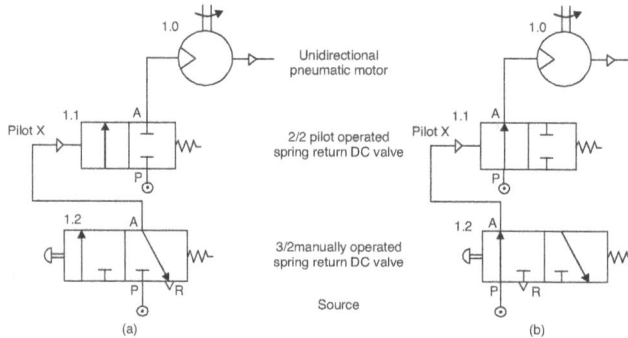

FIGURE 8.5 Circuit Showing Application of 2/2 and 3/2 DC Valves.

The compressed air then enters the pilot line X of the control element 1.1 and actuates the 1.1 DC valve as shown in Figure 8.5 (*b*). When the position of the 1.1 DC valve is changed, port P gets connected to port A and compressed air reaches the inlet port of the pneumatic motor. No return line is required in the case of air motors as air is exhausted directly from the outlet port of the motor. Now in order to stop the rotation of motors, the push button of 1.2 DC valve is released and both valves return back to their normally closed position by action of spring force.

CIRCUIT WITH MECHANICAL FEEDBACK

Refer to Figure 8.6. This circuit is designed for the condition that when the manual push button of 1.2 DC valve is pressed, the piston rod extends from position A to position B and then it retracts automatically back to its start position.

Initially the cylinder 1.0 is in a fully retracted position as shown in Figure 8.6. As the operator pushes the manual push button of the 1.2 DC valve, the supply of compressed air goes to the pilot line X and changes the position of the 1.1 DC valve. With the change of position, the compressed air enters the cylinder 1.0 and the piston rod of cylinder 1.0 starts extending

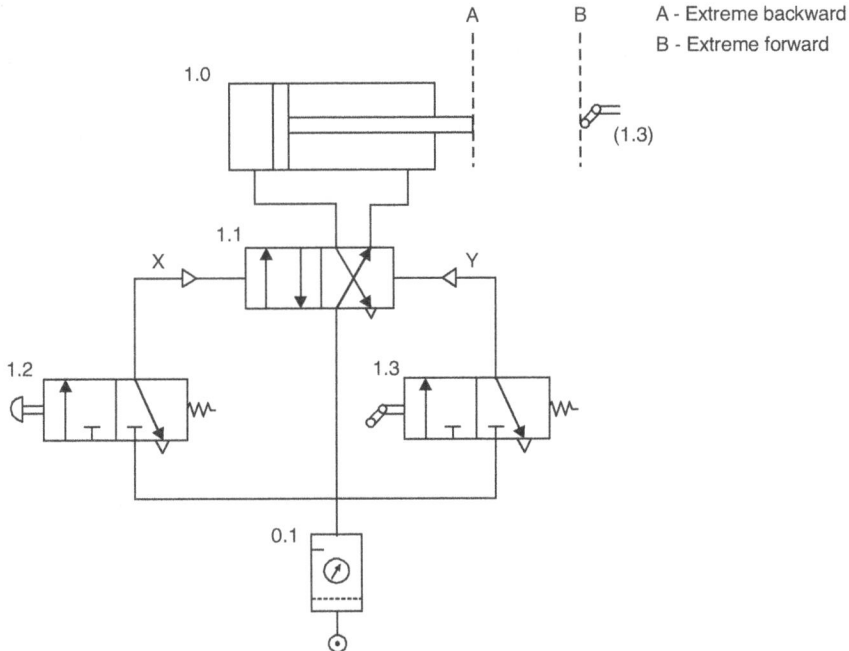

FIGURE 8.6 Circuit with Mechanical Feedback.

from position A. The moment it reaches position B, a roller trip of 1.3 DC valve is actuated which results in inducing compressed air into pilot line Y and further results in changing the position of the 1.1 DC valve. With this, the compressed air enters the cylinder and results in the retraction of the cylinder.

SPEED CONTROL CIRCUITS

Flow control valves are also called speed controllers because they are used in pneumatic circuits to slow down the speed of the cylinder. Non return adjustable throttle valves are commonly used in a pneumatic system, as they can be adjusted to give desired output. A check valve is always used in conjunction with a flow control valve. Depending on the arrangement of the flow control valves, speed control circuits for a double-acting cylinder can be of two types:

▪ Infeed throttling/Throttle in circuit
▪ Outfeed throttling/Throttle out circuit

Infeed throttling

Infeed throttling may be defined as a method to reduce the speed of a pneumatic cylinder in which fluid at the entrance to a double-acting cylinder is passed through control restrictors so as to slow down the speed of the piston in the cylinder whereas no restriction is applied to air exhausting from the cylinder.

Outfeed throttling

Outfeed throttling may be defined as a method to reduce the speed of a pneumatic cylinder in which fluid at the exit of a double-acting cylinder is passed through control restrictors so as to slow down the speed of the piston in the cylinder whereas no restriction is applied to air entering the cylinder.

These methods are further explained in details with the help of circuit diagrams as shown below:

1. Throttle in circuit

Throttle in circuit is shown in Figure 8.7. The circuit consists of a service unit 0.1, a 4/2 DC valve as the control element (1.1), two non return flow control valves (1.01), and (1.02) respectively, and a double-acting cylinder (1.0).

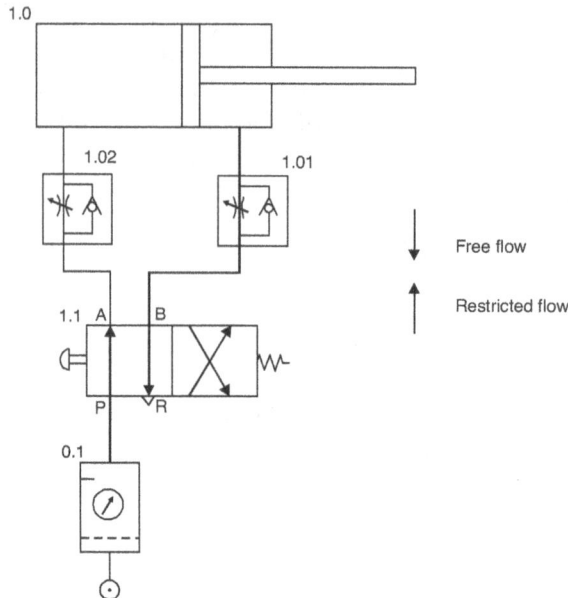

FIGURE 8.7 Throttle In Circuit.

Consider the valve is at its normal position, which is shown in Figure 8.7. The pressure port P is connected to working port A and port B is connected to exhaust port R. The compressed air from the compressor passes through control element 1.1 and further reaches the non return flow control valve 1.02. Here air will have to pass from the variable restrictor because a check valve is mounted on the bypass line. This results in a reduction of flow of compressed air at the entrance of the cylinder (1.0). Hence, the cylinder extends smoothly towards the forward direction. At the same time, compressed air in the rod side of the cylinder is to be exhausted out. The compressed air is bound to pass through the non return flow control valve (1.01). But this time air is vented out from the check valve through the by pass without any restriction. When the manual push button of control element 1.1 is pressed, the position of the 1.1 DC valve changes. This time air is entered through the non return flow control valve (1.01). Again, at the entrance air has to pass through the variable restrictor because of the "no flow" position of the check valve in that direction which results in the slow motion of the piston in a backward direction. Air at other side of the piston is vented out again through the check valve 1.02 and no resistance is encountered while exhausting through the cylinder.

2. *Throttle out circuit*

Throttle out circuit is shown in Figure 8.8. Construction of this circuit is the same as that of the throttle in the circuit. The only difference is that the direction of the check valve is changed in this circuit for providing restriction to air at the outlet.

Consider the valve is at its normal position, which is shown in Figure 8.8. The pressure port P is connected to working port A and port B is connected to exhaust port R. The compressed air from the compressor passes through control element 1.1 and further reaches non return flow control valve 1.02. No restriction is offered in valve 1.02 as the check valve allows the flow of air through it. Air finally enters the cylinder and pushes the piston in a forward direction. At the same time, air has to be exhausted from the rod side of the cylinder and this time air has to pass through the non return flow control valve (1.01). This time the check valve does not allow air to freely pass from it and air passes through restriction. Hence, speed of piston is slowed down as air escapes slowly out of the restrictor. After completion of the forward stroke, the manual push button on control element 1.1 is pushed and the circuit proceeds in the same manner as explained in the case of a forward stroke.

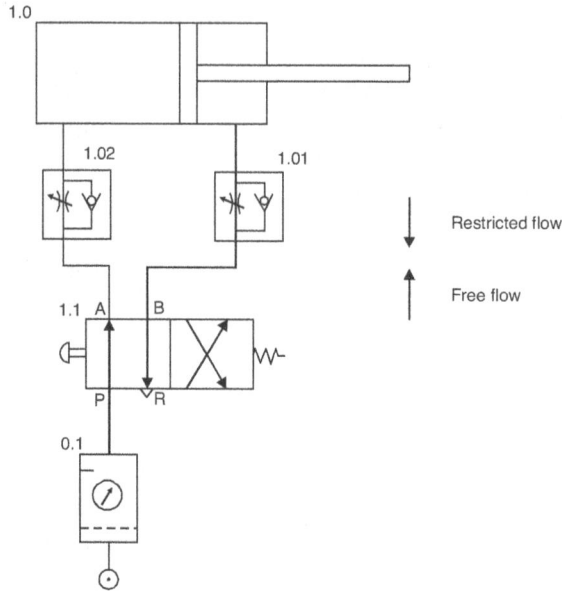

FIGURE 8.8 Throttle Out Circuit.

In pneumatic circuits, generally *outfeed throttling is preferred over infeed throttling* because better control can be attained in the former case. It is also important to add here that flow control valves can be fitted at the exhaust also with 5/2 DC valves as control elements in line. (Refer to Figure 8.9.)

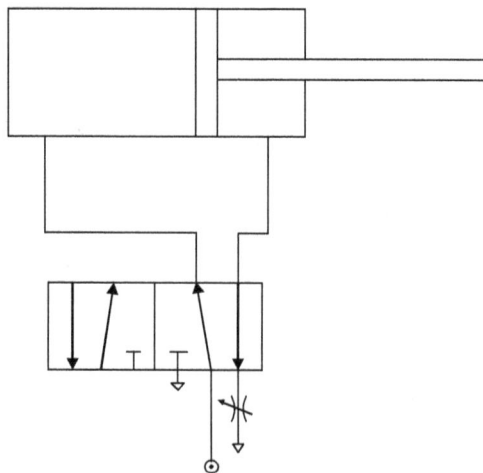

FIGURE 8.9

USE OF FLOW CONTROL VALVE TO CONTROL SINGLE-ACTING CYLINDER

The flow control valves can also be used to control the single-acting cylinder. The circuit shown in Figure 8.10 describes the manner in which the speed is controlled in a single-acting cylinder during both forward and backward strokes by using the flow control valve.

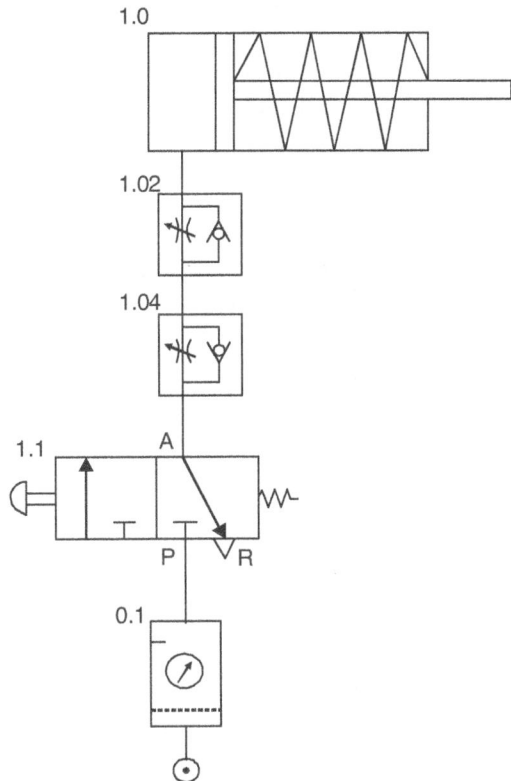

FIGURE 8.10 Control of Single-Acting Cylinder Using Flow Control Valve.

Here, two non return flow control valves 1.02 and 1.04 are used in the series as shown in Figure 8.10. Initially, cylinder 1.0 is in a fully retracted position as shown in the figure. When the push button of 1.1 DC valve is pressed, the pressure port P gets connected to working port A and pressurized fluid, *i.e.*, air enters the cylinder through two non return flow control valves 1.02 and 1.04. The restriction to air is offered by the flow control valve 1.02 during the forward stroke of the cylinder. When the push button of 1.1 DC valve is released, the backward stroke of the cylinder takes place and during the backward stroke the restriction to air is offered by the flow control valve 1.04. So the speed is reduced in both the forward and return stroke of the cylinder.

USE OF QUICK EXHAUST VALVE IN PNEUMATIC CIRCUITS

Flow control valves are used to slow the speed of the pneumatic cylinder. Similarly, the quick exhaust valve is used to exhaust the air out of the cylinder at a faster rate without passing it through the direction control valve thereby increasing the speed of the cylinder. The circuits given below show how the acceleration of the piston in single-acting and double-acting cylinder is achieved using a quick exhaust valve.

Use of a Quick Exhaust Valve in Single-Acting Cylinder

The circuit (refer to Figure 8.11) shows the control of a single-acting cylinder (1.0) by using a 3/2 manually operated spring return DC valve (1.1) and quick exhaust valve (1.01).

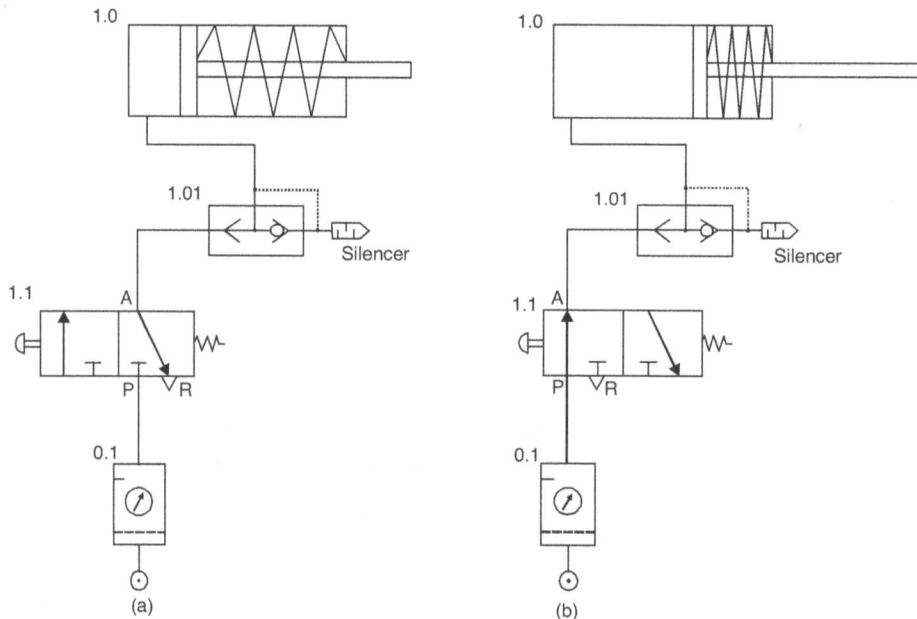

FIGURE 8.11 Control of Single-Acting Cylinder Using Quick Exhaust Valve.

Initially, the cylinder is in a fully retracted position as shown in Figure 8.11 (*a*). Port A is connected to exhaust port R. When the push button of the 1.1 DC valve is pressed, the pressure port P gets connected to working port A and the pressurized fluid, *i.e.*, air starts moving from the 1.1 DC valve to the cylinder 1.0 through the quick exhaust valve 1.01 causing it to move in a forward direction. Figure 8.11 (*b*) shows the fully extended position of the cylinder. To have the backward motion of the cylinder, the push button of 1.1 DC valve is released. When the push button of the 1.1 DC valve is released, the cylinder starts moving backward because of the action of the spring force and the compressed air escapes to the atmosphere through the quick exhaust valve 1.01 without getting exhausted through the 1.1 DC valve. Hence, the speed of the cylinder is increased by exhausting air out of the cylinder at a faster rate. A silencer is installed at the outlet of a quick exhaust valve to reduce the noise of the air going out to the atmosphere.

Use of a Quick Exhaust Valve in a Double-Acting Cylinder

The circuit (refer to Figure 8.12) shows the control of a double-acting cylinder (1.0) by using a 4/2 manually operated spring return DC valve as the control element (1.1) and quick exhaust valve (1.01).

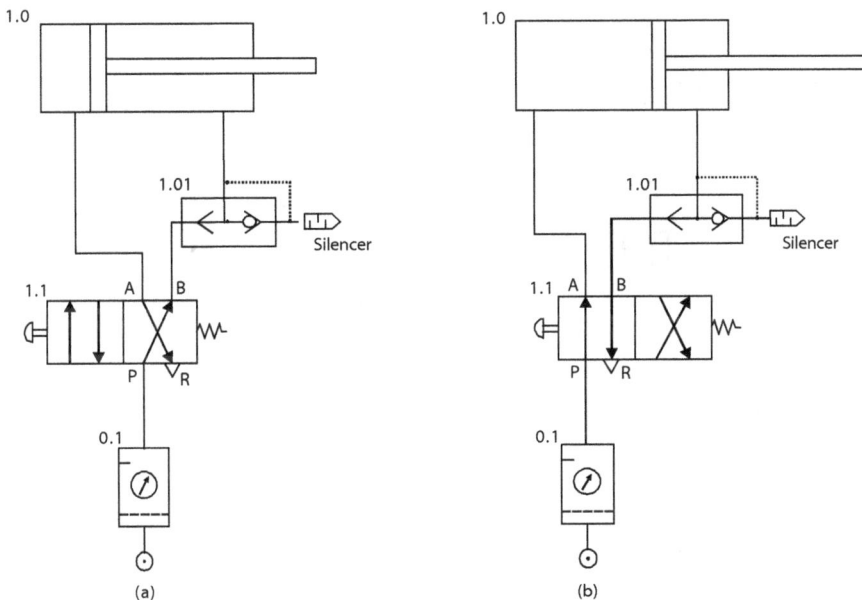

FIGURE 8.12 Control of Double-Acting Cylinder using Quick Exhaust Valve.

Initially, the cylinder 1.0 is in a fully retracted position as shown in Figure 8.12 *(a)*. Pressure port P is connected to working port B and port A is connected to exhaust port R. The pressurized fluid, *i.e.*, air starts moving from control element 1.1 to the cylinder 1.0 through the quick exhaust valve 1.01. When the push button of the 1.1 DC valve is pressed, the pressure port P gets connected to working port A and the pressurized fluid, *i.e.*, air starts moving from control element 1.1 to cylinder 1.0 causing the forward motion of the cylinder. Figure 8.12 *(b)* shows the fully extended position of the cylinder. At exhaust, a quick exhaust valve 1.01 is present which lets the air escape from itself. So air does not go back to the control element 1.1 for the purpose of exhausting. A silencer is installed at the outlet of a quick exhaust valve for reduction in noise of the air going out to the atmosphere. This results in extending the cylinder at a faster rate. It is important to note here that the quick exhaust valve does not play any role during the backward motion (retraction) of the piston.

TIME DELAY CIRCUIT

The time delay circuit as shown in Figure 8.13 is used to delay an already started signal. The circuit consists of a double-acting cylinder 1.0,

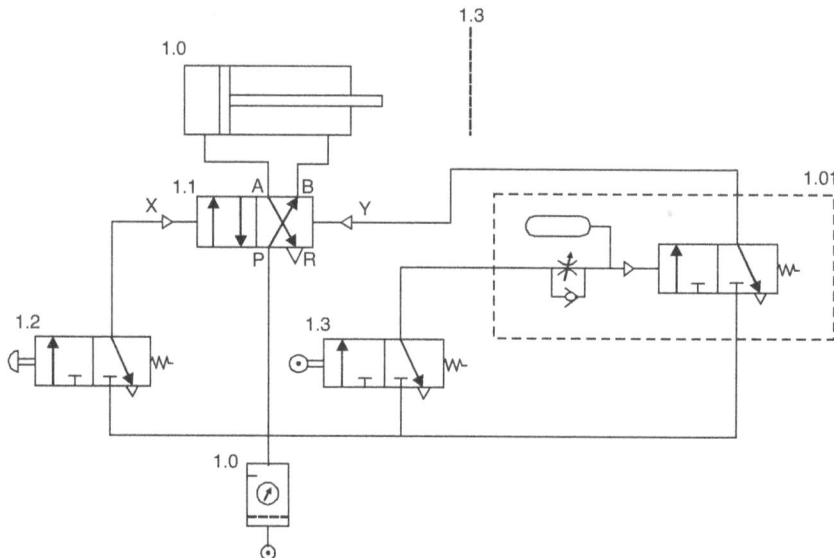

FIGURE 8.13 Time Delay Circuit.

4/2 pilot operated DC valve (1.1), time delay valve (1.01), two 3/2 manual operated and roller button operated spring return DC valves (1.2 and 1.3), FRL unit (0.1), and a compressor.

Initially, supply is given from pressure port P to working port B of a 1.1 DC valve and cylinder 1.0 is in a fully retracted position as shown in Figure 8.13. When the manual push button of the signaling element 1.2 is pressed, the airflow is started in the circuit. Control element 1.1 is actuated by pilot line X. With this, air enters the cylinders and pushes the piston in a forward direction. When the piston rod is completely extended, it actuates the signal valve element 1.3. With this, the compressed air is directed to the time delay valve 1.01. The time delay valve can delay the signal for 5 to 15 seconds depending on the requirement. The signal from the time delay valve 1.01 actuates the pilot line Y of control element 1.1, thereby changing the position of the DC valve and allows fluid to enter the rod side of the cylinder and to discharge from the other side. Time delay valves are used in certain special applications. For example, the task to be preformed by using a double-acting cylinder is bonding of the two work pieces by filling adhesive between them and applying heat and pressure on it for some time. Application of pressure for sometime on the work pieces can be achieved by the above discussed circuit. The cylinder will stop and apply pressure for sometime (5 to 15 seconds depending upon specification of the time delay valve) after extending completely.

CIRCUITS WITH NECESSARY CONDITIONS

These types of circuits are drawn to fulfill a particular condition. The circuits that come under this category are:

- Circuit showing application of the twin pressure valve
- Circuit using application of the shuttle valve

APPLICATION OF THE TWIN PRESSURE VALVE

Figure 8.14 shows the application of the twin pressure valve. This valve requires two equal inputs to give one output. Twin pressure valve is also termed as "pneumatic AND element" because it has the basic logic function of AND gate, *i.e.*, a signal is given at the output only if both inlet signals are present. The circuit diagram is drawn by taking some condition in mind that the double-acting cylinder will start extending only when two jobs are done. These are:

(a) Unloading of processed component, and

(b) Loading of fresh component that is to be processed.

The circuit shown in Figure 8.14 consists of a double-acting cylinder (1.0) as the actuator, a 4/2 pilot operated spring return DC valve (1.1), twin pressure valve (1.01), and two 3/2 push button operated spring return DC valves (1.2 and 1.4).

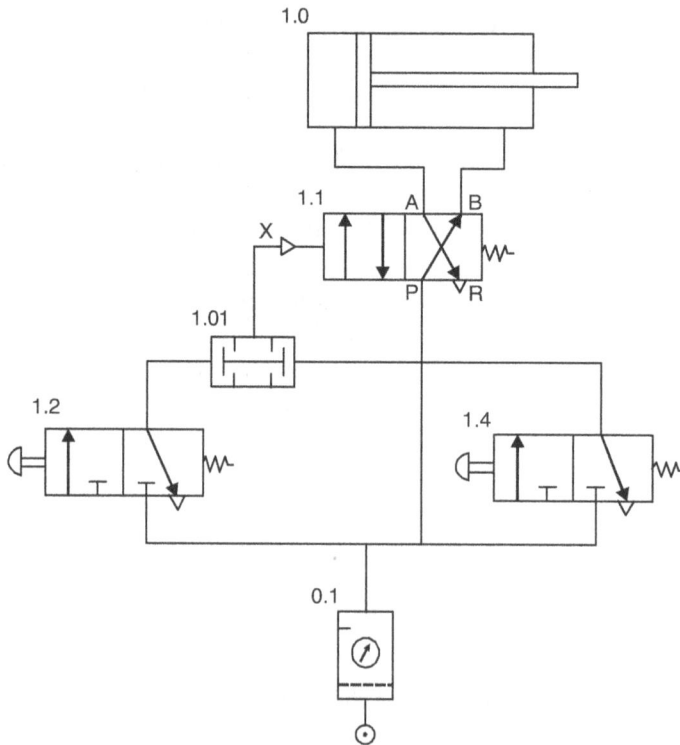

FIGURE 8.14 Circuit Showing Application of Twin Pressure Valve.

In the initial position as shown in the figure, the supply of compressed air is given to control element 1.1. Pressure port P is connected to working port B and port A is connected to exhaust port R. The cylinder is in a fully retracted position. When the two above said conditions are filled, the signaling elements 1.2 and 1.4 are operated by the machine operator and the signal is sent to the pilot line X of control element 1.1 via twin pressure valve 1.01 resulting in a forward motion of the cylinder. It is interesting to note here that until these two switches (1.2 and 1.4) are not pressed, operation of cylinder (on which tool is mounted) cannot be started.

APPLICATION OF THE SHUTTLE VALVE

Figure 8.15 shows the application of the shuttle valve. The shuttle valve always gives one output irrespective of the number of inputs whether 1 or 2, as shown in the circuit diagram. The shuttle valve is also termed as "pneumatic OR element" because it has the basic logic function of OR gate, *i.e.*, a signal is given at the output if a signal is present at either one input or the other or at both inputs. The circuit consists of a double-acting cylinder (1.0) as an actuator, a 4/2 pilot operated spring returned DC valve as control element (1.1), shuttle valve (1.01), and two 3/2 push button operated spring return DC valves (1.2 and 1.4).

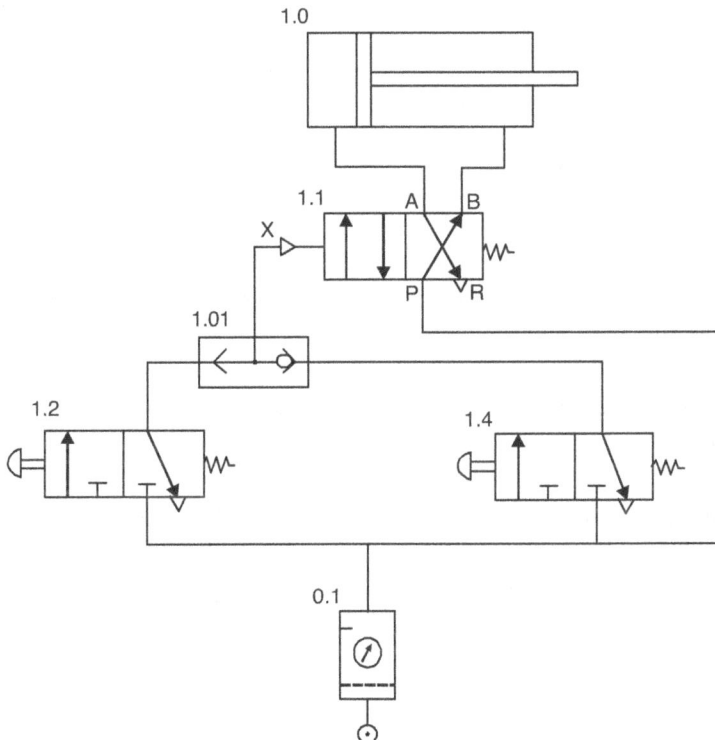

FIGURE 8.15 Circuit Showing Application of the Shuttle Valve.

In the initial position as shown in the figure, the supply of compressed air is given to control element 1.1. Pressure port P is connected to working port B and port A is connected to exhaust port R. The cylinder is in a fully retracted position. The input signaling elements 1.2 and 1.4 perform the same function. Both signaling elements are used to give input to shuttle valve 1.01. By pushing any switch out of 1.2 and 1.4, the compressed air is entered into line and is made to flow through the shuttle valve. From the shuttle valve the compressed air enters the pilot line X of control element 1.1 and changes its position. This results in forward motion of the cylinder. These types of circuits are used to provide a solution to all problems in which 'if' is involved.

HYDRAULIC CIRCUITS

Hydraulic circuits can be defined as graphic representation of the elements of a hydraulic system where these elements are represented by their corresponding symbols and these symbols are sequentially arranged in a specific manner depending upon the nature of output required. The major difference between hydraulic and pneumatic circuits is that hydraulic elements are designed for high pressure (20–400 bar) and pneumatic circuits are designed for low pressure (5–20 bar).

Hydraulic circuits are similar to pneumatic circuits because most of components used in hydraulic and pneumatic systems are similar. There are a few specific requirements of hydraulic circuits which are frequently used in hydraulic systems (example, pressure relief valves, check valves, pressure gauges, etc). The methodology is explained with the help of simple circuits given below.

HYDRAULIC CIRCUIT FOR CONTROL OF SINGLE-ACTING CYLINDER

Hydraulic Circuit for Control of Single-Acting Cylinder using a 2/2 DC Valve

In the circuit shown in Figure 8.16, a single-acting cylinder (1.0) is being controlled with a 2/2 manually operated spring return DC valve (1.1). The required pressure in the line is maintained with the help of a pump and motor assembly. Pressure gauge (0.1) and pressure relief valve (0.3) are provided in the line of the fluid flow. Now, if the pressure rises above the design limit, the pressure relief valve exhausts the excess fluid and ensures

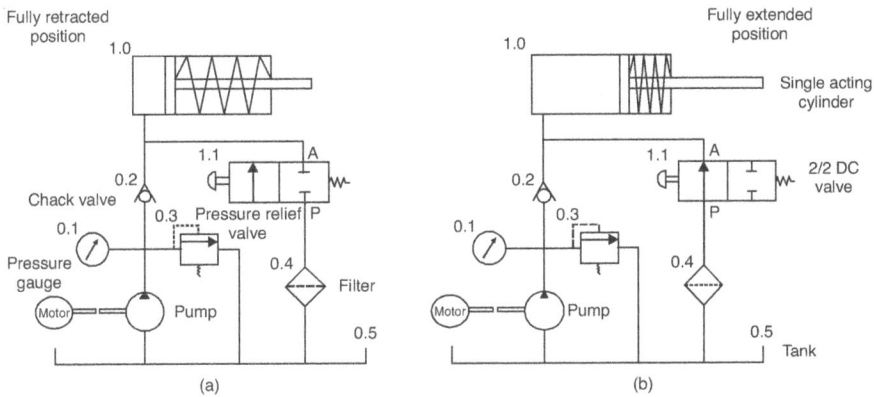

FIGURE 8.16 Control of Single-Acting Cylinder using 2/2 DC Valve.

the safety of the system. Afterwards, a check valve (0.2) is provided to prevent the flow of fluid into the cylinder. The oil is returned to the oil tank by passing it through a hydraulic filter (0.4), which removes any dirt particles present in the oil. In the given circuit, a 2/2 DC valve is used for controlling a single-acting cylinder. Generally, a 3/2 DC valve is used for this purpose but the position of the check valve has made it possible here.

Initially, the cylinder is in a retracted position as shown in Figure 8.16 *(a)*. Pressure port P is not connected to working port A. Pressurized fluid cannot enter into the cylinder either through the check valve or the 1.1 DC valve. When the push button of control element 1.1 is pressed, the pressure port P gets connected to working port A as shown in Figure 8.16 *(b)*. The pressurized fluid starts flowing into the cylinder through control element 1.1, which results in forward motion of the cylinder. As the push button is released, the DC valve comes back to its original position because of the spring action. Pressurized fluid from the cylinder will pass though the check valve into the tank causing its backward motion.

Hydraulic Circuit for Control of Single-Acting Cylinder using 3/2 DC Valve

In the circuit shown in Figure 8.17, a single=acting cylinder (1.0) is being controlled with a 3/2 manually operated spring return DC valve (1.1). The required pressure in the line is maintained with the help of a pump and motor assembly. Pressure gauge (0.1) and pressure relief valve (0.3) are provided in the line of fluid flow. Now, if the pressure rises above the design limit, the pressure relief valve exhausts the excess fluid and ensures the safety of the

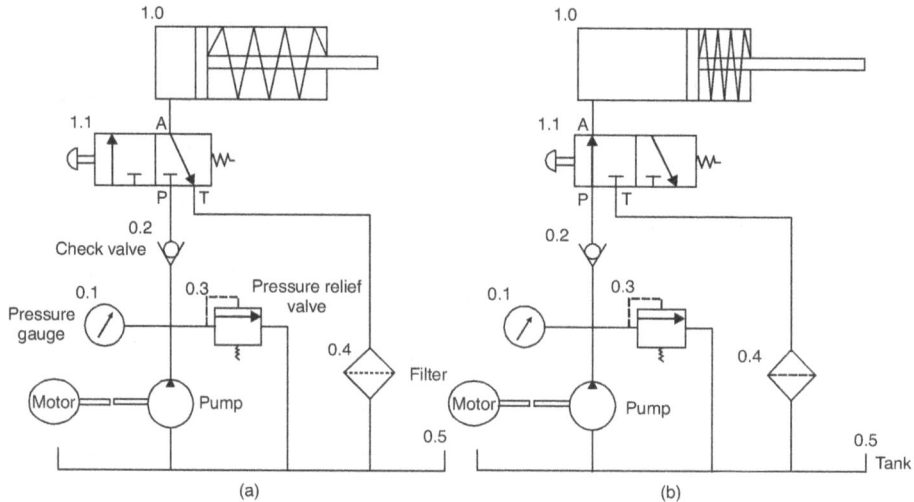

FIGURE 8.17 Control of Single-Acting Cylinder using 3/2 DC Valve.

system. Afterwards, a check valve (0.2) is provided to prevent the flow of fluid into the cylinder. The oil is returned to the oil tank by passing it through a hydraulic filter (0.4), which removes any dirt particles present in the oil.

Initially, the cylinder is in a retracted position as shown in Figure 8.17 (a). Working port A is connected to exhaust port T. Pressurized fluid cannot enter the cylinder. When the push button of control element 1.1 is pressed, the pressure port P gets connected to working port A as shown in Figure 8.17 (b). The pressurized fluid starts flowing into the cylinder through control element 1.1, which results in the forward motion of the cylinder. As the push button is released, the cycle repeats again.

HYDRAULIC CIRCUIT FOR CONTROL OF DOUBLE-ACTING CYLINDER

The basic hydraulic circuit for control of a double-acting cylinder is shown in Figure 8.18. Double-acting cylinders are those in which movement of the piston and piston rod is obtained by fluid entering from one port and simultaneously exhausting it from another port. For this purpose, a 4/2 DC valve is at least required. Here a double-acting cylinder (1.0) is being controlled by a 4/2 push button operated, spring return DC valve (1.1). A fixed displacement pump delivers pressurized fluid to control element 1.1 via check valve (0.2) and pressure relief valve (0.3). The oil is returned to the oil tank by passing it through a

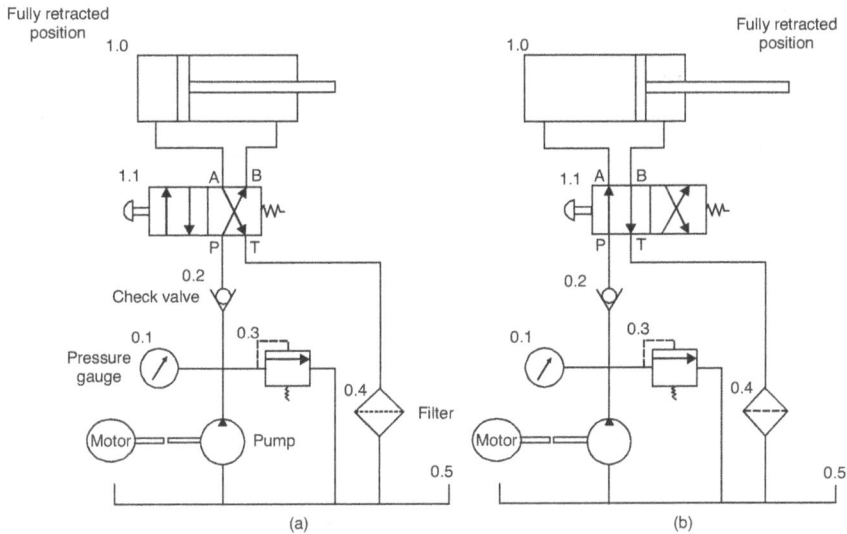

FIGURE 8.18 Control of Double-Acting Cylinder using 3/2 DC Valve.

hydraulic filter, which removes any dirt particles present in the oil. The circuit shows the manual control over the forward and backward motion of the cylinder. In the circuit shown below, there is no return spring in the cylinder for retraction of the piston rod, so both forward and backward motions of the cylinder are controlled by pressurized fluid and a 1.1 DC valve.

Initially, the cylinder 1.0 is in a fully retracted position as shown in Figure 8.18 (a). Pressure port P is connected to working port B and port A is connected to exhaust port T. When the push button of the 1.1 DC valve is pressed, the position of the 1.1 DC valve changes. The pressure port P gets connected to working port A and the pressurized fluid starts flowing from the 1.1 DC valve to the cylinder causing it to move in a forward direction. Figure 8.18 (b) shows the fully extended position of the cylinder. To have the backward motion of the cylinder, the push button of the 4/2 DC valve is released. Because of the action of the spring, the DC valve attains its previous position. The cycle repeats again.

In this circuit diagram, the motion of the piston and piston rod can be stopped only at two extreme positions of the cylinder but it is also possible to stop the piston and piston rod at any extreme position in the cylinder. This can be achieved by using 3 position DC valves. These valves not only serve the above said purpose but also have unique characteristics that are discussed in the following paragraphs.

CIRCUITS USING 3 POSITION VALVES

Closed Center Circuit

This circuit is called a closed center circuit because a close centered 4/3 DC valve is used as a control element. By closed center, it means that all the ports at neutral position of the valve are departed from each other (refer to Figure 8.19). When the valve is brought to its neutral position, the pressurized fluid sent to the double-acting cylinder is unable to come back. When this position is attained, the pressurized fluid gets blocked on the rod side and other side of the cylinder. Due to this, the piston immediately becomes motionless. For example, let's assume initially that the piston is extending or retracting. But once the DC valve is brought to a neutral closed center position, the piston will stop immediately.

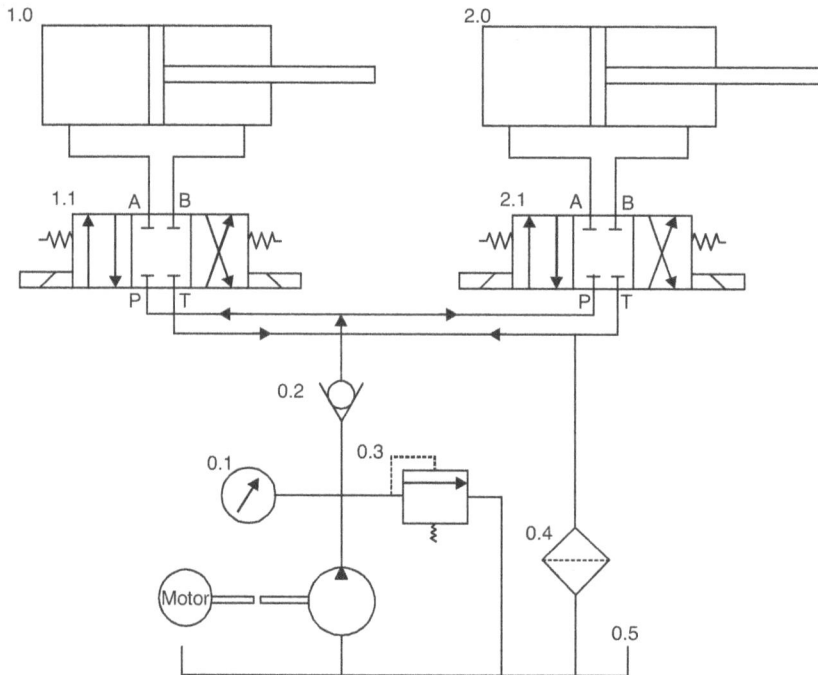

FIGURE 8.19 Closed Center Circuit.

Open Center Circuit

An open circuit is one in which the open center DC valve is used as the control element. This circuit (refer to Figure 8.20) has an advantage

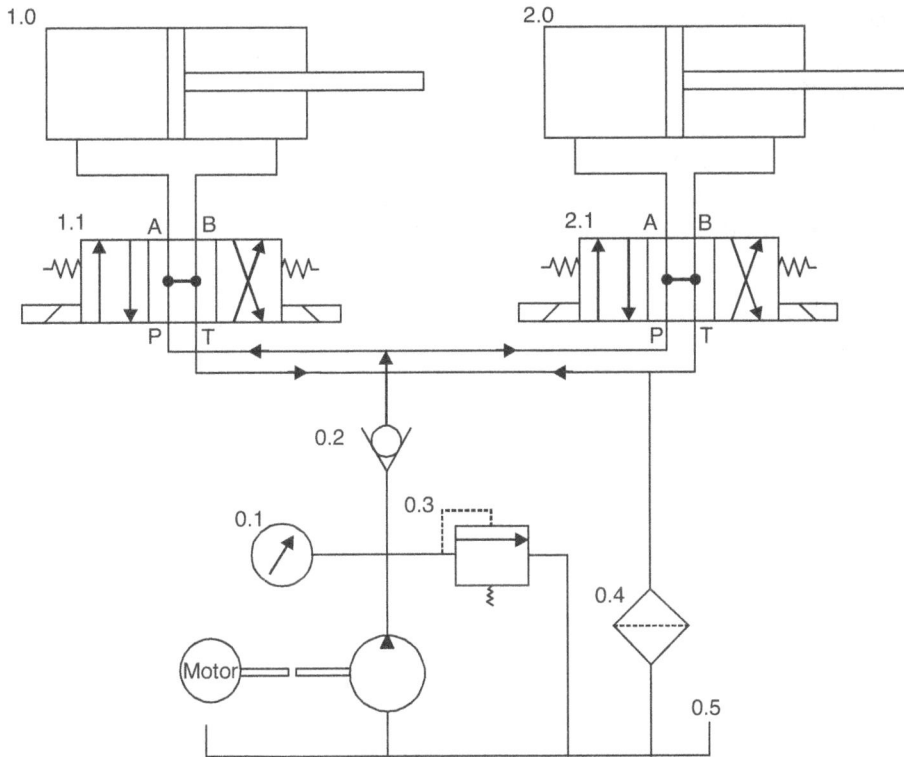

FIGURE 8.20 Open Center Circuit.

that the center position of the DC valve allows fluid (in the cylinder) to escape from the cylinder at a low pressure ensuring lesser heat generation. In this circuit, fluid is raised with the help of a hydraulic pump and is sent to control elements (4/3 solenoid operated spring return DC valves). The two extreme position of valves, *i.e.*, on and off are responsible for retraction and extension of the cylinder and the neutral position results in connecting pressure port P with working port A and exhaust port T with working port B. Thus, the fluid from both sides of the cylinder escapes back to the tank. This position is brought only when the pump is switched off and there must be a low-pressure region set in the hydraulic line.

There are two drawbacks associated with this circuit. First, all the cylinders have to start and stop at the same time as the ports are connected all the time during center position. Secondly, the load cannot be locked in neutral position. This may lead to an accident.

Tandem Center Circuit

In a tandem center circuit as shown in Figure 8.21, a tandem center 4/3 DC valve is used as a controlling element. In this circuit, a pump is not required to be switched off. Also, the movement of a double-acting cylinder can be stopped at any position between two extreme positions of the cylinders. But this valve enjoys an advantage over a closed center circuit, that the pressurized fluid from port P is directly supplied to port T thereby ensuring the flow from one subsystem to another whereas in closed center circuit flow is divided into two lines. Pressure losses and heat generation are also less. The fluid that had gone to both sides of a double-acting cylinder is blocked there only, thereby locking the piston and piston rod at that position. The pressurized fluid blocked in the cylinder may cause the problem of oil spillage.

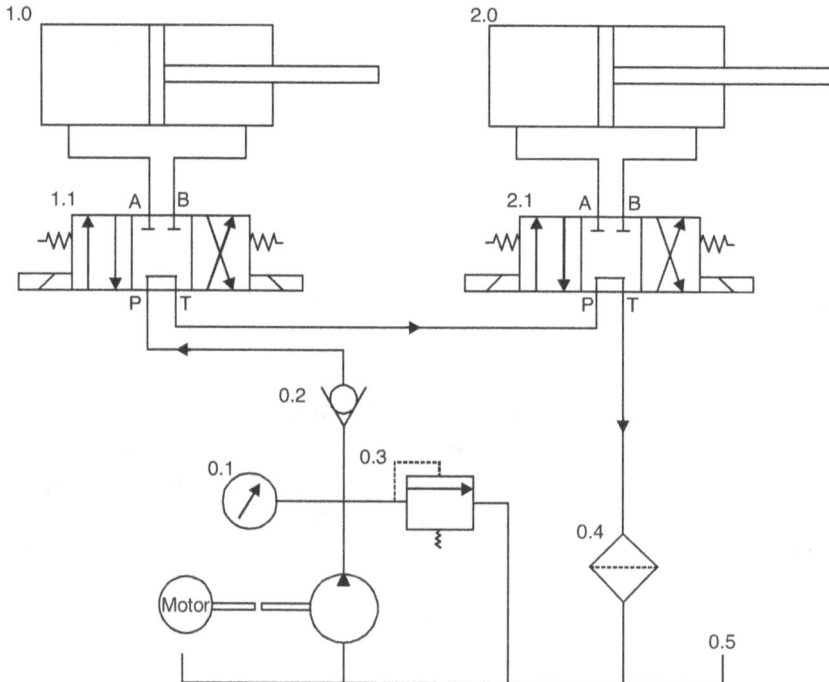

FIGURE 8.21 Tandem Center Circuit.

SPEED CONTROL IN HYDRAULIC CIRCUITS

Speed of hydraulic actuators is controlled by changing flow rates. These flow rates can be varied either by fixed restrictors or flow control valves.

In case of fixed restrictors or orifices, flow rate is dependent on the resistance offered by load, *i.e.*, the speed couldn't be accurately controlled. If a constant speed is required irrespective of the resistance, than the flow control valves are to be used. Flow control valves have further two configurations. *The simple adjustable throttle valve* provides speed regulation in both return and forward strokes of double-acting cylinder whereas *adjustable throttle valves with a by pass check valve* provides speed regulation only in one direction and they are called non return flow control valves.

Figure 8.22 shows the location of an adjustable throttle valve in a hydraulic circuit. The throttle valve (0.6) is mounted at the inlet of DC valves. As a result, the pressure rises in the inlet line that is exhausted from the pressure relief valve. In this way, the double-acting cylinder get lesser fluid on the cap end side of the cylinder. Therefore, the piston will extend at a slower rate. The same process is repeated in the return stroke.

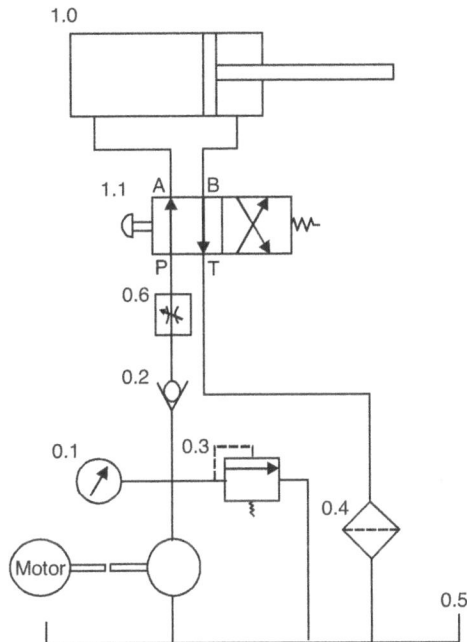

FIGURE 8.22 Use of Adjustable Throttle Valve.

Although the use of adjustable throttle valves are not recommended, this is common practice in hydraulics. The flow control valve is always accompanied by a by pass check valve. The direction of the check valve in a hydraulic line decides the flow to be regulated at what time whether at the inlet or at the exhaust port from the cylinders. Speed control circuits are of two types:

- Meter in circuit
- Meter out circuit

Meter In Circuit

In meter in circuit, the fluid has to pass through the flow control valve before entering the cylinder, *i.e.*, speed of fluid is reduced when it has to enter the cylinder. Figure 8.23 shows the details of meter in circuit. The circuit consists of a 4/2 DC valve as the control element (1.1), two non return flow control valves (1.01) and (1.02), respectively, and a double-acting cylinder (1.0).

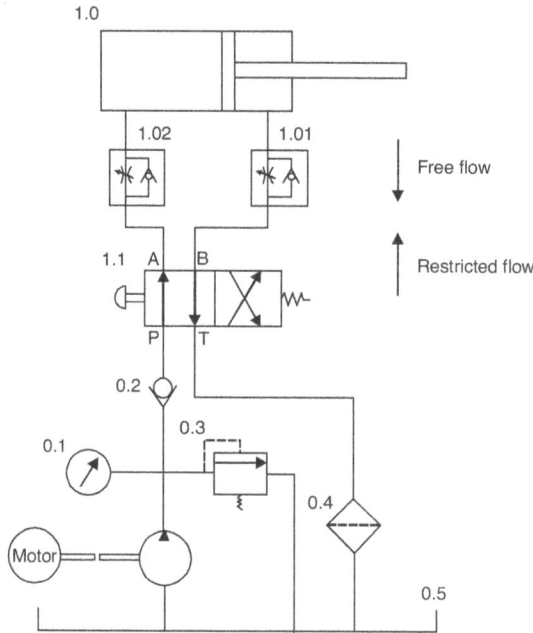

FIGURE 8.23 Meter In Circuit.

Pressurized fluid from the pump is entered in the non return flow control valve 1.02 through control element 1.1. As the flow control valve is placed in the primary line and the check valve is closed in this direction, so the fluid has to pass through restriction, which results in the lowering speed of the actuator during the extension/forward stroke. The fluid to be exhausted out of the cylinder is bound to pass through the flow control valve

1.01 without any restriction. When the push button of control element 1.1 is pressed, its position changes. This time pressurized fluid is entered through the non return flow control valve (1.01). Again, at the inlet, the fluid has to pass through the variable restrictor because of the "no flow" position of the check valve in that direction which results in the slow motion of the piston in a backward direction.

Meter in circuit is not recommended, except in few cases, because the pressure of hydraulic fluid which was built up in order to push the piston of the cylinder is reduced before even entering the hydraulic cylinder. This arrangement is suitable for precise and low-pressure systems because a better speed control can be attained.

Meter Out Circuit

In the meter out circuit, the fluid has to pass through the non return flow control valve after escaping from the cylinder, *i.e.*, speed of fluid is reduced when it is coming out of the cylinder. Construction of this circuit shown in Figure 8.24 is the same as that of meter in circuit. The only difference is that the direction of the check valve is changed in this circuit for providing restriction to fluid at the outlet.

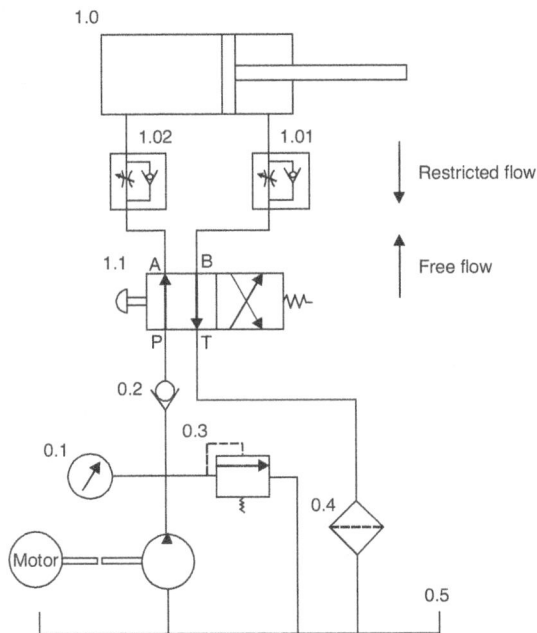

FIGURE 8.24 Meter Out Circuit.

Consider the valve at its normal position, which is shown in Figure 8.24. The pressure port P is connected to working port A and port B is connected to exhaust port T. The pressurized fluid passing from control element 1.1 reaches the non return flow control valve 1.02. No restriction is offered in valve 1.02 as the check valve allows the flow of fluid through it. The fluid enters the cylinder and pushes the piston in a forward direction. At the same time, the fluid has to be exhausted from the rod side of the cylinder and this time the fluid has to pass through the non return flow control valve (1.01). This time check valve does not allow fluid to pass freely from it and fluid passes through restriction. Hence, the speed of the piston is slowed as the fluid escapes slowly out of the restrictor. After completion of the forward stroke, the manual push button on control element 1.1 is pressed and the circuit proceeds in the same manner as explained in the case of the forward stroke.

Meter out circuit is preferred in hydraulics because the pressure of entering fluid is not reduced here as is done in the case of meter in circuit. The whole pressure energy is converted into mechanical energy. Meter out circuits can also be used to control speed in the case of hydraulic motors whereas meter in circuits are not suitable to perform this task. In meter out circuit, heat generation due to the flow control valve does not affect the system, i.e., actuator, because fluid approaches the flow control valve after the actuator.

BLEED OFF CIRCUIT

In this circuit, a small quantity of pressurized fluid from the pump is sent directly to the tank. This is done in order to slow down the speed of the cylinder in both directions, *i.e.*, during the forward and backward stroke. A flow control valve is used in the by pass line for regulating the flow. These circuits are used only to slow down the speed of the cylinder. As the flow control valve is placed before the DC valve in the circuit, speed can be reduced only to some limited value.

Bleed off circuit is explained with the help of a simple circuit diagram shown in Figure 8.25 for controlling a double-acting cylinder. For ease of readers, the circuit is explained in three steps taking all positions of the DC valve. Here, a double-acting cylinder (1.0) is being controlled by a 4/3 solenoid operated spring return DC valve as control element (1.1). The required pressure in the line is maintained with the help of a motor and pump assembly. A pressure relief valve (0.3) is also provided. Flow control valve (0.2) is used in by pass line for regulating the flow.

Step 1: OFF or neutral position of circuit

Initially, as shown in Figure 8.25 *(a)*, the control element is in its normally closed position. All the ports, P, A, B, and T, are blocked to each other. During OFF position when the circuit is not in use all the coils, *i.e.*, X and Y are de-energized. At this stage, pressure in both sides of the cylinder, *i.e.*, cap end and rod end sides are equal.

Step 2: Forward stoke

The forward motion of the cylinder can be achieved by energizing the solenoid X. When the solenoid X gets energized, the control element 1.1 changes its position. Pressure port P gets connected to working port A and port B gets connected to port T as shown in Figure 8.25 *(b)*. A small quantity of pressurized fluid, which is going from P to A, is sent to the tank through flow control valve 0.2. This slows down the speed of the cylinder during the forward stroke. The solenoid gets de-energized, then due to the action of the spring force, the 1.1 DC valve comes back to its original position, *i.e.*, normally closed position.

Step 3: Backward stroke

Now, when the solenoid Y is energized, the 1.1 DC valve changes its position. Pressure port P gets connected to working port B and port A gets connected to port T as shown in Figure 8.25 *(c)*. This results in slowing down the speed of the cylinder during the backward stroke as a small quantity of pressurized fluid, which is going from P to B, is sent to the tank through flow control valve 0.2. The solenoid gets de-energized, then due to the action of the spring force, the 1.1 DC valve comes back to its original position, *i.e.*, normally closed position.

REGENERATIVE CIRCUIT

Regenerative circuit is used when there is a requirement of rapid extension of the cylinder against load. This requirement of forward motion with push force can be achieved by using a regenerative circuit. Regenerative circuit is explained with the help of a simple circuit diagram shown in Figure 8.26 for a controlling double-acting cylinder. Here, a double-acting cylinder (1.0) is being controlled by a 4/3 solenoid operated spring return DC valve as control element (1.1). A 3/2 solenoid operated spring return DC valve (1.3) is placed in between the cylinder and control element 1.1. The required pressure in the line is maintained with the help of a motor and pump assembly. A pressure relief valve (0.3) is also provided. For the ease of readers, this circuit is explained in three steps taking all positions of DC valve.

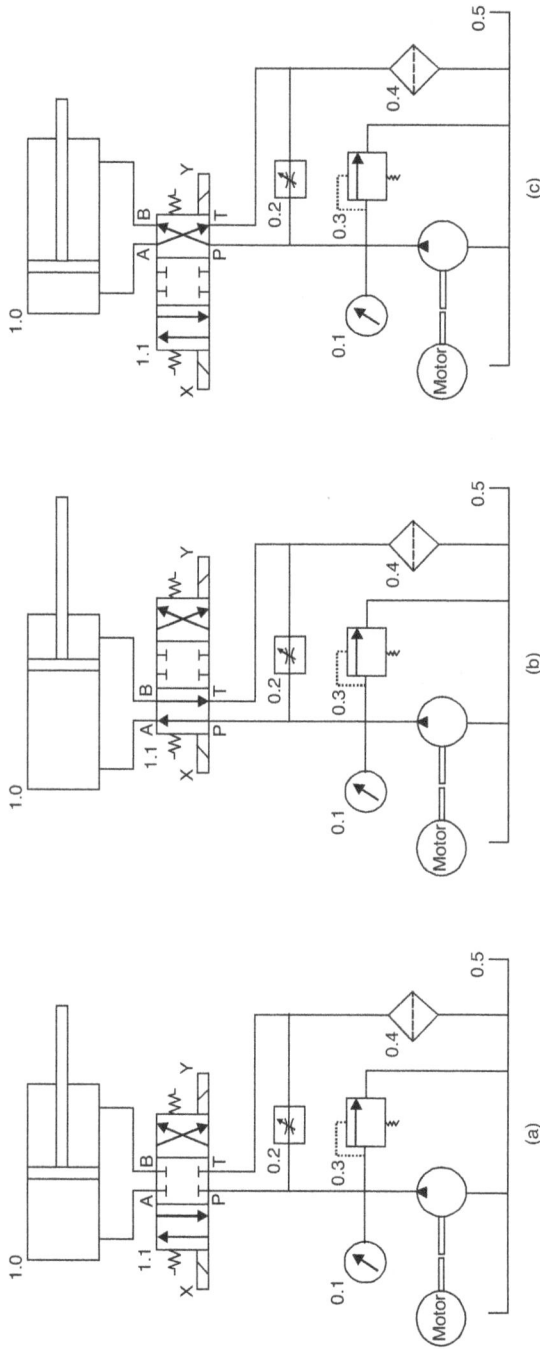

FIGURE 8.25 Bleed Off Circuit.

Step 1: OFF or neutral position of circuit

Initially, as shown in Figure 8.26 *(a)*, the control element is in its normally closed position. All the ports, P, A, B, and T, are blocked to each other. During the OFF position when the circuit is not in use all the coils, *i.e.*, X, Y and Z, are de-energized. At this stage, pressure in both sides of the cylinder, *i.e.*, cap end and rod end side are equal.

Step 2: Forward stroke

The forward motion of the cylinder can be achieved by energizing the solenoid X. When the solenoid gets energized, the 1.1 DC valve changes its position. Pressure port P gets connected to working port A and port B gets connected to port T as shown in Figure 8.26 *(b)*. The pressurized fluid coming out of cylinder 1.0 is passed through 1.3 DC valve where it gets mixed with the fluid coming from the pump. The fluid from the pump as well as from the rod side of the cylinder is directed to the cap end side of the cylinder. This induces a fast forward motion in the piston and piston rod end assembly.

Step 3: Backward stroke

During the backward stroke of the cylinder, the regenerative circuit is not required because rapid movement of cylinder is only required during the forward stroke. So, solenoids Y and Z are energized to deactivate the regenerative circuit and to change the position of control element, respectively. When solenoids Y and Z are energized, the pressure port P gets connected to working port B and the pressurized fluid enters the rod side of the cylinder through a 1.3 DC valve, which results in the backward motion of the cylinder as shown in Figure 8.26 *(c)*. The backward motion of the cylinder will be at normal speed because fluid coming out of the cylinder during the backward stroke is not sent back to the pressure line.

CIRCUIT SHOWING APPLICATION OF COUNTER BALANCE VALVE

The counter balance valve is used to prevent the free fall of the cylinder under its own weight or due to the load attached to it. The application of the counter balance valve is explained with the help of a simple circuit diagram as shown in Figure 8.27 for the controlling double-acting cylinder. Here, a double-acting cylinder (1.0) is being controlled by a 4/3 solenoid operated spring return DC valve as control element (1.1). The counter balance valve (1.01) is placed in between the cylinder and control element. The required pres-

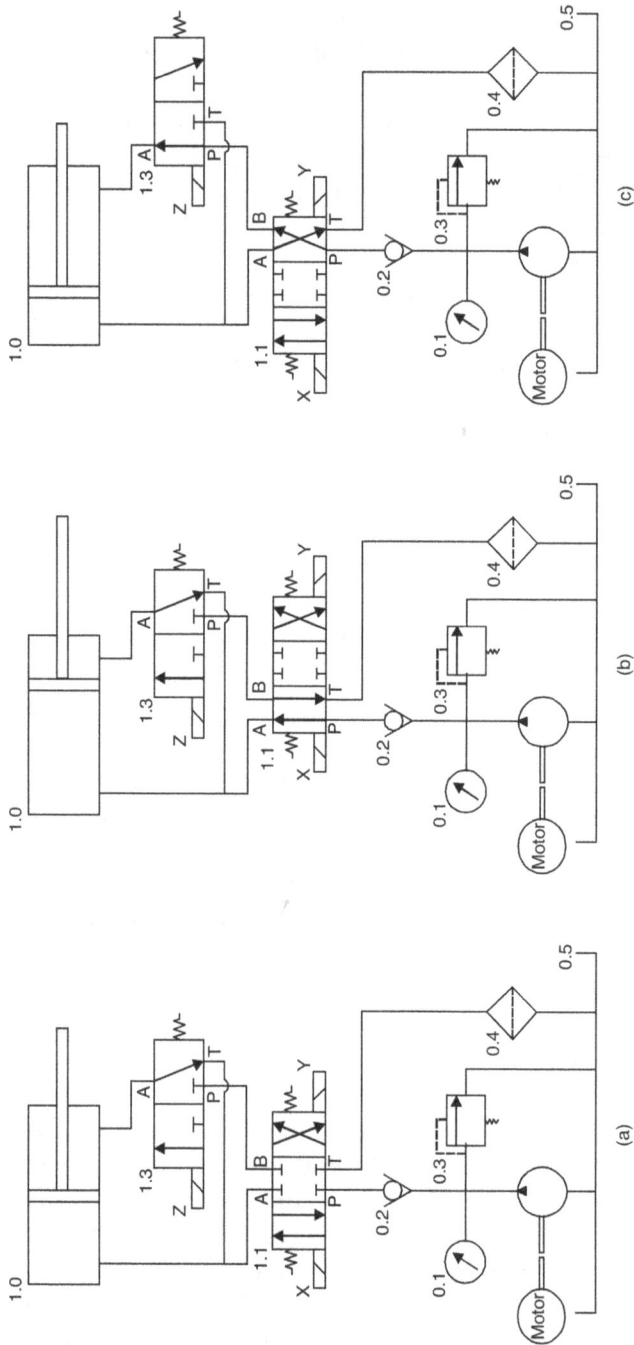

FIGURE 8.26 Regenerative Circuit.

sure in the line is maintained with the help of a motor and pump assembly. A pressure relief valve (0.3) is also provided. For ease of readers, the circuit is explained in three steps taking all positions of the DC valve.

Step 1: OFF or neutral position of circuit

Initially, as shown in Figure 8.27 (*a*), the control element is in its normally closed position. All the ports, P, A, B, and T, are blocked to each other. During the OFF position, when the circuit is not in use, all the coils, *i.e.*, X and Y are de-energized. At this stage, pressure in both sides of the cylinder, *i.e.*, cap end and rod end sides are equal.

Step 2: Upward stroke of cylinder

The upward motion of the cylinder can be achieved by energizing the solenoid X. When the solenoid X gets energized, pressure port P gets connected to working port A and port B gets connected to port T as shown in Figure 8.27 (*b*). Because the counter balance valve 1.01 is in its closed position, the pressurized fluid going from port P to port A passes through the check valve and lifts the piston in an upward direction. It is the tendency of the piston to fall because of the load attached to it. Here comes the use of the counterbalance valve. The pressure setting of the counter balance valve is done at higher limits as compared to the pressure exerted by the load attached to the piston on the fluid. So because of a higher pressure setting in the counter balance valve, the piston remains held in vertical position.

Step 3: Downward stroke of cylinder

When the solenoid Y gets energized, pressure port P gets connected to working port B and port A gets connected to port T as shown in Figure 8.27 (*c*). The pressurized fluid going from port P to port B exerts force on the piston which overcomes the pressure setting of the counter balance valve, so the counter balance valve gets open and the piston starts moving downward as shown in Figure 8.27 (*c*).

SEQUENCING CIRCUIT

Sequencing means control of more than one cylinder to move in the desired sequence during forward and return strokes. Sequencing circuit is explained by taking a simple sequence of two cylinders. For example, there are two cylinders, 1 and 2, and it is required that when the start button is pressed, the piston of cylinder 1 extends and when it is fully extended, the

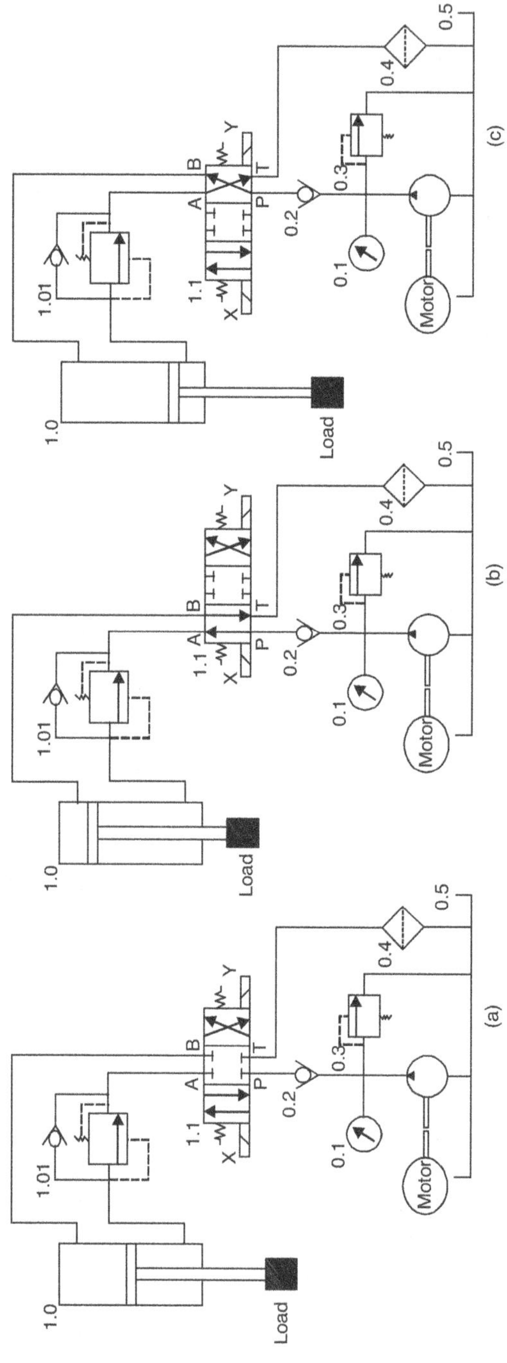

FIGURE 8.27 Application of Counter Balance Valve.

piston of cylinder 2 extends. When both the cylinders are in an extended position, it is required that the piston of cylinder 1 retracts, and when the piston of cylinder 1 is in a fully retracted position, the piston of cylinder 2 retracts. The sequencing circuit for the above sequence is shown in Figure 8.28.

The circuit is divided into groups based on the number of actuators. First actuator 1.0 is driven by a double pilot operated 4/2 DC valve (1.1). Two roller-operated spring returned DC valves 2.2 and 2.3 are placed at inner and outer extreme position of cylinder 2.0, respectively. One roller-operated spring returned 3/2 DC valve (2.4) is placed at the outer extreme position of cylinder 1.0.

Initially, both the cylinders are in a fully retracted position as shown in the figure. The 2.2 DC valve is initially in an actuated position. The pressurized fluid coming from 2.2 DC valve actuates the pilot signal X of control element 1.1. The pressurized fluid enters the cylinder 1.0. This causes the forward motion of cylinder 1.0. This forward motion leads to the actuation of a 2.4 DC

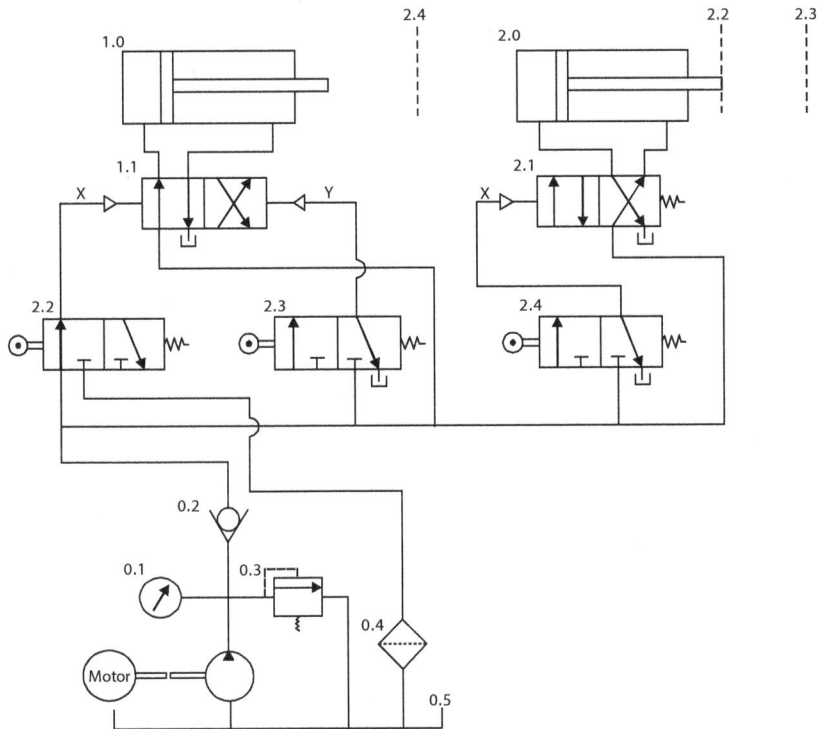

FIGURE 8.28 Sequence Control Circuit.

valve. The position of control element 2.1 changes with the actuation of the 2.4 DC valve. The pressurized fluid coming from control element 2.1 enters the cylinder 2.0. This causes the forward motion of cylinder 2.0. The forward motion leads to the actuation of the 2.3 DC valve.

The actuation of the 2.3 DC valve changes the position of control element 1.1, which leads to the backward motion of cylinder 1.0. Because the 2.1 DC valve is spring returned, it causes the backward motion of cylinder 2.0. This completes the required sequence of two cylinders.

Example of Sequencing Circuit

The circuit (refer to Figure 8.29) shows the sequencing of two cylinders to perform the drill operation on a workpiece. The sequence of cylinders to be accomplished is as given below:

1. Cylinder 1 clamps the workpiece.
2. Cylinder 2 performs drilling operation on workpiece.
3. Cylinder 2 returns back.
4. Cylinder1 unclamps the workpiece.

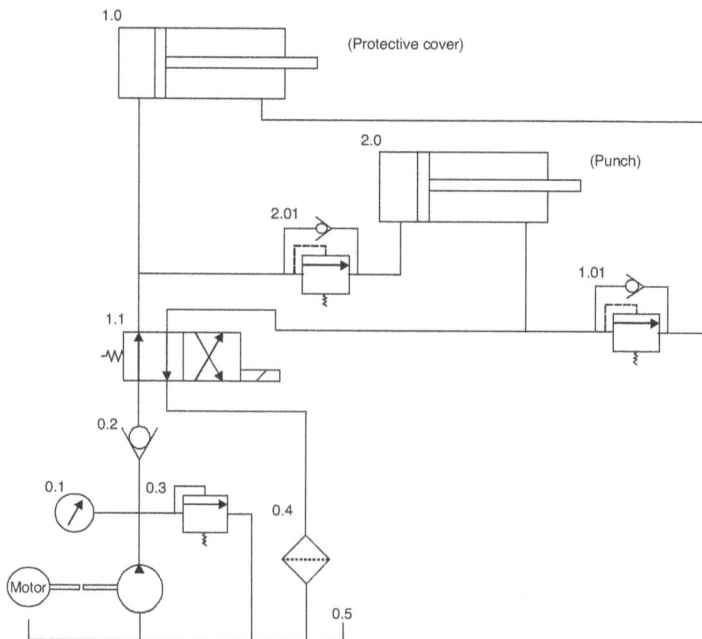

FIGURE 8.29

Initially both cylinders 1.0 and 2.0 are in a fully retracted position as shown in Figure 8.29. Pressurized fluid from the pump enters the cylinder 1.0 through 1.1 DC valve. This causes forward motion of cylinder 1.0, which results in the clamping operation. The pressurized fluid gets exhausted from cylinder 1.0 through the check valve of 1.01 sequence valve. The fully extended position of cylinder 1.0 results in opening of sequence valve 2.01. Now pressurized fluid starts flowing into cylinder 2.0 through the sequence valve 2.01, which results in the drilling operation. When the solenoid is energized, the pressurized fluid starts flowing directly into cylinder 2.0 resulting in its backward motion, which further results in retraction of cylinder 1.0. Hence, the above sequence is achieved.

PRESSURE REDUCTION CIRCUIT

A pressure reducing valve is used in a supply line joining two subsystems. Suppose the supply from the pump taken at 10-bar pressure is sent directly to the actuator (1.0) via the DC valve. In order to drive the actuator (2.0) in a low-pressure circuit, fluid has to pass through a pressure reducing valve where its pressure is reduced to 5-bar and then sent to actuator 2.0 via the DC valve. (Refer to Figure 8.30.)

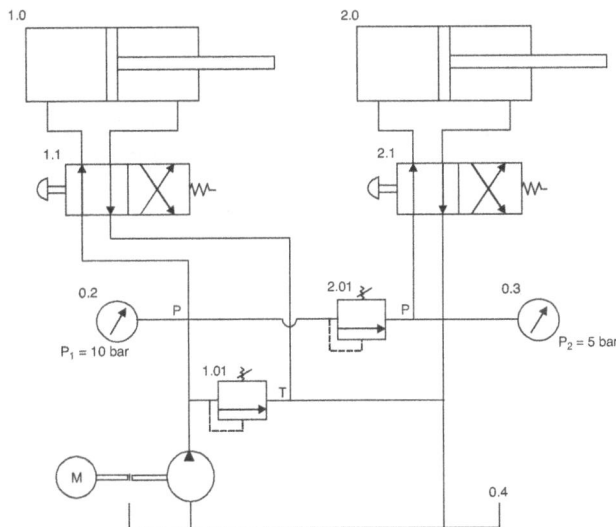

FIGURE 8.30 Pressure Reduction Circuit.

PROBLEMS IN CIRCUIT DESIGN

Problem 1: *Design a pneumatic system having cylinder flow control valves, pressure regulating valves, FRL unit, etc. for operating the cylinder once in 100 seconds with a dwell time of 15 seconds in the retracted as well as extended extremes.*

Solution: (Refer to Figure 8.31.) For satisfying the conditions given in the above problem, non return flow control valves (1.01 and 1.02) and time delay valves (1.2 and 1.3) are used in the circuit. In this circuit, compressed air from the source is sent to the 2/2 solenoid operated DC valve through a FRL unit. When the solenoid of the 2/2 DC valve is energized, the valve opens and the pressurized fluid, *i.e.,* air is sent to the rest of the system. Limit switches (3/2 DC valves) 1.4 and 1.5 are placed at both extremes of the cylinder. Limit switches are actuated automatically as the piston rod moves between two extreme positions

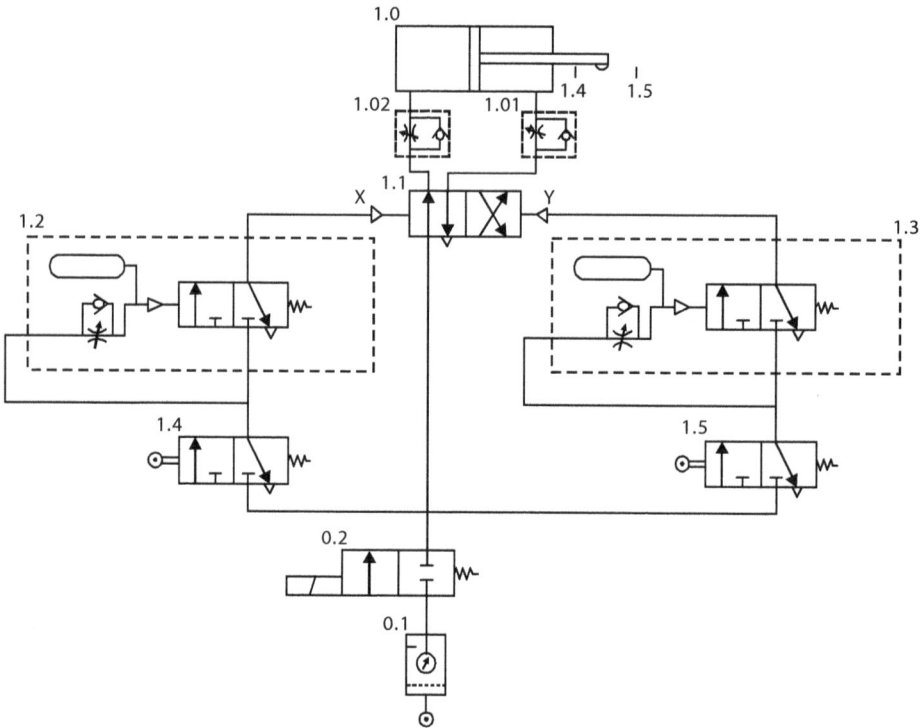

FIGURE 8.31

and pilot signals X and Y are sent alternatively via the time delay valves. Limit switch (1.5) and time delay valve (1.3) are responsible for the backward stroke of the cylinder and limit switch (1.4) and the time delay valve (1.2) are responsible for the forward stroke of the piston. Time delay valves are for a time delay of 15 seconds. Further, flow control valves 1.01 and 1.02 are provided so that forward and backward movement is obtained in 100 seconds.

Problem 2: *A steel door is installed in an auditorium. This door may be opened or closed by ON and OFF switches operated either from outside or inside as shown in Figure 8.32. Design a pneumatic circuit and explain it works.*

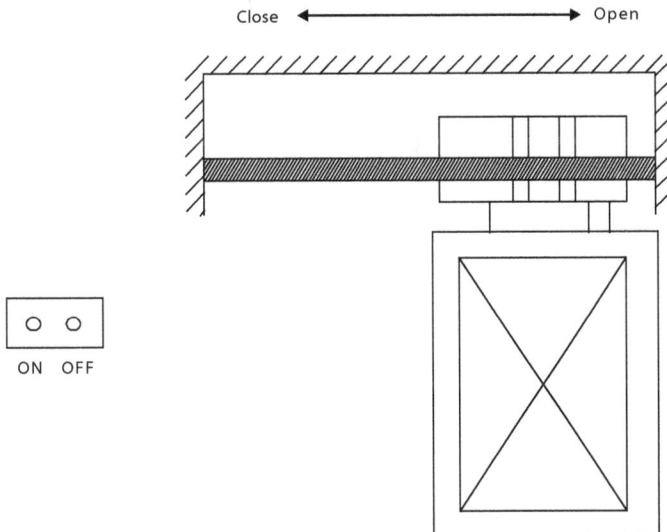

FIGURE 8.32

Solution: (Refer to Figure 8.33.) In this problem, two switches have to be provided for opening the door and two for closing the door, *i.e.*, one ON and one OFF switch on both sides of the wall. For these two OR gates/ shuttle valves (1.01, 1.02) are used. Valve 1.01 is connected with two 3/2 push button DC valves (1.3 and 1.5). These two valves perform the function of manually operated OFF switches. Similarly, valve 1.02 is connected with two 3/2 push button DC valves (1.2 and 1.4) and these valves perform the function of manually operated ON switches.

There will be one shuttle valve for each set of ON switches and another for each set of OFF switches. The whole door assembly is surmounted on a through rod double-acting cylinder with a fixed rod (1.0). In this type of cylinder, the rod is fixed between two ends and the cylinder body is free to move forward and backward by entering and exhausting compressed air in two ends alternately. The double-acting cylinder is driven by means of compressed air and the 4/3 closed center DC valve (1.1) is used in the system to control the direction of the fluid. The two shuttle valves (1.01, 1.02), as discussed above, provide the pilot signal (depending on which switch is pressed) to the 4/3 DC valve (1.1) and the door will move in that particular direction.

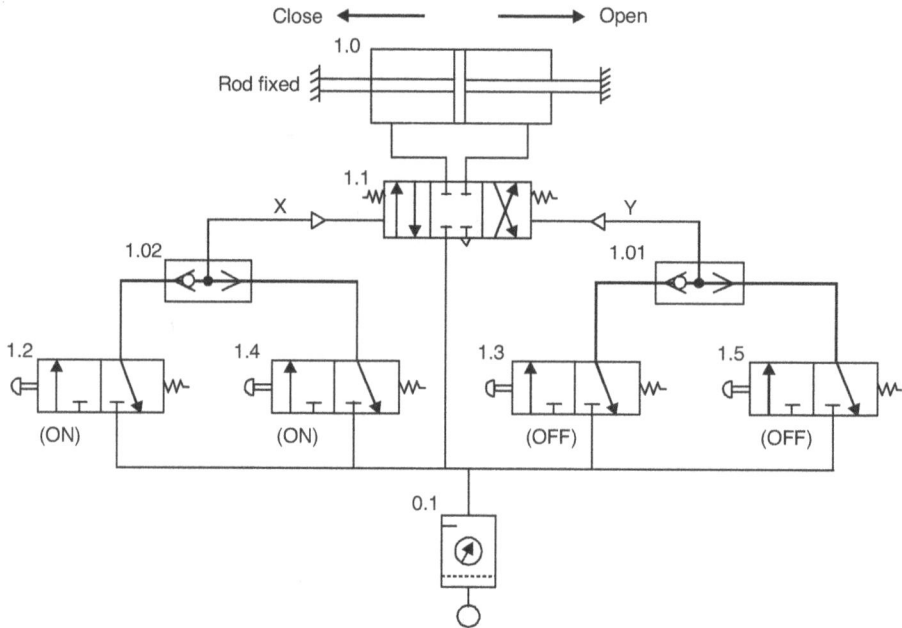

FIGURE 8.33

Problem 3: *Components are to be supplied from a gravity magazine to workstation by using a double-acting cylinder as shown in Figure 8.34. Feeding starts when the push button is pressed. The piston returns automatically to start the process again. Design a pneumatic circuit for the above problem and explain its construction and how it works.*

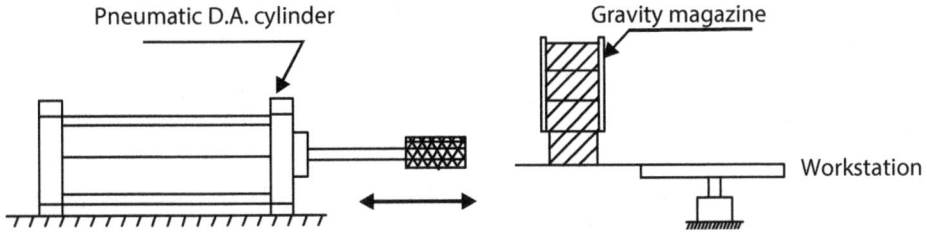

FIGURE 8.34

Solution: As shown in Figure 8.35 forward and backward motion of the cylinder is obtained by changing the position of the 4/2 DC valve (1.1). The system starts by operating the 1.2 DC valve. When the manual push button of the 1.2 DC valve is pressed, the supply of compressed air is sent to the pilot signal X of control element 1.1, which results in the forward motion of cylinder 1.0. On reaching the extreme outward end, the piston rod pushes the roller switch operated (3/2) DC valve (1.3) and due to which the piston returns automatically.

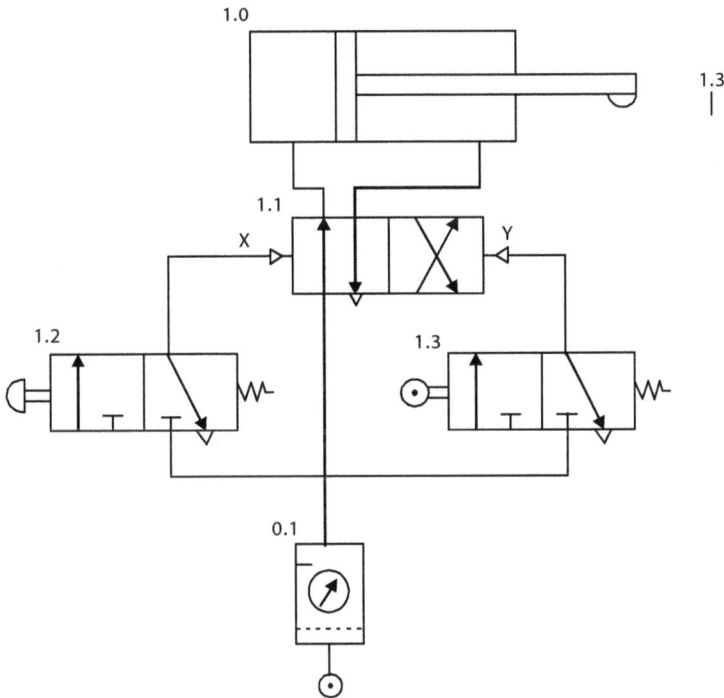

FIGURE 8.35

Problem 4: *Design a hydraulic circuit to control the motion of a double-acting cylinder in such a way that the forward movement of the piston rod can be possible with three different speeds (lower than normal speed) but the piston comes back with its normal speed. The piston can be stopped at any position between two extreme positions too.*

Solution: (Refer to Figure 8.36.) Fluid is made to pass through the hydraulic filter and check valve to the 4/3 DC valve (closed center). A and B are the working ports of the control valve (1.1). Working port A is connected to a hydraulic subsystem consisting of three 2/2 DC valves (1.01, 1.02, 1.03), all of them having a separate variable flow control valve (1.05, 1.06, 1.07) which is responsible for changing the velocity of fluid to a particular set level for which it is designed. Also, there is a check valve (1.04) in the line whose function is to by pass the discharged fluid from the cylinder to the reservoir during return stroke.

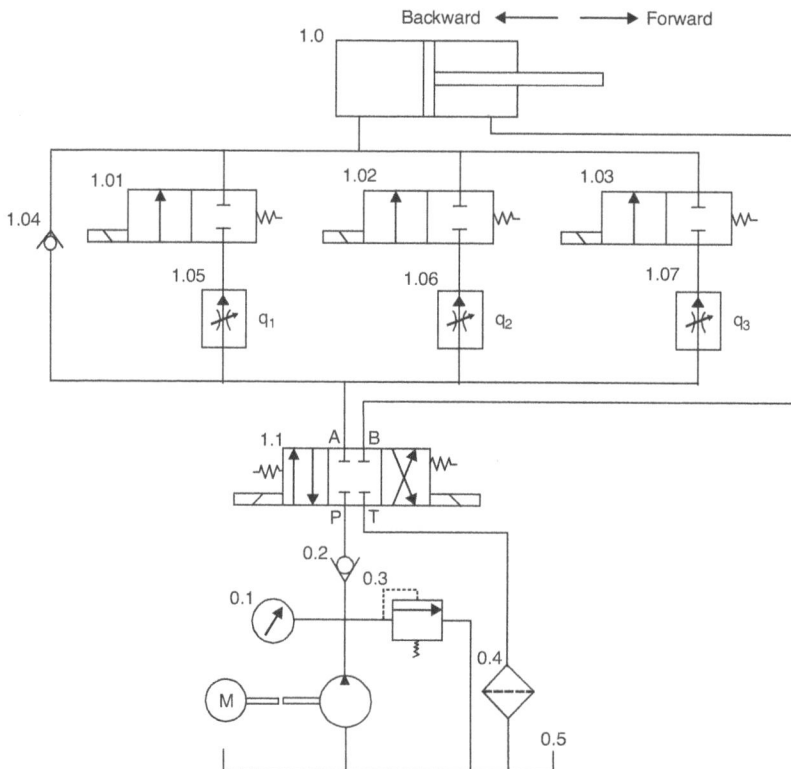

FIGURE 8.36

The output from this subsystem depends on valves 1.01, 1.02, and 1.03 that are solenoid operated. The required speed is obtained by energizing and de-energizing the respective solenoids of valves (1.01, 1.02, and 1.03). During the return stroke, there is no such arrangement provided after working port B and the cylinder comes with a faster speed. In order to satisfy the condition of stopping the piston of the double-acting cylinder at any position between two extremes, we use the 4/3 closed center DC valve. It satisfies the above-mentioned condition.

Problem 5: *In some manufacturing operations, it is required that the movement of the ram of a machine tool should be fast but sometimes it should be slow, both during forward and backward stroke. In case of electricity failure, the reciprocating motion of the ram becomes free and operation can be carried out manually. Design a hydraulic circuit keeping the above conditions in mind.*

(Note: A constant speed pump is used in this system.)

Solution: (Refer to Figure 8.37.) A 4/3 float-centered valve (1.1) is used to control the motion of the double-acting cylinder. The working ports A and B are connected to exhaust port T in a neutral position that discharges all the fluid to the tank and the piston is free to move in the cylinder.

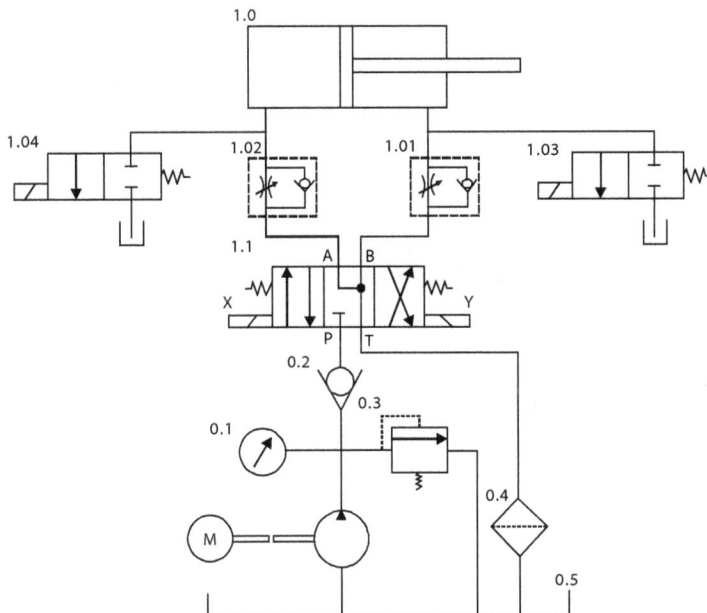

FIGURE 8.37

Two non return flow control valves (1.01 and 1.02) are used to slow down the speed of the cylinder but when it is not required to slowdown the speed, the fluid is by passed directly from the 2/2 DC valves namely 1.03 and 1.04 by simply energizing the electromagnets of valve 1.03. This will directly flush the discharged fluid to the tank, hence causing no restriction to the forward and backward movement of the ram. Throttle out non return flow control valve is used because it is preferred with constant speed pumps.

EXERCISES

1. What are the basic differences between pneumatic and hydraulic circuit diagrams?

2. Explain the method of designating components in a hydraulic or pneumatic circuit with the help of a sketch.

3. List the basic types of hydraulic circuits.

4. Explain the construction, working, and performance characteristics of fluidic elements.

5. Identify the basic elements used in a hydraulic circuit.

6. Differentiate between throttle in and throttle out speed control circuits in pneumatics with the help of a sketch.

7. Draw a basic block of a circuit showing the reservoir, accessories, pressure relief valve, and the pump and tank lines.

8. Draw a pneumatic circuit to achieve the continuous reciprocating motion of a single-acting spring return pneumatic actuator.

9. Explain the function of the quick exhaust valve with the help of a sketch.

10. Differentiate between an open center and closed center circuit.

11. Draw a circuit diagram to control the hydraulic double-acting cylinder with the condition that the cylinder can be stopped anywhere between the extreme positions.

12. Draw a pneumatic circuit to operate a double-acting pneumatic actuator such that the forward and backward motions of the actuator can be affected with varying speeds.

13. Why is speed control less accurate and also difficult to obtain the intermediate positioning of the cylinder in the case of a pneumatic system design?

14. Sketch a bleed-off circuit.

15. Why is it necessary to use the pressure relief valve in hydraulic circuits?

16. Draw a double-handed pneumatic safety circuit for a clamping operation and explain it works. Compare this circuit with a single-handed circuit for the same purpose.

17. Draw and discuss the pneumatic circuit for speed control of a double-acting cylinder both in forward as well as in backward motion.

18. What are the pressures ranges in hydraulic and pneumatic circuits?

19. Draw pressure reduction circuit and explain how it works.

20. Draw the schematic of a hydraulic circuit for the following sequence: Extend a cylinder, provide a dwell, and then retract the cylinder.

21. Explain the sequence control circuit with the help of a sketch.

22. Draw the pneumatic circuits for the following:
    - To control the motion of a single-acting cylinder.
    - To control the motion of a double-acting cylinder.
    - To control the motion of a motor.

CHAPTER 9

PNEUMATIC LOGIC CIRCUITS

INTRODUCTION

Pneumatic systems are well suited to process automation, as they not only provide the necessary force to accomplish the task but also perform control functions. Pneumatic self-contained circuits are possible by which the process can be completely automated. This can be achieved by using simple fluidic elements in a logical manner. The use of special purpose electronic equipment is popular because these are considered the best media to provide control on pneumatic power. Its main functions are to generate a signal, stop and start a process, and provide feedback, etc., but nowadays pneumatic controls has developed to such an extent that it can perform all the control related functions alone. Because the problems related to pneumatic controls are sometimes complex, certain techniques are developed to solve the problems of circuit design that add simplicity to the process of circuit diagram. These techniques are explained in this chapter with some examples.

CONTROL SYSTEM

Control system can be defined as a mechanism by virtue of which any quantity of interest in equipment can be maintained or altered in accordance to one's desire. Pneumatic systems as discussed already can be a control system when air is used as controlling media. Like other control systems, pneumatic control systems can be classified as:

1. Open loop control systems
2. Closed loop control systems

OPEN LOOP CONTROL SYSTEM

In an open loop control system, control function is independent of output.

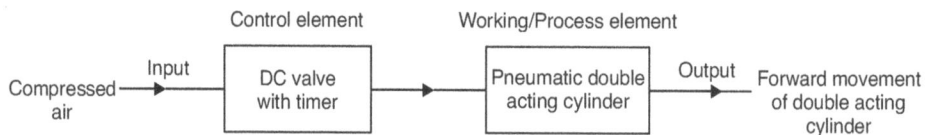

FIGURE 9.1 Open Loop Control System.

A pneumatic open loop control system is shown in Figure 9.1 in which compressed air is taken as the input to the system. A DC valve is used as the control element, which changes its position after fixed time intervals. The flow of air is directed to the pneumatic cylinder and as an output the double-acting pneumatic cylinder extends outward. Here one cannot determine when end position of cylinder will be reached and also for backward stroke of cylinder operator has to push another button. Open loop control systems are simple systems, not troubled with problems of instability. Their ability to provide desire and accurate output depends on the stability, calibration, and quality of the control element. Output once generated cannot be altered again in these systems.

CLOSED LOOP CONTROL SYSTEM

Closed loop control systems, on the other hand, are called feedback systems. In these systems, control function is dependent on output (Refer to Figure 9.2).

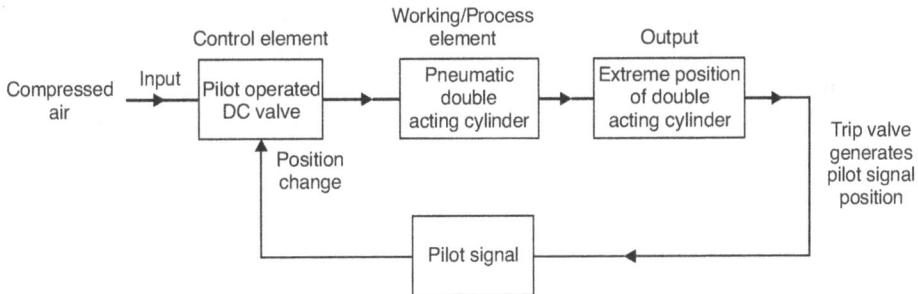

FIGURE 9.2 Closed Loop Control System.

In the above pneumatic control system, a pilot operated DC valve is used as the control element which directs the flow of compressed air to the appropriate side of the double-acting cylinder (working element) and, as an output, the double-acting cylinder extends forward. At the moment the piston rod reaches extreme position, it hits the trip valve placed there and actuates a pilot signal, which changes the position of the DC valve. Hence, the process is controlled, *i.e.*, air is allowed to enter one side of the cylinder until the piston reaches its extreme position, and after that, the same air is directed to the other side of the piston. This ensures the automatic forward and backward motion of the piston. Closed loop control systems are complex and costly but their output is accurate. Pneumatic closed loop systems are considerably cheaper because of the low cost of pneumatic equipment as well as their reliable operation.

CIRCUIT DESIGN METHODS

There are two primary methods of circuit design:

▪ Trail and error
▪ Methodological design

Trial and Error Methods

Trial and error methods are also called intuitive methods because these methods do not have any methodological approach but they are dependent on past experience and knowledge of the designer. Trial and error methods require more time for designing complex circuits. One can easily realize that there can be a number of solutions to one design problem. It is also important to note that design with this method is dependent on personal influences of the designer such as ability, mood, knowledge of pneumatic systems, etc. and it also relies on whether one gives more weight to the least expensive solution or the more reliable solution.

Methodological Design

On the other hand, methodological design follows a certain set of rules and instructions to arrive at the optimal solution. This method consists of a precisely defined method and the designer's influence is less, whereas some theoretical knowledge about the technique that is being followed is required. A methodological circuit design takes lesser time and provides a reliable solution. Because of these reasons these methods are being used in industry. Control is always independent of personal influences of designer. It is also worth noting that more components are required while designing a circuit diagram by using these techniques than in circuits designed by intuitive methods. The additional cost of the components is compensated by saving time in completing a project as well as reduced maintenance costs.

At this stage, when both the methods have their own advantages and disadvantages, it is difficult to decide which method is to be used. Again, it is worth noting that regardless of the method or technique used to design a circuit diagram, one should have acquired sound knowledge about pneumatic valves, switches, fluidic elements, and means of actuation.

Some of the commonly used techniques under heading of methodological design are:

- Cascade design
- Algebric method (using Boolean Algebra)
- Graphical method (K-mapping)

MOTION SEQUENCE REPRESENTATION

An automatic system constitutes working elements and control elements. It is required that the motion schedule and the start and stop conditions of the elements in a system must be known and interpreted clearly. Understanding what is to be done is very important for final results. So it becomes important to show the sequence of operation of various elements in a pneumatic system in the form of a graph. There are two types of diagrams, which are used to represent functional sequences:

- Motion diagrams
- Control diagrams

Motion diagrams are the graphical representation of all the conditions of working elements (elements which do not take part in controlling). Similarly, *control diagrams* are those which provide information about control elements only. These are explained below.

MOTION DIAGRAMS

Motion diagrams are of following types:

- Position step diagram
- Position time diagram

Position Step Diagram

The position step diagram shown in Figure 9.3 is drawn only by considering the two positions of a double-acting pneumatic cylinder.

Position A is retracted or the rear position of the piston rod and position B represents the extended or forward position of the piston rod.

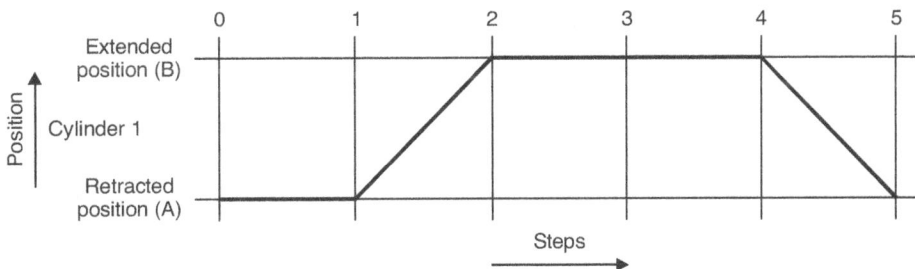

FIGURE 9.3 Position Step Diagram.

Position is plotted against steps (change in position of any component). Steps are plotted on the X-axis and position is plotted on the Y-axis. On actuating or operating any particular switch/element in the system, there are two possibilities. The position of the piston rod either changes (for example from extreme inward to extreme outward) or remains the same. Only these two cases can be shown with this diagram. Position step diagram for cylinder 1 is shown in Figure 9.3 and is explained in Table 9.1.

Table 9.1

Step	Cylinder Position
0-1	Retracted
1-2	Extended
2-3	Extended
3-4	Extended
4-5	Retracted

Example: *Draw the position step diagram for the example shown in Figure 9.4 for which the series of operations to be carried out are:*

1. Cylinder A clamps the workpiece.

2. Machine operation is performed on workpiece.

3. Cylinder A proceeds backward. Cylinder B proceeds forward.

4. After pushing the workpiece on the conveyor, cylinder B travels back.

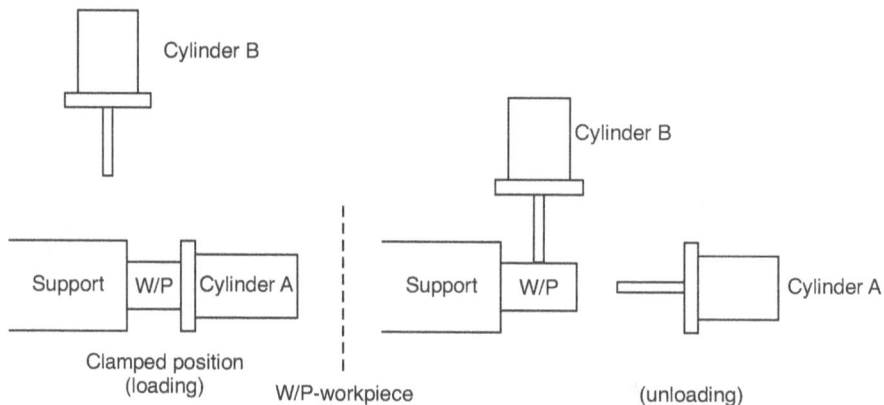

FIGURE 9.4

Solution:

Sequence of Operations: The sequence of operations of two cylinders for the above said example is shown in Table 9.2.

Table 9.2

Step	Cylinder A	Cylinder B
0-1	Extends	——
1-2	——	——
2-3	Retract	——
3-4	——	Extends
4-5	——	Retract

Position Step Diagram: The position step diagram for the above example is shown in Figure 9.5.

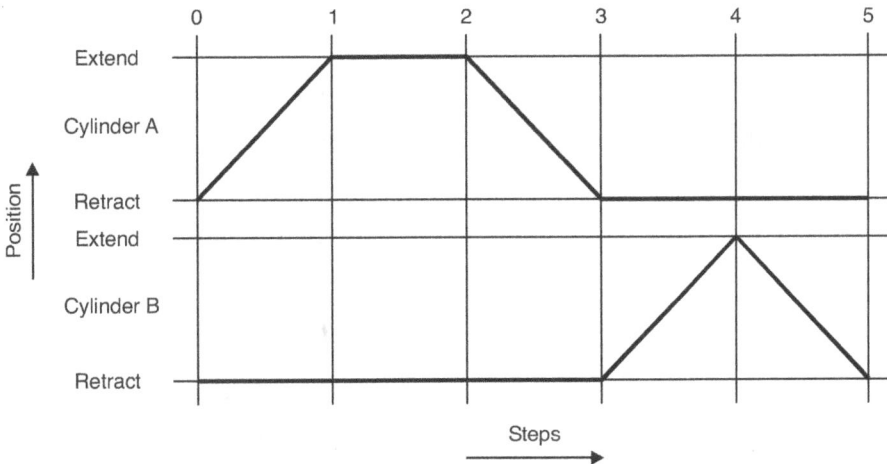

FIGURE 9.5 Position Step Diagram

Position Time Diagram

A position time diagram is similar to of the position step diagram. Here, cylinder positions are plotted with respect to time. The displacement step diagram is simple to understand and draw, on other hand, the position time diagram overlaps and operating speeds can be better shown. The position time diagram for the above example is shown in Figure 9.6.

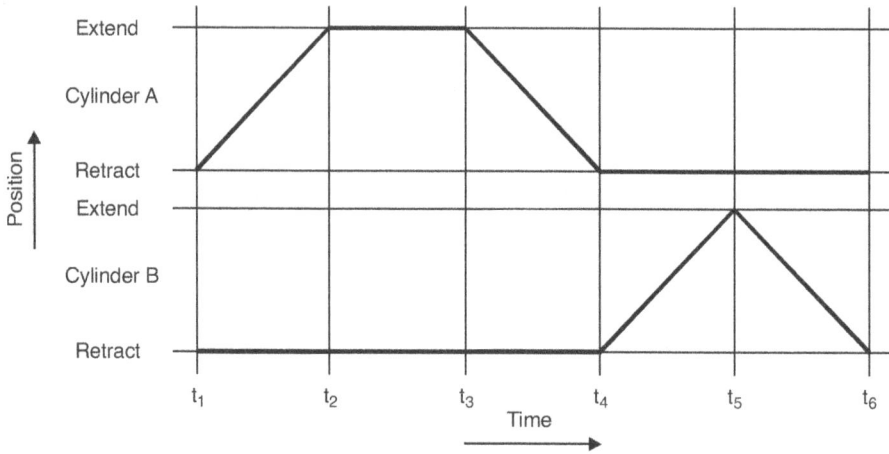

FIGURE 9.6 Position Time Diagram.

CONTROL DIAGRAM

Control diagrams are drawn for a particular control element. They simply display the sequence of actions conducted by a particular control element. It may be called a condition step diagram.

The diagram shown in Figure 9.7 is drawn for an electromagnet which constitutes a spool valve. The flow of fluid is started when electromagnets are energized and the flow of fluid stops when the coil is de-energized.

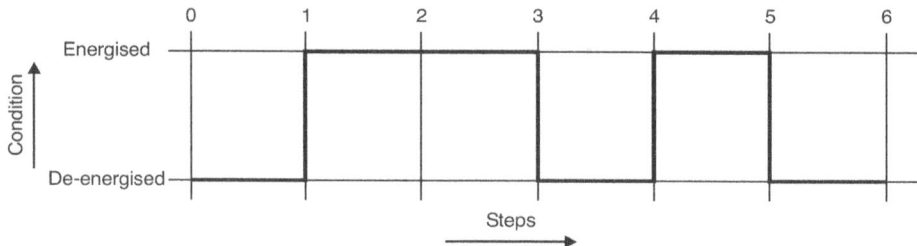

FIGURE 9.7 Control Diagram.

CASCADE DESIGN

Cascading is a methodological approach to the problem of pneumatic circuit design. As already discussed, intuitive techniques of circuit design are not suitable for complex circuits. This technique not only reduces the complexity in design but also reduce the size and number of components used in circuits.

Cascading means "in series." In this method, the sequence of pneumatic cylinders is controlled by using various types of signaling elements. These signaling elements are, of course, driven by the forward and backward strokes of the cylinders but the air supply to pilot lines is delivered through a cascade system. In a cascade system, the forward and backward motions of the pneumatic cylinders are classified into various groups. Then, those particular groups of movements are controlled with components in the cascade system. The cascade system consists of group selector valves, bus bar lines, and pilot lines. Bus bar lines are basically pneumatic energy lines, which pass from plant. These are used to supply pneumatic energy to pneumatic systems.

STEPS INVOLVED IN CASCADE DESIGN

The steps involved in drawing a circuit using the cascade method are the following:

1. Cylinders are named as A, B, C, etc. depending on their number.

2. Cylinder advanced movement is designated by "+" sign and cylinder backward movement is designated by "-" sign. So the forward motion of cylinder A is denoted by the sign A⁺ and backward motion by A⁻.

3. The given sequence is written with "+" and "-" sign depending upon the problem.

4. The step position diagram for the given sequence is drawn.

5. The given sequence of cylinders is divided into groups. Groups are formed by ensuring that the letters A or B should come only once in one group regardless of the sign, *i.e.*, it may be +ve or –ve.

6. Each group is assigned "bus bars" or "compressed air lines." The number of bus bar lines is equal to the number of groups formed. The bus bar line is responsible for the supply of compressed air in one group only.

7. Each cylinder is provided with a 4/2 pilot operated DC valve. The number of DC valves is equal to the number of cylinders.

8. Limit switches/valves are positioned at either end and are actuated by piston rod. These identify the extension and retraction of the cylinders. Limit

switches consist of mechanically actuated electrical contacts. The contacts open or close when some machine component reaches a certain position (*i.e.*, limit) and actuates the switch. The limit switches are denoted by suffix "o" and "1." Suffix "o" means the return stroke of the cylinder and suffix "1" means the forward stroke of the cylinder. For each cylinder, there are two limit switches. Each bus bar line supplies compressed air to the limit switches within the group. These are normally 3/2 DC valves.

9. Group selector valves act as an energy source for the limit switches used for controlling the motion of cylinders. These are 4/2 pilot operated DC valves. As a rule of thumb, group selector switches are always equal to "the number of bus bar lines or number of groups minus 1."

10. The circuit is drawn by keeping the given sequence in mind. To know and learn how the cascade system is drawn for a particular problem, one must be familiar with the components of the cascade system. In the following paragraphs, the method of cascading is explained step by step by showing various examples. Various components of the cascade system are also explained.

SIGN CONVENTIONS

Certain sign conventions have to be adopted to denote forward and backward motions of cylinders. These are given below:

Sign conventions

Cylinder advanced movement is designated by: + ve(positive) sign

Cylinder backward movement is designated by: –ve (negative) sign

Cylinders can be named A, B, C, D, etc. .... depending on their number

So we can denote the forward motion of cylinder A by the sign A^+ and backward motion by A^-.

SEQUENCING

Sequencing may be defined as the process of putting things in the right order. It is the prime step in circuit design depending on the problem assigned. One can arrange the forward and backward motions of all the cylinders in the circuit. Position step diagrams can further show this sequence graphically.

EXAMPLES

Example 1: *Design a pneumatic logic circuit by using the cascade method for the following sequence of cylinders: In a two cylinder circuit, cylinder A extends in the first step, B extends in the second step, cylinder A retracts in the third step, and cylinder B retracts in the fourth step.*

Solution:

Drawing a pneumatic circuit by using the cascade method involves the following steps:

Step 1. Sequence of Motions

The sequence of motions for the above problem is

A B A B

The motion that occurs in the same steps can be written in the same column.

Step 2. Position Step Diagram

For the above sequence of movements, the position step diagram is shown in Figure 9.8.

Step 3. Grouping

The next step is to divide the sequence of cylinder motions into groups. Groups are formed by ensuring that the letters A or B should come only

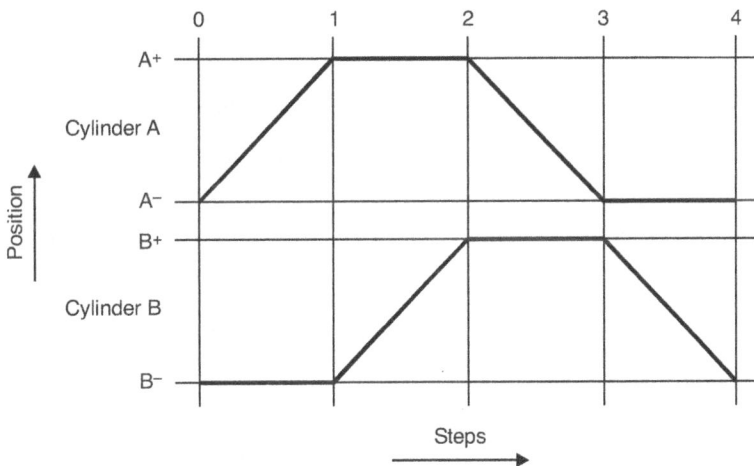

FIGURE 9.8 Position Step Diagram.

once in one group regardless of the sign, *i.e.* , it may be +ve or –ve. For example, grouping for the sequence written above is:

$$\left| \text{A} \quad \text{B} \right| \left| \text{A} \quad \text{B} \right|$$
$$\qquad G_1 \qquad G_2$$

Groups are denoted by letter G. So, the groups are G_1 and G_2.

Step 4. Bus Bar Lines

The next step is to draw the cascade system. For this "bus bars" or "compressed air lines" are drawn. The number of bus bar lines are equal to the number of groups formed. This is because each bus bar line is responsible for the supply of compressed air in one group only. For the above example, these lines are drawn as shown in Figure 9.9.

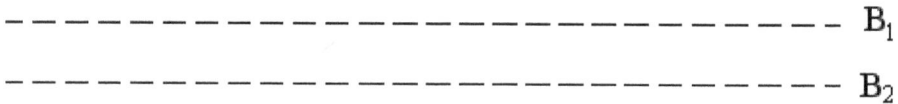

$$- \quad \mathbf{B_1}$$

$$- \quad \mathbf{B_2}$$

FIGURE 9.9 Bus Bar Lines.

Step 5. Group Selector Switches

Group selector switches are basically 4/2 DC valves (pilot operated). As a rule of thumb, group selector switches are always equal to "the number of bus bar lines –1."

Or mathematically: if B_1, B_2, B_3,........ ,B_n implies bus bar lines in a cascade system,

Number of group selector switches in cascade system = n-1

For above example, the arrangement of the group selector valves is shown in Figure 9.10.

Group selector valves act as an energy source for the limit switches used for controlling the motion of the cylinders. Group selector valves are placed in a cascade system by keeping in mind that only one bus bar line is given compressed air supply at a time to ensure the proper functioning of the circuit.

Currently, the supply is being given to bus bar line B_1 as shown in Figure 9.10 and rest of the lines are empty. When the requirement in line B_1 is over, pilot signal S_2 is generated which changes the position of GS1.

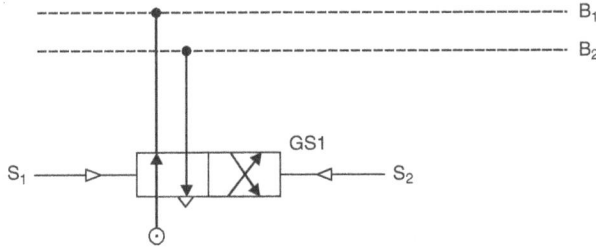

FIGURE 9.10

Now, air supply is directed to bus bar line B_2. At last, when the above process is to be repeated, *i.e.*, when the air supply is required at bus bar B_1, the pilot signal S_1 is given to GS1. Now again the supply is restored to bus bar line B_1. The above cycle is repeated again.

Step 6. Drawing a Circuit

After drawing the cascade system for a circuit, all the elements of the circuit are drawn according to the requirements. The number of cylinders, controlling elements, limit switches, pressure control valves, flow control valves, etc. are determined from the given problem. All the required elements of the circuit for above example are shown in Figure 9.11.

The next step is to make the cylinder move in a sequence as determined above. For ease of readers, the circuit is drawn in 3 steps:

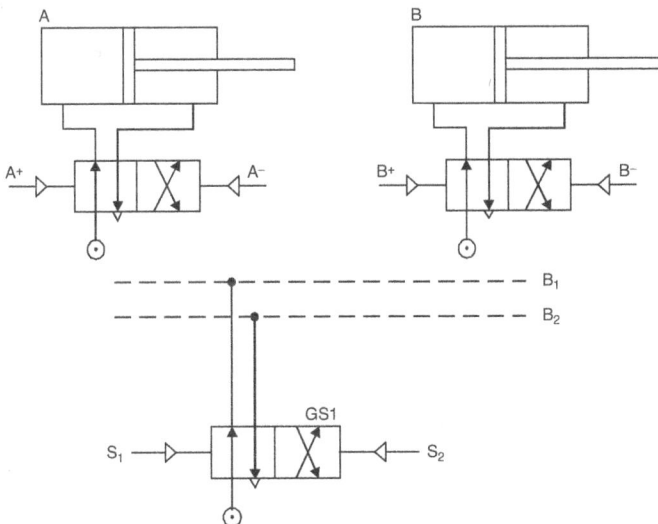

FIGURE 9.11

Step 1: In the first step, only the first group is taking compressed air from bus bar line B_1. The sequence of the motion for the first group is:

$$\left|\,A\quad B\,\right|$$

The pneumatic circuit for the sequence of the first group is shown in Figure 9.12.

In this figure, group selector valve GS1 is at its initial position by supplying compressed air to bus bar B1. The compressed air will be supplied to the control elements in such a way that the sequence of motions of group 1 should be accomplished. The air after activating the respective pilot line also rushes to limit switches (3/2 DC valves). These signaling elements only allow the air to pass, when the piston rod reaches the extreme desired position. For example, 1.1 DC valve is actuated by the pilot line. The cylinder 1.0 starts moving forward. When outer extreme position A^+ is reached, only then the limit switch a1 is actuated and air is sent to the next valve for completion of motion sequence B^+. Thus, the sequence $A^+\ B^+$ is achieved.

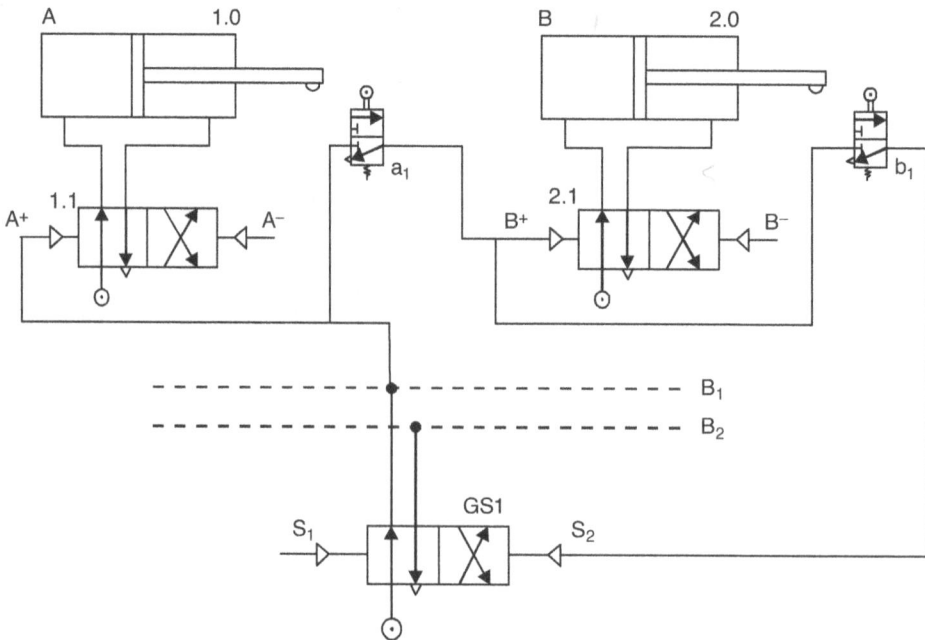

FIGURE 9.12

Step 2: In step 2, the signal coming from limit switch b_1 of step 1 acts as a pilot signal S_2 for GS1. Now bus bar line B_2 is selected. Sequence of motion for group 2 is

$$|A \quad B|$$

The pneumatic circuit for the sequence of second group is shown in Figure 9.13.

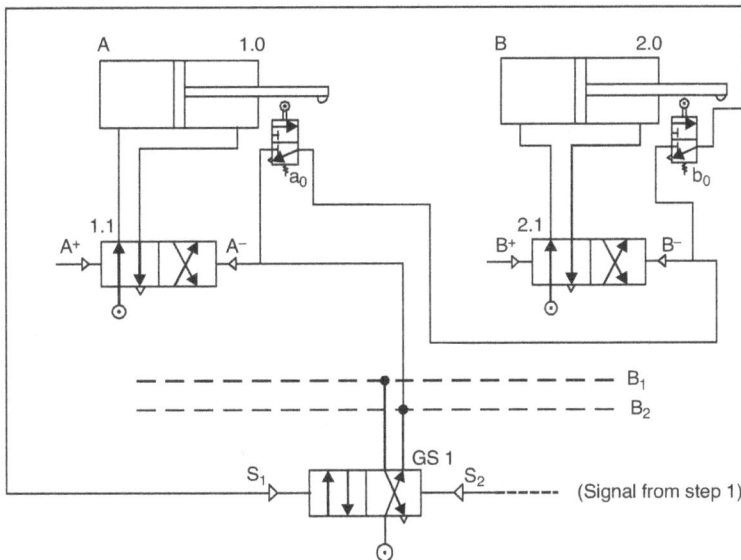

FIGURE 9.13

Compressed air taken from bus bar line B_2 is used to apply pilot pressure in attaining A- position. The cylinder A starts moving backward. When the extreme position A- is reached only then the limit switch a_0 is actuated and air is sent to 2.1 DC valve for completion of B- sequence. Thus, the sequence **A- B-** is achieved.

Step 3: In this step, the circuits of step 1 and 2 are combined together. The final circuit is shown in Figure 9.14.

The backward motion of cylinder 2.0, *i.e.*, B- leads to the actuation of b_0. Now the compressed air from limit switch b_0 will act as a pilot signal S_1 for GS1 and resets it to its original position which again selects bus bar line B1. Thus, the sequence **A+ B+ A- B-** is achieved. Now, the cycle is again ready to start.

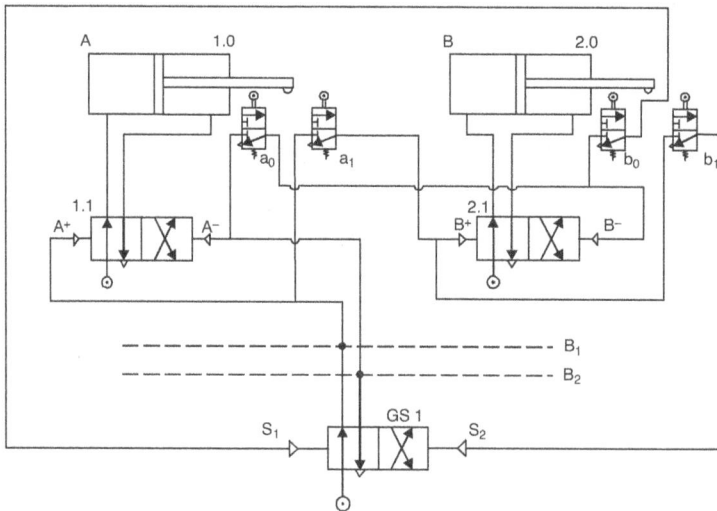

FIGURE 9.14

Note: The circuit shown in Figure 9.14 will not stop at any moment once it gets activated. It keeps running continuously. To make the circuit control, a push button operated spring returned 3/2 DC valve is added. This makes the circuit manually controlled which is shown in Figure 9.15.

Example 2: *The drilling operation is to be performed on a workpiece. The sequence of motion of the cylinders are:*

1. *Cylinder 1 clamps the workpiece.*
2. *Cylinder 2 performs a drilling operation on the workpiece.*
3. *Cylinder 2 returns back.*
4. *Cylinder 1 unclamps the workpiece.*

For the above said sequence, draw a pneumatic circuit.

Solution:

The sequence of motion for the above example is

A   B   B   A

The pneumatic circuit for the above sequence is shown in Figure 9.16.

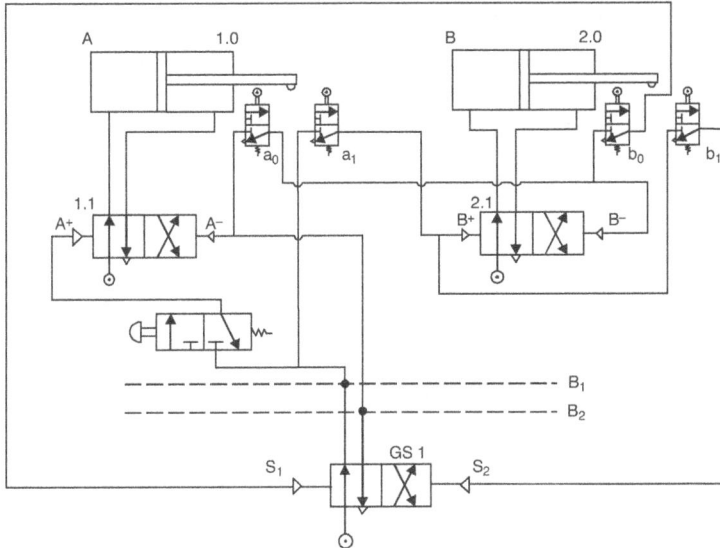

FIGURE 9.15

Initially, both the cylinders are in the retracting position as shown in Figure 9.16. When the push button of the 1.2 DC valve is pressed, the compressed air starts flowing into cylinder 1.0 through control element 1.1. This results in the forward motion of cylinder 1.0, which leads to the actuation of the 2.2 DC valve. Now, the compressed air starts flowing into cylinder 2.0 through control element 2.1, which leads to the forward motion of cylinder 2.0. This forward motion of cylinder 2.0 leads to the actuation of the 2.3 DC valve. The 2.3 DC valve sends a pilot signal to the control element 2.1. As the pilot signal is also coming from the previous operation, $i.e.$, B$^+$, so both the pilot signals (one from the 2.2 DC valve and second from the 2.3 DC valve) are available to control element 2.1. The extended position of cylinder A is actuating the 2.2 DC valve and, on the other side, the extended position of cylinder B is actuating the 2.3 DC valve. Both these leads to locking of the system. This is called a signal overlapping problem as the signal is coming from both sides. To overcome the problem of overlapping, cascading is used.

Use of cascading for the above example

The steps involved in drawing the circuit by cascading method for the above example are:

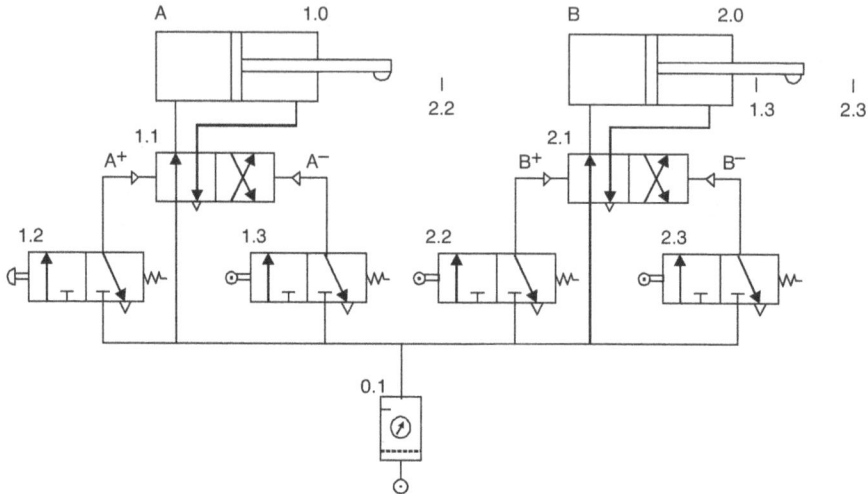

FIGURE 9.16

Step 1. Sequence of Motions

The sequence of motions for the above said problem is

A   B   B   A

The motion that occurs in same steps can be written in same column.

Step 2. Position Step Diagram

For the above sequence of movements, the position step diagram is shown in Figure 9.17.

Step 3. Grouping

The next step is to divide the sequence of the cylinder motions into groups. Groups are formed by ensuring that the letters A or B should come only once in one group regardless of the , *i.e.*, it may be +ve or –ve. For example, grouping for the sequence written above is:

$$|A \quad B \; \| A \quad B \; |$$
$$\quad G_1 \qquad G_2$$

Groups are denoted by letter G. So, the groups are G_1 and G_2.

Step 4. Bus Bar Lines

The next step is to draw the cascade system. For this "bus bars" or "compressed air lines" are drawn. The number of bus bar lines are equal to the

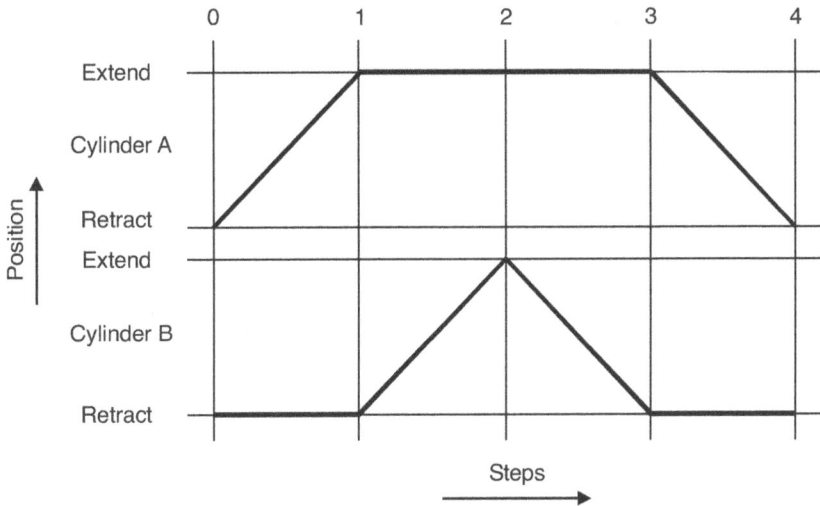

FIGURE 9.17

number of groups formed. This is because each bus bar line is responsible for the supply of compressed air in one group only. For the above example, these lines are drawn as shown in Figure 9.18.

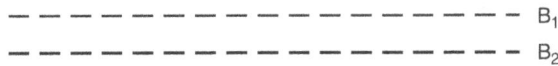

FIGURE 9.18

Step 5. Group Selector Switches

Group selector switches are basically 4/2 DC valves (pilot operated). As a rule of thumb, group selector switches are always equal to "the number of bus bar lines − 1."

Or mathematically: if B_1, B_2, B_3,…….., B_n implies bus bar lines in a cascade system,

Number of group selector switches in cascade system = $n − 1$

For the above example, the arrangement of group selector valves is shown in Figure 9.19.

Group selector valves act as an energy source for limit switches used for controlling the motion of cylinders. Group selector valves are placed in the cascade system by keeping in mind that only one bus bar line at a time is given compressed air supply to ensure proper functioning of circuit.

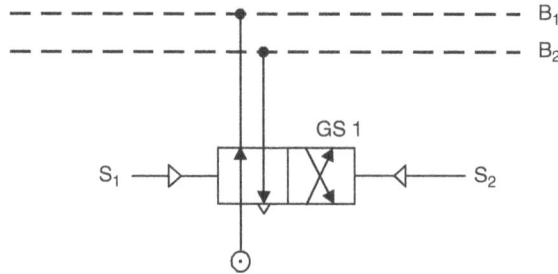

FIGURE 9.19

Refer to Figure 9.19. Currently, the supply is being given to the bus bar line B_1 and all the rest of the lines are empty. When the requirement in line B_1 is over, pilot signal S_2 is generated which changes the position of GS1. Now air supply is directed to bus bar line B_2. At last, when the above process is to be repeated, *i.e.*, when the air supply is required at bus bar B_1, the pilot signal S1 is given to GS1. Now again the supply is restored to bus bar line B_1. The above cycle is repeated again.

Step 6. Drawing a Circuit

After drawing the cascade system for a circuit, all the elements of the circuit are drawn according to these requirements. The number of cylinders, controlling elements, limit switches, pressure control valves, flow control valves, etc. are determined from the problem given. All the required elements of the circuit for the above example are shown in Figure 9.20.

The next step is to make the cylinder move in a sequence as determined above. For ease of readers, the circuit is drawn in 3 steps:

Step 1: In the first step, only the first group is taking compressed air from bus bar line B_1. The sequence of the motion for first group is:

$$|A \ B \ |$$

The pneumatic circuit for the sequence of the first group is shown in Figure 9.21.

In this figure, group selector valve GS1 is at its initial position by supplying compressed air to bus bar B1. The compressed air will be supplied to the control elements in such a way that the sequence of motions of group 1

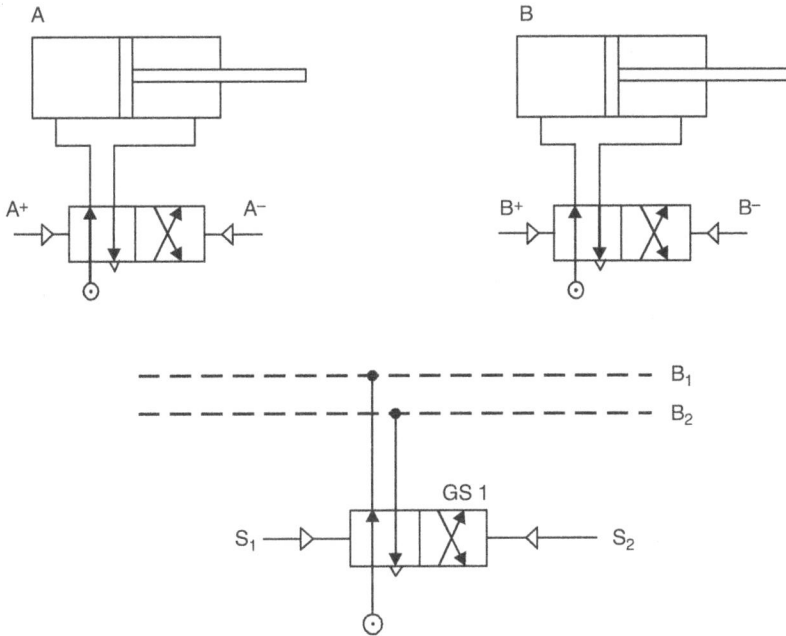

FIGURE 9.20

should be accomplished. The air after activating the respective pilot line also rushes to the limit switches (3/2 DC valves). These signaling elements only allow the air to pass, when the piston rod reaches the extreme desired position. For example, the 1.1 DC valve is actuated by the pilot line. The cylinder 1.0 starts moving forward. When outer extreme position A⁺ is reached, only then the limit switch a1 is actuated and air is sent to the next valve for completion of motion sequence B⁺. Thus, the sequence **A⁺ B⁺** is achieved.

Step 2: In step 2, the signal coming from the limit switch b1 of step 1 acts as a pilot signal S_2 for GS1. Now bus bar line B_2 is selected. Sequence of motion for group 2 is

$$| B \quad A \: |$$

The pneumatic circuit for the sequence of the second group is shown in Figure 9.22.

Compressed air taken from bus bar line B_2 is used to apply pilot pressure in attaining B⁻ position. The cylinder B starts moving backward. When the extreme position B⁻ is reached, only then the limit switch b_o is actuated and air is sent to the 1.1 DC valve for completion of the A⁻ sequence. Thus, the sequence **B⁻ A⁻** is achieved.

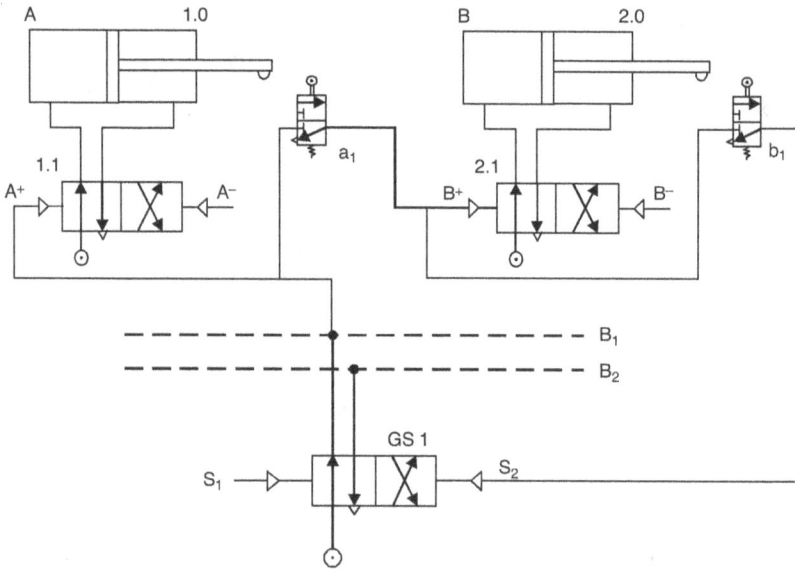

FIGURE 9.21

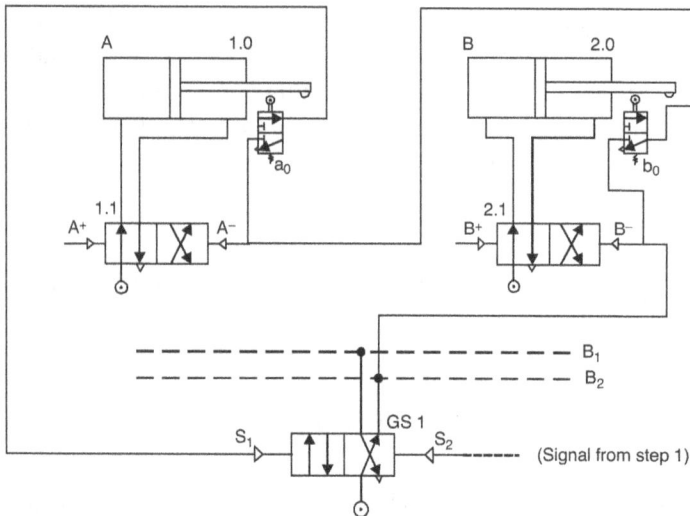

FIGURE 9.22

Step 3: In this step, the circuits of step 1 and 2 are combined together. The final circuit is shown in Figure 9.23.

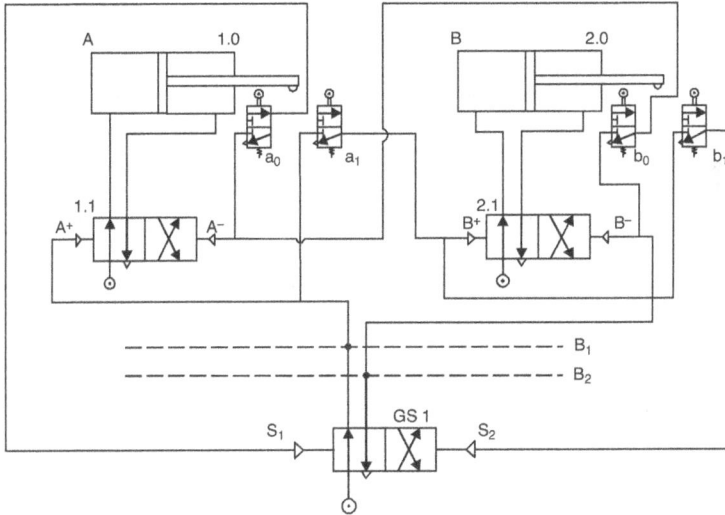

FIGURE 9.23

The backward motion of cylinder 1.0, *i.e.*, A⁻ leads to actuation of a_0. Now the compressed air from limit switch a_0 will act as pilot signal S_1 for GS1 and resets it to its original position, which again selects bus bar line B_1. Now the cycle is again ready to start. Thus, the sequence **A⁺ B⁺ B⁻ A⁻** is achieved. Hence, with the use of cascading, the problem of signal overlapping is avoided.

Example 3: *Movement diagram for a 3 cylinder (A, B, and C) pneumatic system is given in Figure 9.24. Draw the pneumatic circuit diagram for the above.*

Solution: The steps involved in drawing the circuit by the cascading method for the above said example are:

Step 1. Sequence of Motions

The sequence for the given position step diagram is:

A B C   A

B

C

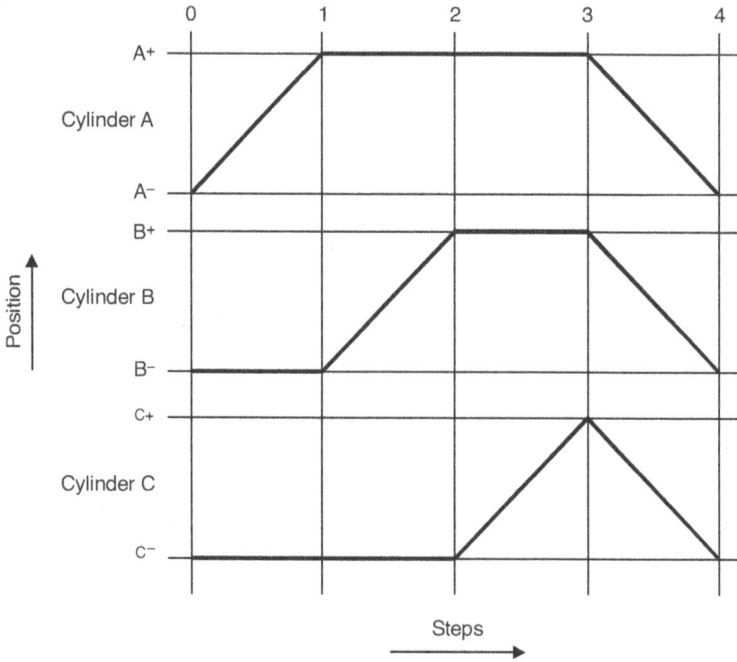

FIGURE 9.24

Step 2. Grouping

The next step is to divide the sequence of cylinder motions into groups. Groups are formed by ensuring that the letters A, B, or C should come only once in one group regardless of the sign, *i.e.* , it may be +ve or –ve. For example, grouping for the sequence written above is:

$$\left| A \ \ B \ \ C \right| \quad \left| \begin{array}{c} A \\ B \\ C \end{array} \right|$$

$$G_1 \qquad \quad G_2$$

Groups are denoted by letter G. So the groups are G_1 and G_2.

Step 3. Bus Bar Lines

The next step is to draw the cascade system. For this "bus bars" or "compressed air lines" are drawn. The number of bus bar lines is equal to number of groups formed. This is because each bus bar line is responsible for the supply of compressed air in one group only. For the above example, these lines are drawn as shown in Figure 9.25.

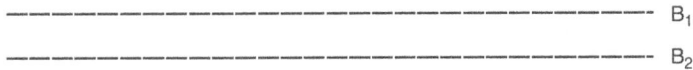

FIGURE 9.25

Step 4. Group Selector Switches

Group selector switches are basically 4/2 DC valves (pilot operated). As a rule of thumb, group selector switches are always equal to "the number of bus bar lines − 1."

Or mathematically: if B_1, B_2, B_3,…….. ,B_n implies bus bar lines in a cascade system

Number of group selector switches in cascade system = n-1

For the above example, the arrangement of the group selector switches valves is shown in Figure 9.26.

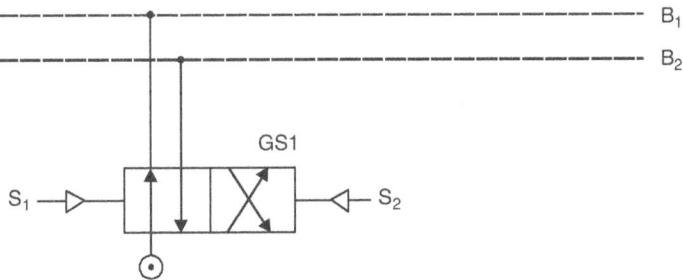

FIGURE 9.26

Group selector valves act as the energy source for the limit switches used for controlling the motion of the cylinders. Group selector valves are placed in the cascade system by keeping in mind that only one bus bar line is given compressed air supply at a time to ensure proper functioning of circuit.

Refer to Figure 9.26. Currently, the supply is being given to bus bar line B_1 and the rest of all the lines are empty. When the requirement in line 'B_1' is over, pilot signal S_2 is generated which changes the position of GS1. Now, air supply is directed to bus bar line B_2. At last, when the above process is to be repeated, *i.e.* , the air supply is required at bus bar B_1, the pilot signal S_1 is given to GS1. Now again the supply is restored to bus bar line B_1. The above cycle is repeated again.

Step 5. Drawing a Circuit

After drawing the cascade system for a circuit, all the elements of the circuit are drawn according to these requirements. The number of cylinders, controlling elements, limit switches, pressure control valves, flow control valves, etc. are determined from the problem given. All the required elements of the circuit for the example are shown in Figure 9.27.

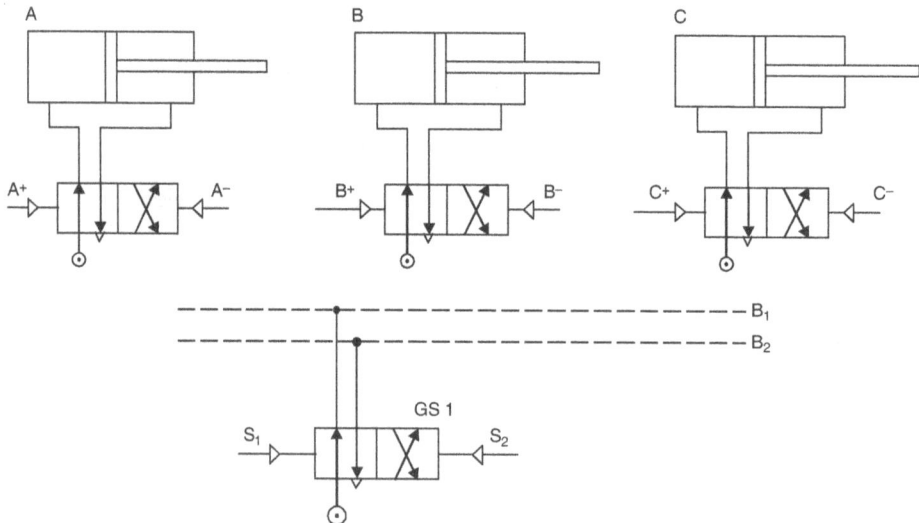

FIGURE 9.27

The next step is to make the cylinder move in a sequence as determined above. For ease of readers, the circuit is drawn in 3 steps:

Step 1: In the first step, only the first group is taking compressed air from bus bar line B_1. The sequence of the motion for the first group is:

$$|A \quad B \quad C \ |$$

The pneumatic circuit for the sequence of the first group is shown in Figure 9.28.

In this figure, group selector valve is at its initial position by supplying compressed air to bus bar line B1. The compressed air will be supplied to the control elements in such a way that the sequence of motions of group 1 should be accomplished. The air after activating the respective pilot line also rushes to the limit switches (3/2 DC valves). These signaling elements only allow the air to pass, when the piston rod reaches the extreme desired position. For example, the 1.1 DC valve is actuated by the pilot line. The cylinder 1.0 starts moving forward. When the outer extreme position A^+ is reached, only then the limit switch a_1 is actuated and air is placed to the next valve for the completion of motion sequence. Thus, the sequence A^+ B^+ C^+ is achieved.

Step 2: In step 2, the signal coming from the limit switch c_1 of step 1 acts as a pilot signal S_2 for GS1. Now, bus bar line B_2 is selected. The sequence of motion for group 2 is

$$\left| \begin{matrix} A \\ B \\ C \end{matrix} \right|$$

The pneumatic circuit for the sequence of second group is shown in Figure 9.29.

Compressed air supply taken from bus bar line B_2 is used to apply pilot pressure in attaining A^- B^- C^- simultaneously. The sequence A^- B^- C^- is taking place in a single step, *i.e.* , backward motion of all the cylinders is taking place simultaneously in step 2. The group selector switch GS1 will change its position only after the completion of A^- B^- C^- in a single step.

When the cylinders 1.0 and 2.0 are in a fully retracted position, both the limit switches a_0 and b_0 gets actuated. Their supply is routed to two input ports of AND gate I, *i.e.* , a twin pressure valve. The output of AND gate I is routed to the input of AND gate II. The other input, which is given to AND gate II is coming from limit switch c_0, which gets actuated after completion of C^-, *i.e.*, the backward stroke of cylinder 3.0. The output of AND gate II acts as a pilot signal S_1 for GS1.

Step 3: In this step the circuits of step 1 and 2 are combined together. The final circuit is shown in Figure 9.30. The output of AND gate II acts as a pilot signal S_1. the group selector switch GS1 resets GS1 to its original position and the cycle is repeated gain.

Example 4: *In a 4 cylinder circuit, cylinder A extends in the first step, cylinder B extends in the second step, cylinder C extends in the third step*

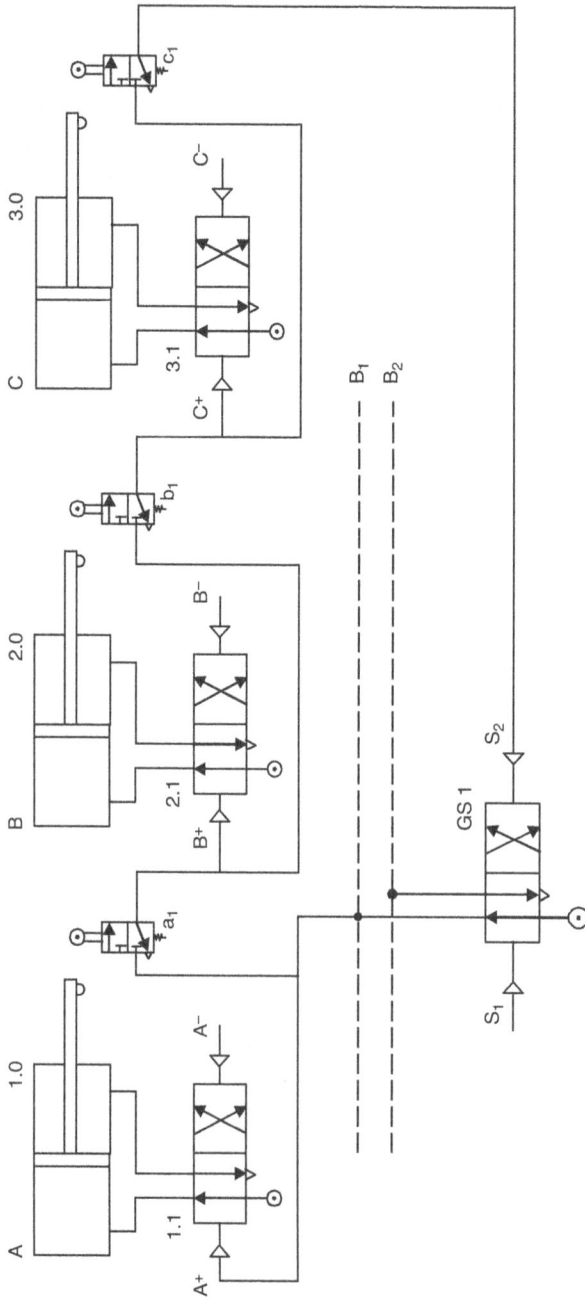

FIGURE 9.28

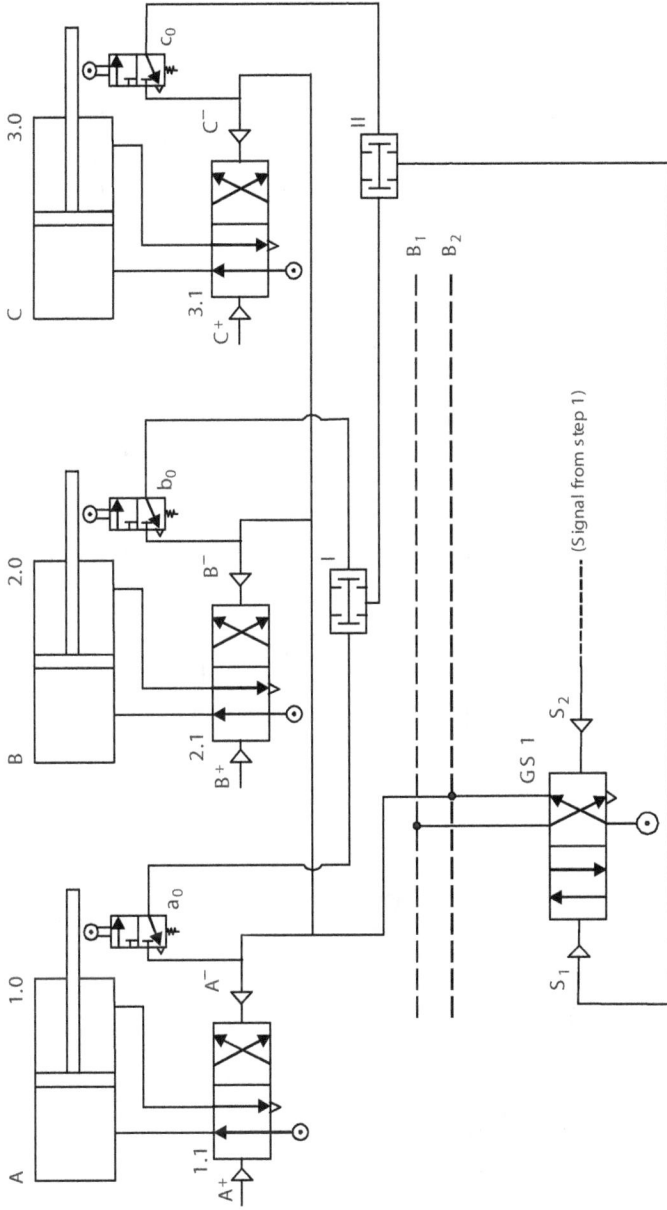

FIGURE 9.29

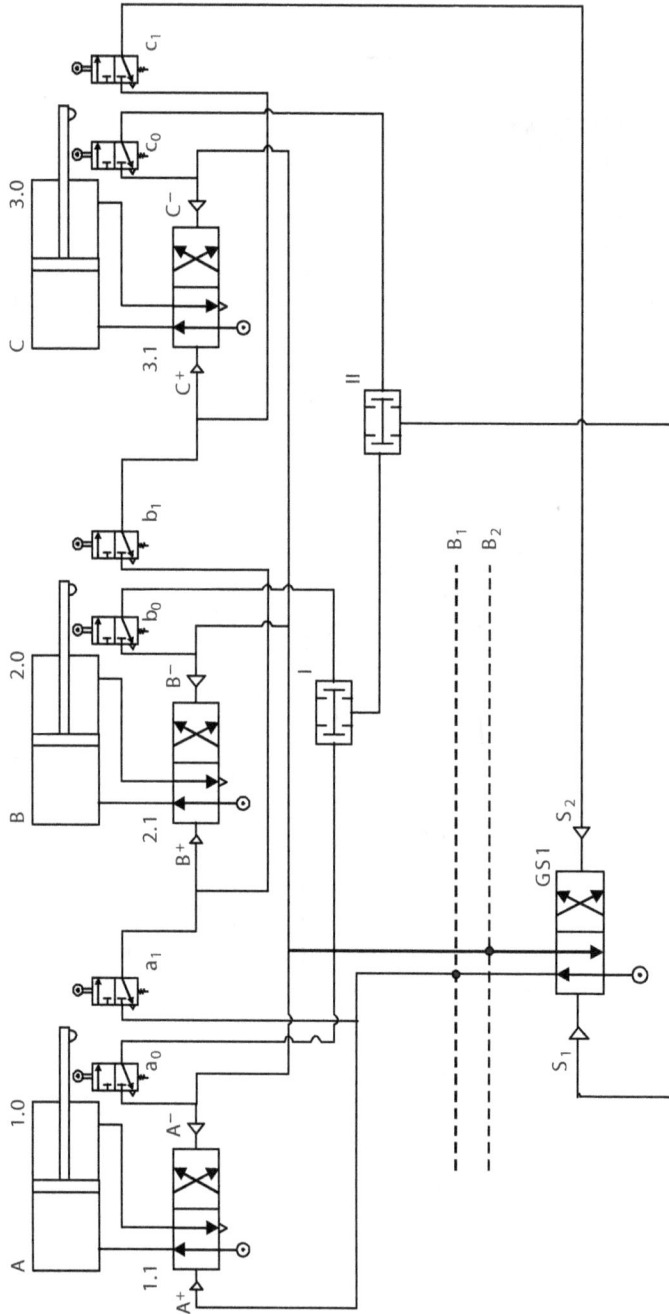

FIGURE 9.30

and cylinder A retracts in the fourth step and at the same time cylinder D extends in the fourth step. Similarly, cylinder D and cylinder B retracts in the next step and cylinder C retracts in the sixth step. For the above sequence of cylinders, design a pneumatic circuit by using cascading.

Solution: Drawing a pneumatic circuit by using cascade method involves the following steps.

Step 1. Sequence of Motions

The sequence of motions for the above said problem is

$$A \quad B \quad C \quad A \quad D \quad C$$
$$D \quad B$$

Step 2. Position Step Diagram

For the above sequence of movements, the position step diagram is shown in Figure 9.31.

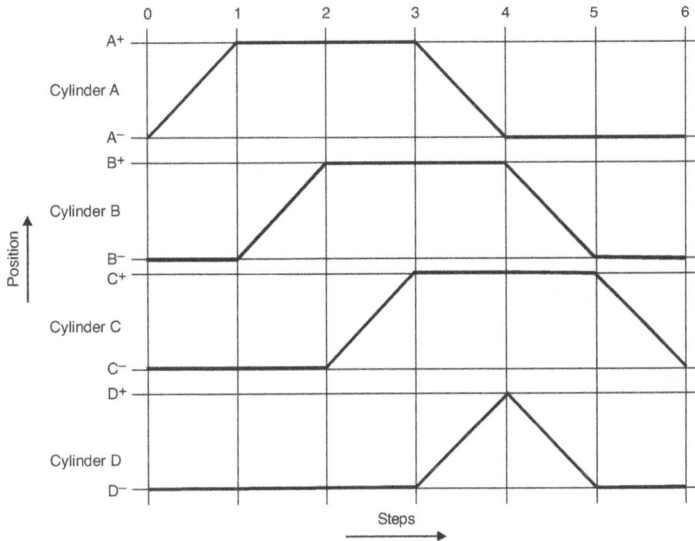

FIGURE 9.31

Step 3. Grouping

The next step is to divide the sequence of cylinder motions into groups. Groups are formed by ensuring that the letters A B C or D should come only once in one group regardless of the sign, *i.e.* , it may be +ve or –ve. For example, grouping for the sequence written above is:

$$\left| A \ \ B \ \ C \ \right| \begin{matrix} A \\ D \end{matrix} \left| \left| \begin{matrix} D \ \ C \\ B \end{matrix} \right| \right.$$

$$G_1 \qquad G_2 \ \ G_3$$

Groups are denoted by letter G, so the groups are G1, G2, and G3.

Step 4. Bus Bar Lines

The next step is to draw the "bus bars" or "compressed air lines." The numbers of bus bar lines are equal to the number of groups formed. This is because the bus bar line is responsible for the supply of compressed air in one group only. For the above example, these lines are drawn as shown in Figure 9.32.

-- B_1
-- B_2
-- B_3

FIGURE 9.32

Step 5. Group Selector Switches

Group selector switches are basically 4/2 DC valves (pilot operated). As a rule of thumb, group selector switches are always equal to "the number of bus bar lines - 1"

Or mathematically: if B_1, B_2, B_3,…….., B_n implies the bus bar lines in a cascade system

Group selector switches = n – 1

For the above example, the arrangement of the group selector valves is shown in Figure 9.33.

Refer to Figure 9.33. Currently, the supply is being given to bus bar line B_1 and the rest of the lines are empty. When the requirement in line 'B_1' is over, pilot signal S_1 is generated which changes the position of GS1. Now the air supply is directed to the bus bar line B_2. Similarly, when it is required to activate the line B_3, the pilot signal S_2 will be generated and compressed air goes to bus bar line B_3 thereby emptying all the other lines. At last, when the above process is to be repeated, *i.e.*, the air supply is required at the bus bar line B_1, the pilot signal S_3 is given to GS2 without changing the position of GS1 and the supply is restored to bus bar line B_1. The above cycle is repeated again.

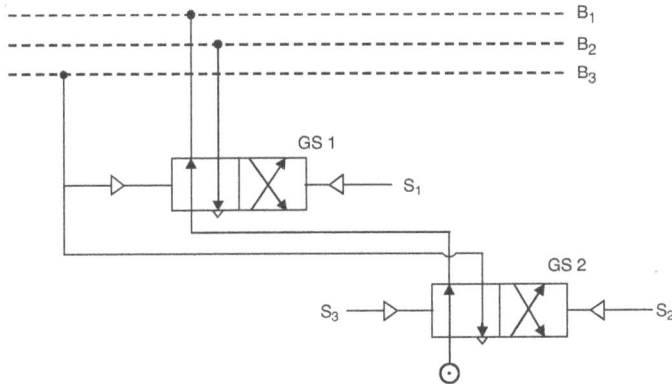

FIGURE 9.33

Step 6 . Drawing a Circuit

After drawing the cascade system for a circuit, all of the elements of the circuit are drawn according to these requirements. The number of cylinders, controlling elements, limit switches, pressure control valves, flow control valves, etc. are determined from the problem given. All of the required elements of the circuit for the above example are shown in Figure 9.34.

The next step is to make the cylinder move in a sequence as determined above. For ease of the readers, the circuit is drawn in 3 steps:

Step 1: In the first step, only the first group is taking compressed air from bus bar line B_1. The sequence of the motion for the first group is:

$$\left| A \quad B \quad C \right|$$

The pneumatic circuit for the sequence of the first group is shown in Figure 9.35.

In this figure, the group selector valve is at its initial position by supplying the compressed air to bus bar B_1. The compressed air will be supplied to the control elements in such a way that the sequence of motions of group 1 should be accomplished. The air after activating the respective pilot line also rushes to the limit switches (3/2 DC valves). These signaling elements only allow the air to pass, when the piston rod reaches the extreme desired position. For example, 1.1 DC valve is actuated by the pilot line. The cylinder 1.0 starts moving forward. When the outer extreme position A^+ is reached, only then the trip valve a_1 is actuated and air is sent to next valve for completion of motion sequence B^+. When the outer extreme position B^+ is reached, only then the trip valve b_1 is actuated and air is sent to the next valve for the completion

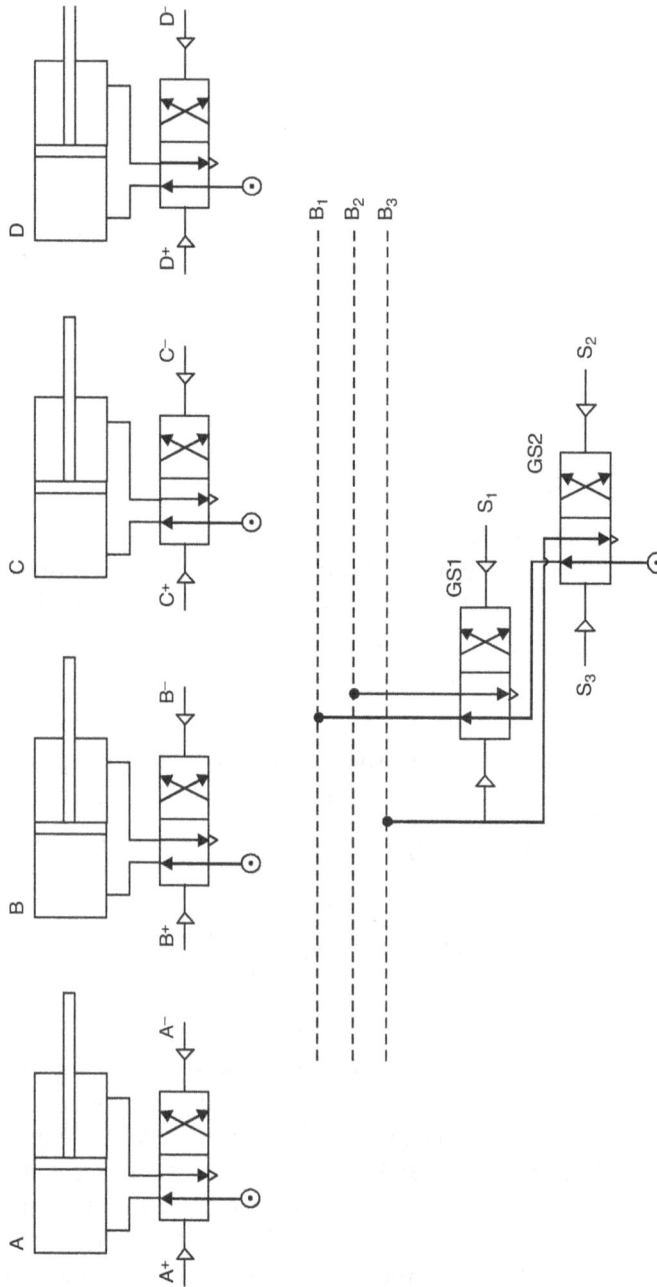

FIGURE 9.34

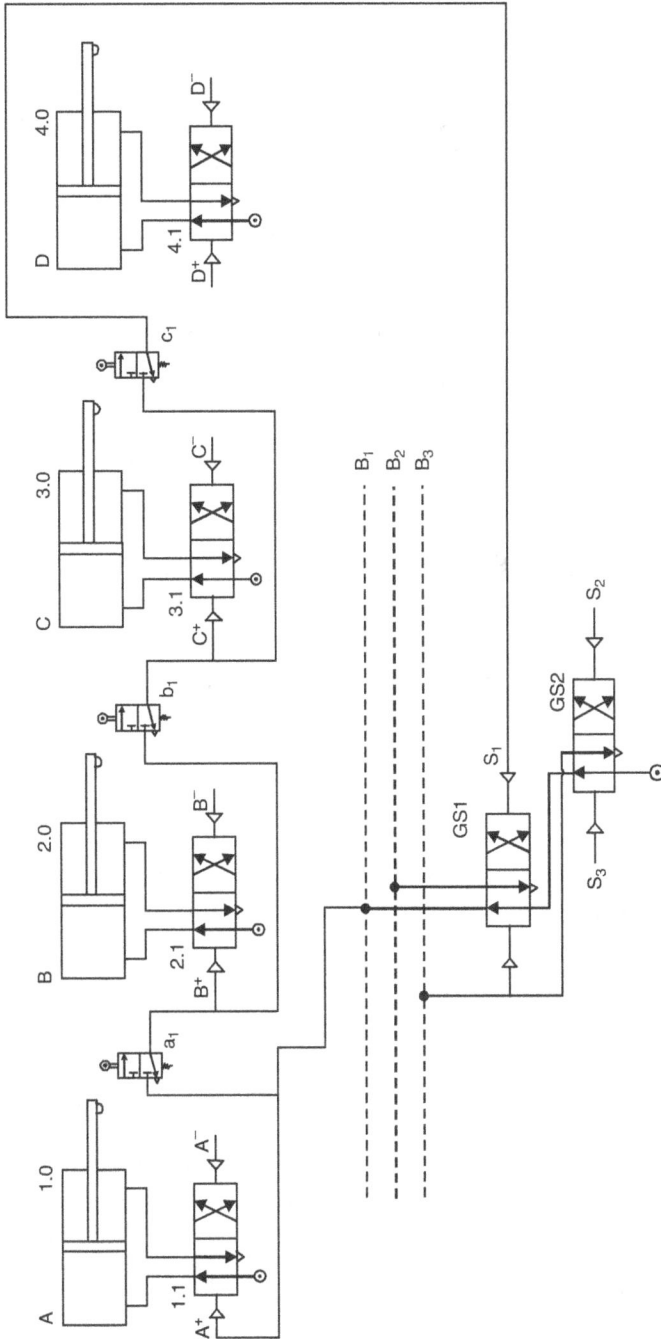

FIGURE 9.35

of motion sequence C^+. When the outer extreme position C^+ is reached, only then the trip valve c_1 is actuated and air is sent to the group selector switch GS1 which acts as a pilot signal S_1 and selects the bus bar line B_2.

Step 2: In step 2, the signal coming from limit switch c_1 of step 1 acts as a pilot signal S_1 for GS1. Now bus bar line B_2 is selected. The sequence of motion for group 2 is

$$\begin{vmatrix} A \\ D \end{vmatrix}$$

The pneumatic circuit for the sequence of the second group is shown in Figure 9.36.

Compressed air taken from bus bar line B_2 is used to apply pilot pressure in attaining A^- and D^+ position. It is required that the group selector switch GS2 is actuated by the pilot signal S_2 only when both positions A^- and D^+ are attained. For this, the outputs from limit switches a_0 and d_1 are given to twin pressure valve I and the common output is used as a pilot signal S_2 for GS2, so that both tasks should be accomplished before starting the new task.

Step 3: In the third step, GS2 is actuated as in last step, *i.e.*, step 2, pilot signal S_2 has changed the position of group selector switch GS2. Supply is taken from bus bar line B_3 and the following motions are accomplished.

$$\begin{vmatrix} D & C \\ B \end{vmatrix}$$

Compressed air taken from bus bar line B_3 is used to apply pilot pressure in attaining D^-, B^-, and C^- positions, where D^- and B^- are accomplished simultaneously and C^- is started after finishing the former motions. The pneumatic circuit for the sequence of the third group is shown in Figure 9.37.

Again, the twin pressure valve is used in this case. Cylinder D and B are retracted back by inducing pressure in the respective pilot line. Then, the outputs from trip valves d_0 and b_0 are directed to twin pressure valve II, so as to ensure the accomplishment of these motions before the starting of motion C^-. Finally, output from twin pressure valve II is directed to the pilot line of 3.1 DC valve to attain the C^- position and output of the trip valve C_0 is sent to the pilot line S_3 of group selector switch GS2. The pilot signal S_3 again resets the valves GS1 and GS2 to their previous positions and cycle is again ready to start.

Figure 9.38 shows the accomplishment of all the motions simultaneously.

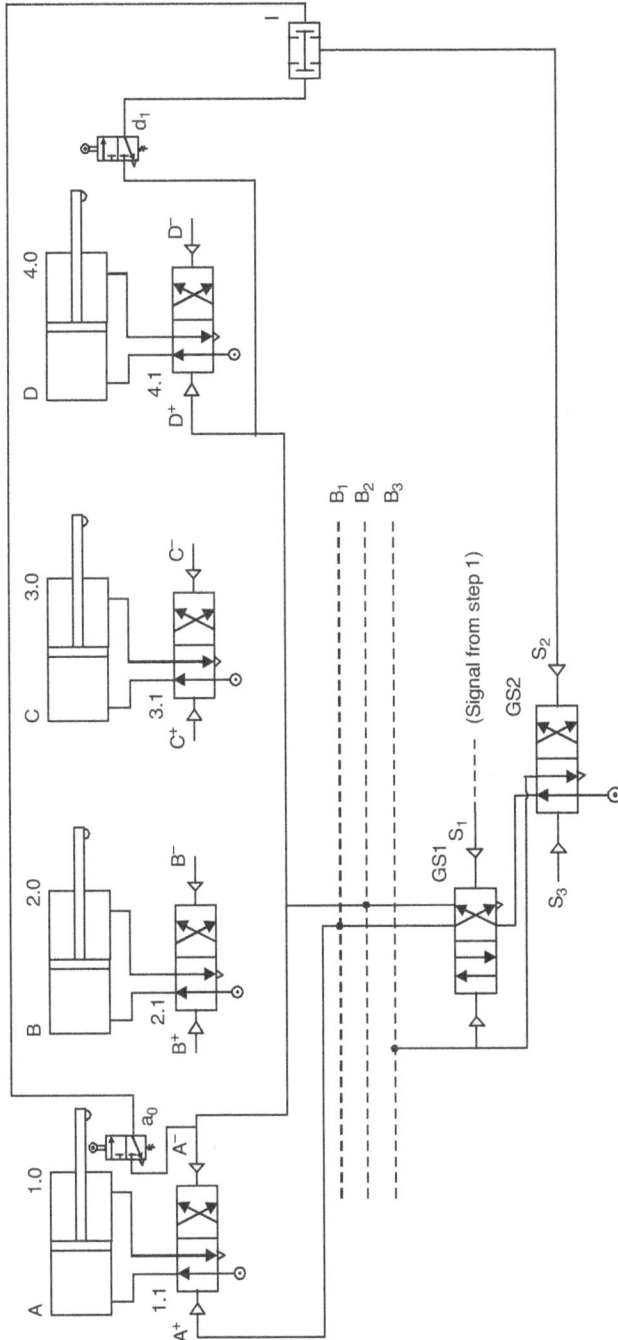

FIGURE 9.36

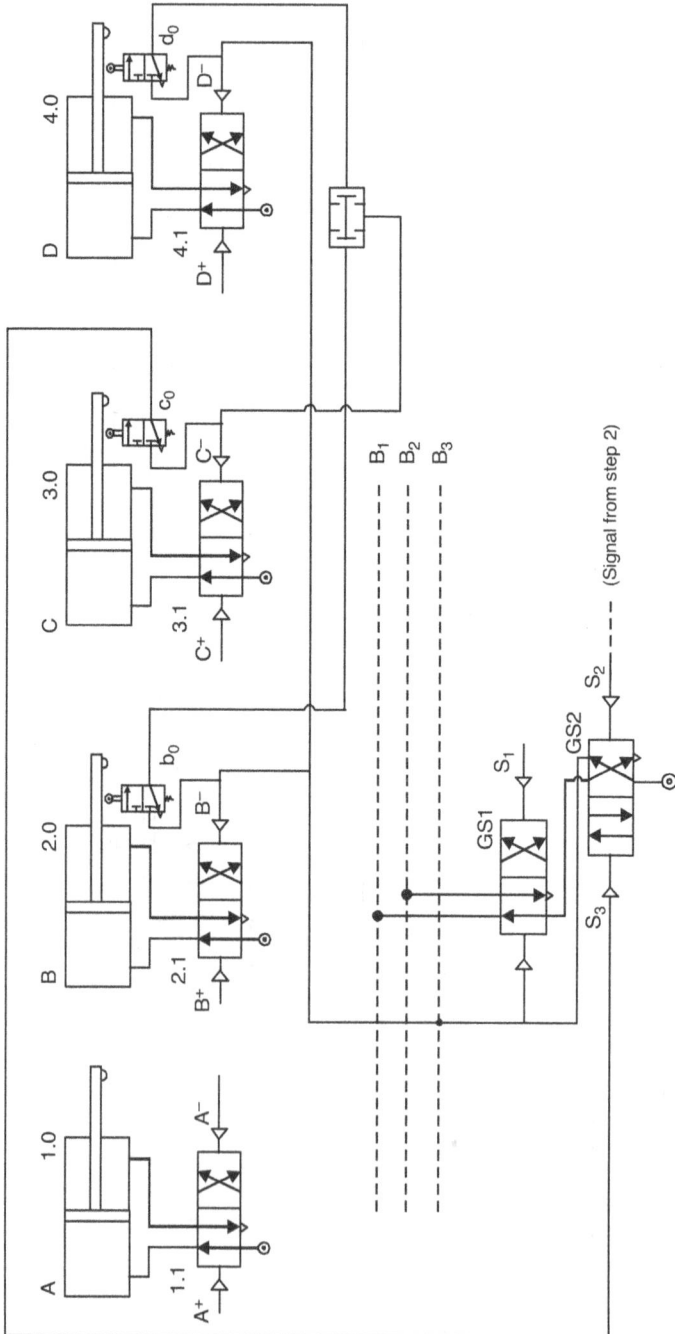

FIGURE 9.37

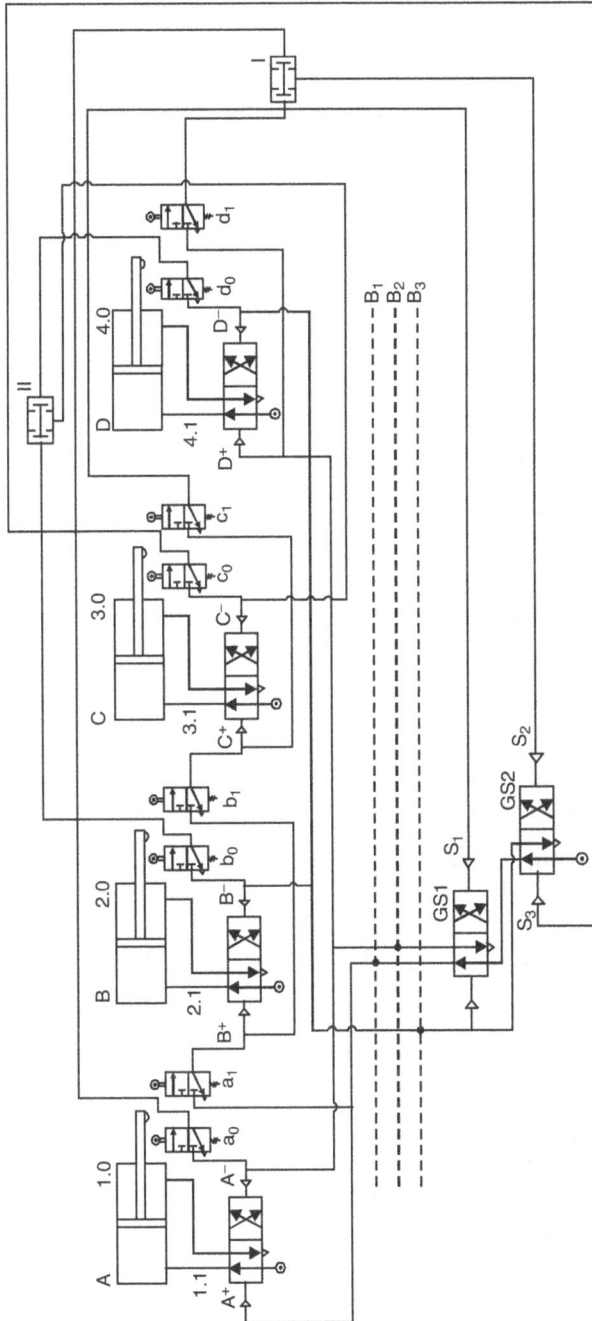

FIGURE 9.38

EXERCISES

1. Define pneumatic logic control circuits.

2. Give the step-by-step procedure for the design of pneumatic logic control circuits. List the various components of pneumatic logic control circuits.

3. Differentiate between the open and closed loop control system.

4. What are the methods of circuit design?

5. Discuss the step-wise procedure for the design of the pneumatic logic circuit for the given sequence of operation. Illustrate the procedure by taking any simple example.

6. What is the need of pneumatic logic circuits?

7. What is an open loop control system? Give two examples.

8. What is the function of using cascade?

9. What is a closed loop control system? Give two examples.

10. Explain the function of trip valve?

11. What rules are to be followed while deciding the number of bus bar lines and group selector switches in cascading?

12. Differentiate between the motion and control diagram.

13. Explain the importance of the position step diagram.

14. List the advantages of cascading over intuitive method.

15. A pneumatic circuit is to be designed for the following sequence:

   - Camping a job and maintaining its position while clamping.
   - Moving the tool for machining.
   - Returning the tool.
   - Unclamping the job.

   Sketch the movement diagram and explain the complete circuit.

16. Design a pneumatic valve circuit to give the sequence A- followed by B+ and then simultaneously followed by A- and B-.

17. If the movement diagram for a 2 cylinder (A&B) pneumatic system is given in figure, draw the pneumatic circuit diagram for the same.

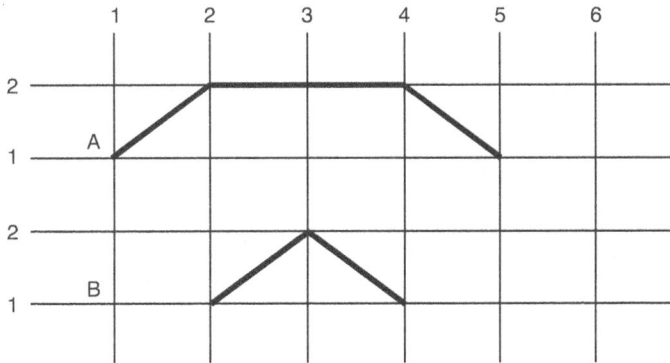

CHAPTER 10

FLUIDICS

INTRODUCTION

Fluidics, a branch of engineering and technology, is concerned with the development of equivalents of various electronic circuits using movements of fluid rather than movements of electric charge. The basic devices used in fluidics are specially designed valves that can be arranged to act as amplifiers and logic circuits. The principal advantage of fluidic systems is that they can be designed to tolerate conditions under which electronic systems could not possibly operate. For example, a fluidic system could operate in the exhaust of a rocket, using the exhaust as its working fluid. Fluidic systems are also advantageous where the system output is to be a flow of fluid, as in an automobile carburetor.

BOOLEAN ALGEBRA

Boolean algebra is a mathematical system for formulating logical statements with symbols so that problems can be solved in a manner to ordinary algebra. In short, Boolean algebra is the mathematics of digital systems.

The most obvious way to simplify Boolean expressions is to manipulate them in the same way normal algebraic expressions are manipulated. With regards to logic relations in digital forms, a set of rules for symbolic manipulation is needed in order to solve for the unknowns. A set of rules formulated by the English mathematician, George Boole, describes certain propositions whose outcome would be either true or false. With regard to digital logic, these rules are used to describe circuits whose state can be either, 1 (true) or 0 (false). The basic rules for Boolean addition and multiplication are shown below:

Addition Rules	Multiplication Rules
$0 + 0 = 0$	$0.0 = 0$
$0 + 1 = 1$	$0.1 = 0$
$1 + 0 = 1$	$1.0 = 0$
$1 + 1 = 1$	$1.1 = 1$

LAWS OF BOOLEAN ALGEBRA

Just like in regular algebra, Boolean algebra has postulates and identities. These laws can be used to reduce expressions or put expressions into a more desirable form. Using just the basic postulates such as Commutative laws, Distributive laws, Identity and Inverse laws, everything else can be derived. Note that every law has two expressions, (a) and (b). Some of the basic laws are:

1. **Commutative Law**

   a $A + B = B + A$

   b $A.B = B.A$

2. **Distributive Law**

   a $A.(B + C) = A.B + A.C$

   b $A + B.C = (A + B).(A + C)$

3. **Identity Law**

   a A   0   A

   b A.1   A

4. **Inverse Law**

   a $\overline{A}$ A   1

   b $\overline{A}$.A   0

Other identities can be derived from the basic postulates such as:

5. **Laws of Ones and Zeros**

   a A   1   1 Law of Ones

   b A.0   0 Law of Zeros

6. **Associative Laws**

   a A   B   C   A   B   C

   b A. B.C   A.B .C

7. **Demorgan's Theorem**

   a $\overline{(A\ B)}$ $\overline{A}\,\overline{B}$

   b $\overline{(A\,B)}$ $\overline{A}$ $\overline{B}$

TRUTH TABLE

The truth table is a type of mathematical table used in logic to determine whether an expression is true or valid. Truth tables are means of representing the results of a logic function using a table. They are constructed by defining all possible combinations of the inputs to a function, and then calculating the output for each combination in turn. Truth tables are widely used because of the following reasons:

- They are relatively easy to understand, as they do not involve any formulas, yet can precisely describe the result of any Boolean formula.

- In the abbreviated form, they are succinct descriptions of Boolean operations, and are widely used in the data sheets of electronic logic devices for this reason.

- Difficult operations, such as simplifying Boolean expressions, can readily be performed by manipulating truth tables, and the abbreviation technique given here is a large part of such simplification.

- They can be used to define a logical formula, without that formula being known, and the formula can then be determined from the truth table.

Construction of a Truth Table

Every Boolean function can be specified as a table. Suppose that the function has n arguments, and then as each has two possible values, there are only 2^n possible argument combinations, which can be listed in full. For each list entry, the function value can then be added, forming the *truth table* of the function. The truth table is a complete and unambiguous definition of the function, because it gives the function value in every possible case. It shows the output states for every possible combination of input states. The symbols 0 (false) and 1 (true) are usually used in truth tables.

LOGIC GATES

A logic gate is a device (whether it is electrical, electronic, or mechanical is irrelevant), which has inputs and outputs. The logic states of the inputs determine the logic states of the outputs. The way in which this is done can be (completely) described using truth tables. All the logic devices will take inputs and produce outputs, which are either ON or OFF. A logic gate is an elementary building block of a *digital circuit*. Most logic gates have two inputs and one output. At any given moment, every terminal is in one of the two *binary* conditions *low* (0) or *high* (1), represented by different voltage levels. There are seven basic logic gates: AND, OR, XOR, NOT, NAND, NOR, and XNOR.

1. The *AND gate* is so named because, if 0 is called "false" and 1 is called "true," the gate acts in the same way as the logical "and" operator. The following illustration and table show the circuit symbol and logic combinations for an

AND gate. (In the symbol, the input terminals are at the left and the output terminal is at the right.) The output is "true" when both inputs are "true." Otherwise, the output is "false."

AND gate

Input 1	Input 2	Output
0	0	0
0	1	0
1	0	0
1	1	1

2. The *OR gate* gets its name from the fact that it behaves after the fashion of the logical inclusive "or." The output is "true" if either or both of the inputs are "true." If both inputs are "false," then the output is "false."

OR gate

Input 1	Input 2	Output
0	0	0
0	1	1
1	0	1
1	1	1

3. The *XOR (exclusive-OR) gate* acts in the same way as the logical "either/or." The output is "true" if either, but not both, of the inputs are "true." The output is "false" if both inputs are "false" or if both inputs are "true." Another way of looking at this circuit is to observe that the output is 1 if the inputs are different, but 0 if the inputs are the same.

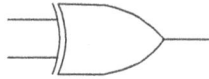

XOR gate

Input 1	Input 2	Output
0	0	0
0	1	1
1	0	1
1	1	0

4. A *logical inverter*, sometimes called a *NOT gate* to differentiate it from other types of electronic inverter devices, has only one input. It reverses the logic state.

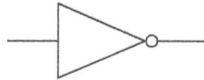

Inverter or NOT gate

Input	Output
1	0
0	1

5. The *NAND gate* operates as an AND gate followed by a NOT gate. It acts in the manner of the logical operation "and" followed by negation. The output is "false" if both inputs are "true." Otherwise, the output is "true."

NAND gate

Input 1	Input 2	Output
0	0	1
0	1	1
1	0	1
1	1	0

6. The *NOR gate* is a combination OR gate followed by an inverter. Its output is "true" if both inputs are "false." Otherwise, the output is "false."

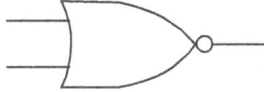

NOR gate

Input 1	Input 2	Output
0	0	1
0	1	0
1	0	0
1	1	0

7. The *XNOR (exclusive-NOR) gate* is a combination XOR gate followed by an inverter. Its output is "true" if the inputs are the same, and "false" if the inputs are different.

XNOR gate

Input 1	Input 2	Output
0	0	1
0	1	0
1	0	0
1	1	1

ORIGIN AND DEVELOPMENT OF FLUIDICS

Fluidics (also known as *Fluidic Logic*) is the use of a fluid or compressible medium to perform analog or digital operations similar to those performed with electronics. The physical basis of fluidics is pneumatics and hydraulics, based on the theoretical foundation of fluid dynamics. The term fluidics is normally used when the devices have no moving parts, so ordinary hydraulic components such as hydraulic cylinders and spool valves are not referred to as fluidic devices.

Fluidics technologies have grown to such an extent that fluids are being used for calculation and data transfer purposes. But in the previous phases of development of fluidics, the electronics systems were predominant in the field of control engineering. The basic development in the field of fluidics started taking place from 1904, when L. Prandtl a German aircraft engineer suggested the solution for a problem of flow separation in diffuser. According to him, by applying suction to the boundary layer, the flow separation can be prevented by applying suction to boundary layer on the diffuser. Prandtl himself devised the first NOT gate. Further in 1916, Nicola Tesla has claimed to be the inventor of first fluidic diode. He has devised a valvular conduit. The patent was granted in 1980. This tube has moving parts but the tube offers resistance to the motion of fluid in one direction and allows the free flow in the outer direction.

Then, in 1938 Henri Coanda, a Russian engineer discovered the theory of 'wall attachment,' Coanda's theory of wall attachment was a major stepping-stone in the development of this field. Later, this effect was used for the development of fluidic logic components. Further in 1962, Ray Auger discovered a fluidic logic element called turbulence amplifier.

Developments in the previous year have led to significant growth in this particular field. Fluid logic elements have gained popularity over last few years as they replaced electronic equipments in the tolerant working environment.

COANDA'S EFFECT

The Coanda Effect was discovered in 1930 by the Romanian aerodynamicist Henri Marie Coanda (1885–1972). It states that the Coanda Effect is the phenomena in which a jet flow attaches itself to a nearby surface and remains attached even when the surface curves away from the initial jet direction. He observed that a steam of air (or fluid) emerging from a nozzle tends to follow a nearby curved surface, if the curvature of the surface or angle the surface makes with the stream is not too sharp.

It is easily demonstrated by holding the back of a spoon vertically under a thin stream of water from a faucet. If one holds the spoon so that it can swing, one will feel it being pulled *towards* the stream of water (Refer to Figure 10.1). The effect has limits: if one uses a sphere instead of a spoon, one will find that the water will only follow a part of the way around.

Further, if the surface is too sharply curved, the water will not follow but will just bend a bit and break away from the surface.

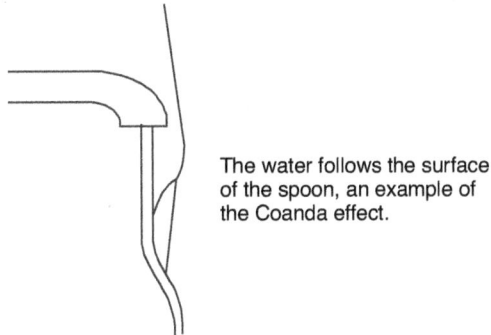

The water follows the surface of the spoon, an example of the Coanda effect.

FIGURE 10.1

Principle of Coanda's Effect

Coanda effect or wall-attachment effect, is the tendency of a moving fluid, either liquid or gas, to attach itself to a surface and flow along it. As a fluid moves across a surface, a certain amount of friction (called "skin friction") occurs between the fluid and the surface, which tends to slow the moving fluid. This resistance to the flow of the fluid pulls the fluid towards the surface, causing it to stick to the surface. Thus, a fluid emerging from a nozzle tends to follow a nearby curved surface even to the point of bending around corners if the curvature of the surface or the angle the surface makes with the stream is not too sharp. This phenomenon has many practical applications in fluidics and aerodynamics. It has important applications in various high-lift devices on aircraft, where air moving over the wing can be "bent down" towards the ground using flaps.

Coanda's Experiments

Coanda's experiments (Refer to Figure 10.2) showed that a sheet of fluid discharged through a slit onto an extended and rounded lip would attach itself to the curved surface and follow its contour. He discovered that a shoulder made of a series of short flat surfaces, each at a specified angle to the one before and each with a certain length would make it possible to bend a jet stream around an 180° arc. Further, he found that the deflected air-stream sucked up air from the surroundings. He found that as the jet flowed around the shoulder, it entrained up to twenty times the amount of air in the original jet.

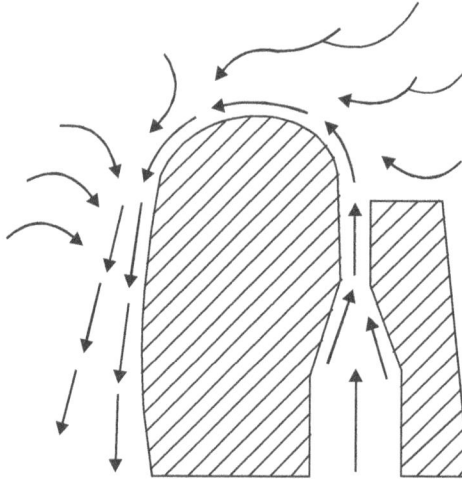

FIGURE 10.2 Coanda's Experiment.

TESLA'S VALVULAR CONDUIT

Tesla claimed that the valvular conduit (Refer to Figure 10.3) devised by him composed of a closed passage having recesses in its walls so formed as to permit a fluid to pass freely from it in the direction of flow but fluid will be subjected to a rapid reversal of direction when entered from another or opposite direction, thereby interposing friction and mass resistance.

This device resembles the today's check valve but this kind of conduit has no moving parts. This is why Tesla's tube has been given the name of fluid diode of 1916.

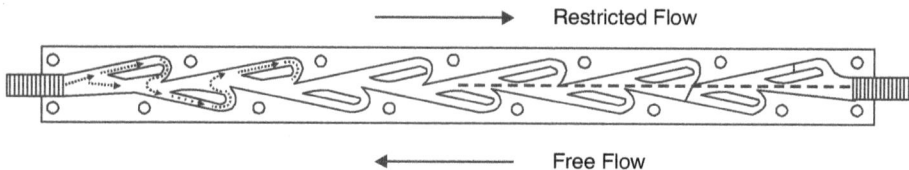

FIGURE 10.3 Tesla's Valvular Conduit.

FLUIDIC DEVICES

Fluidic components are the simple components devised on the basis of some well-known principles and phenomenon. Fluidic devices are simple to design, repair, and use, and one can intend them as a substitute for electronics for their use in fluid control systems. Fluidic devices can be classified as:

- Digital fluidic devices
- Analog fluidic devices

Digital fluidic devices are those which has two outputs. There will be flow from either one output or the other output. The flow never splits in two outputs. In digital fluidic devices, there is no question of more output or less output, they just give signal in the form of "Output" or "No output." Examples of such devices are bi-stable flip flops, logic, etc.

Analog fluidic devices are those, which can vary the output from low to high. Here the output will increase or decrease as a function of control input. Examples of such devices are fluidic position sensors, fluidic vortex amplifier, wall-attachment amplifier, fluidic oscillator, etc.

Fluidic devices can be further divided into three categories:

- Fluidic logic devices
- Fluidic sensors
- Fluidic amplifiers

FLUIDIC LOGIC DEVICES

Generally, all fluidic devices are based on Coanda's theory of wall attachment. This phenomenon has already been explained earlier in this chapter. Based on the theories, a few simple devices have been patented by various agencies in last few decades. Most popular amongst these are bi-stable flip-flop, AND gate, OR-NOR gate, etc. Some of these important fluidic logic devices are explained in the following text.

Bi-stable Flip Flop

Bi-stable flip flop is a most common fluidic logic element (Refer to Figure 10.4). It works on the principle of Coanda effect. Its general configuration consists of five ports out of which one is the supply port; two control jet ports and two output ports. Supply is always present on the input port whereas output is dependent on the control jets, *i.e.*, when fluid is passed from control jet C_1 there will be an output at O_1 and, similarly, when there is supply at control jet C_2 the output will be at O_2.

It is also important to know here the difference between mono-stable and bi-stable.

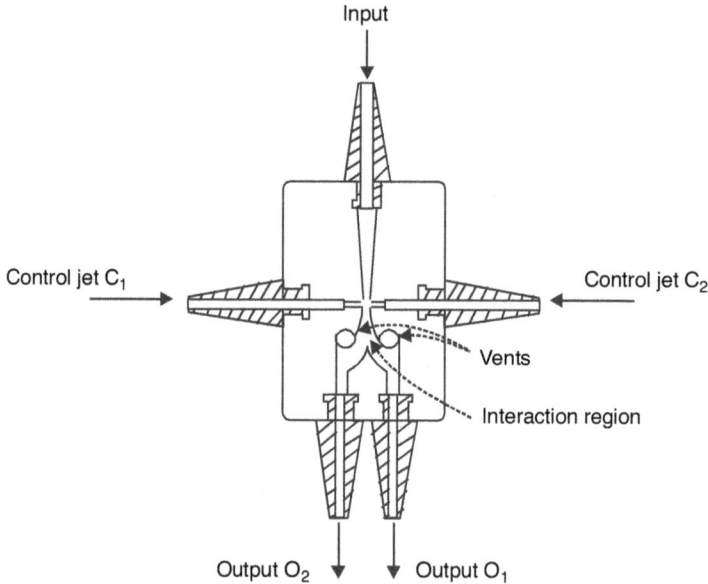

FIGURE 10.4 Bi-stable Flip Flop.

Difference between Mono-stable and Bi-stable Elements

Mono-stable means when the device has one stable state, *i.e.*, in the absence of signal it will always remain in that particular state only. The second state can only be achieved in the presence of an applied signal, for example, a 3/2 manually operated spring return DC valve is a mono-stable device.

On the other hand, *bi-stable* means a device is stable in any one position out of its two possible positions; whether the signal is applied or not, for example, a 4/2 pilot operated DC valve is a bi-stable device.

In this way, by using the two control jets alternatively, this fluidic logic element can be used to perform memory functions. The output from a bi-stable flip-flop can be used as a pilot signal for actuating various valves with a low pressure actuating element.

Fluidic AND Gate

Figure 10.5 shows a fluidic AND gate. Supply A and supply B show input ports from where fluid is entered. Further, there is one output vent Y and two drain vents named 1 and 2. If the fluid stream is present in supply port A and absent in supply port B, the fluid is drained out from drain vent 2. Similarly, if the fluid stream is present in supply port B but absent in supply port A, fluid is drained through drain vent 1. This means there will be no output or zero output in the above two cases.

The only condition for the output at Y is the presence of fluid streams of equal strength at the two supply ports. Operation of this fluid logic component also depends on Coanda's wall-attachment theory.

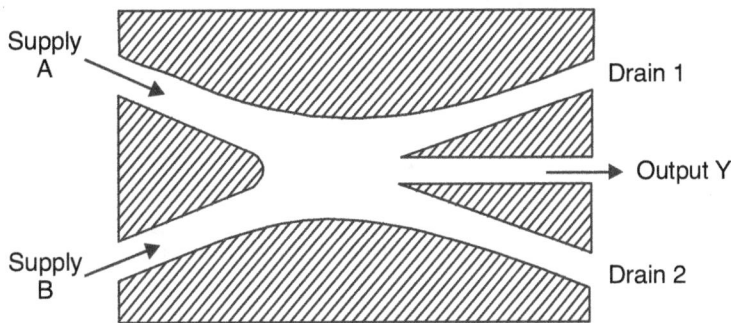

FIGURE 10.5 Fluidic AND Gate.

Fluidic OR-NOR Gate

Figure 10.6 shows a fluidic OR-NOR gate. This fluidic device consists of supply port A from which fluid is supplied all the time. There are two control ports, 'a' and 'b' and two output ports, Y and Y1. Y output represents the OR gate and Y1 output represents the NOR gate. In the absence of both the control jets 'a' and 'b,' supply A is passed out through the gate via output Y1. It implies that when there is no input ('a' or 'b'), there is an output. This is the logic characteristic of the NOR gate and Y1 is considered as a NOR output. On the other hand, when a fluid stream is present at either control jet 'a' or 'b' or at both 'a' and 'b,' there will be an output at Y. It implies that even with one input there is an output. This is the logic characteristic of the OR gate and Y output is considered an OR output. This complete device is called an OR-NOR fluidic gate.

FLUIDIC SENSORS

Fluidic sensors are primarily used for detecting the presence of objects. These sensors are designed to provide a signal in the form of fluid jet to announce the presence of an object. Fluidic sensors are generally of very

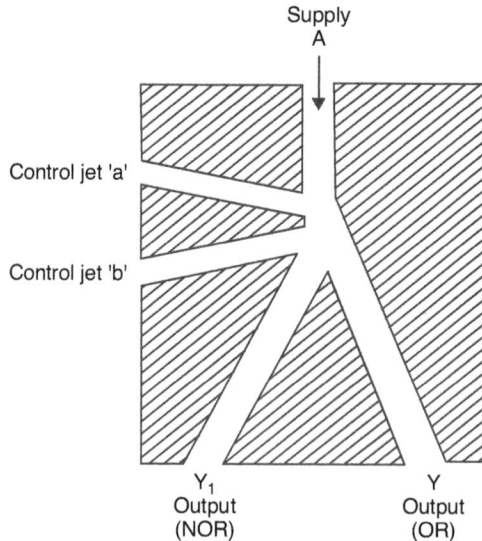

FIGURE 10.6 Fluidic OR-Nor Gate.

small size. This is to reduce the air consumption. Similarly, their other design aspects like the diameter of nozzles producing jets and the size of inlet. Outlet orifices are also important from the view of minimizing air consumption. The output of fluidic sensors can be directly used as a control signal for pneumatic logic circuits. But fluidic control signals require amplification when used in such pneumatic and hydraulic devices. A few important fluidic sensors are explained in the following text.

Cone Jet Sensor

Figure 10.7 shows a cone jet sensor. A cone jet sensor is used to sense the positions of an object even when the object is placed at a considerable distance from it. A cone jet sensor has two supply nozzles connected with supply port 'A' and 'B,' respectively, from which fluid is projected onto the object to be located and between the two nozzles there is cylindrical passage. This cylindrical passage is connected with output orifice. Fluid, which comes back after hitting the object, is directed to the cylindrical passage and further, passed from output orifice 'Y.' When the object is at some distance from the cone jet sensor, there is low pressure output at 'Y,' whereas if the object comes closer to the cone jet sensor, the high pressure output is observed at output orifice 'Y.' Thus, output pressure is indicative of the current position of the object.

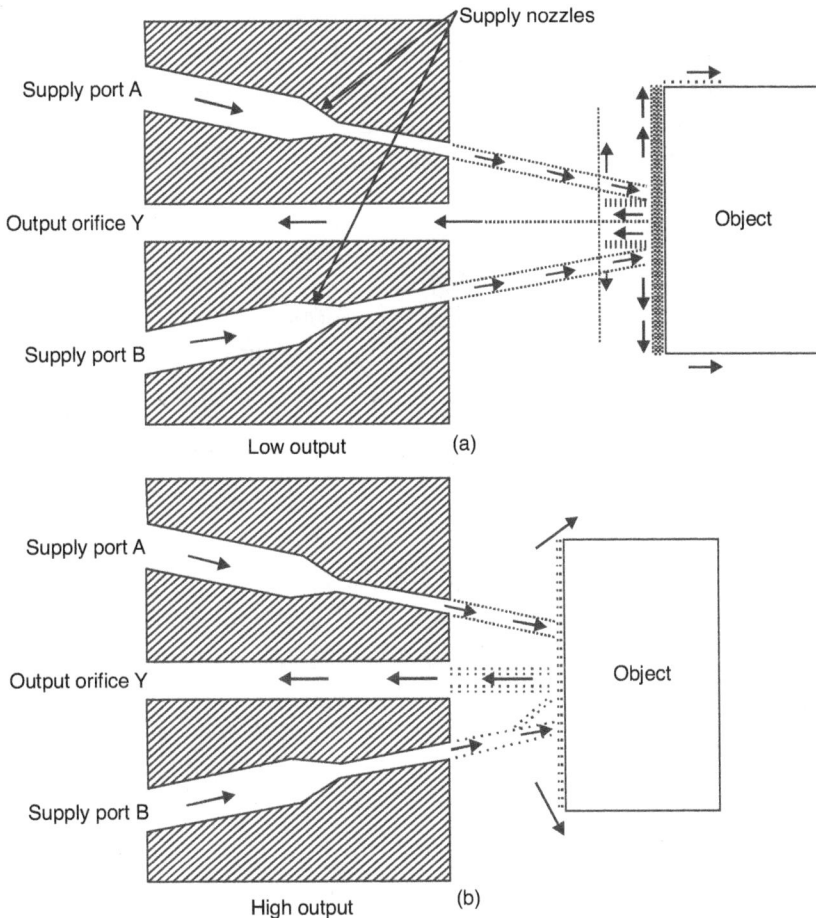

FIGURE 10.7 Cone Jet Sensor.

Interruptible Jet Sensor

The principle of interruptible jet sensor is the same as discussed above for the cone jet sensor. Figure 10.8 shows an interruptible jet sensor. Low-pressure air flowing in laminar mode is supplied from the nozzle. The air has to pass across a gap between the issuing nozzle and the collector. The jet of air passes uninterrupted if the object does not obstruct the path. An interruptible jet sensor consists of a nozzle and a collector. There is a considerable gap between both. From the nozzle, fluid is supplied and is directed to the collector. If the path between A and Y is not blocked by the object or component, then the laminar flow remains continuous and the stream of air issuing from the nozzle provides a signal to the collector in the form of output Y. On the other hand, if the object blocks the supply air jet A, then the flow will become turbulent and output Y will be very small. This decreased output can be sufficiently used to actuate a gate, which is incorporated in the device.

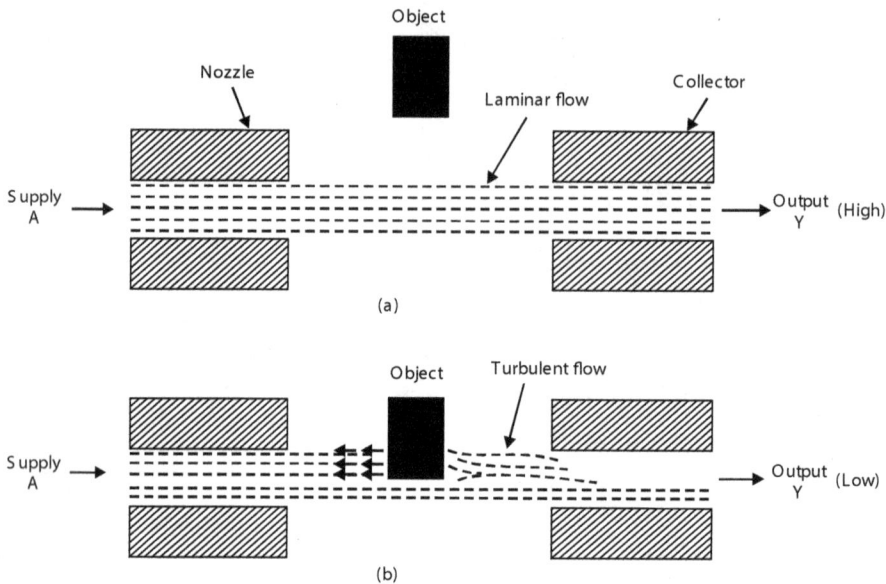

FIGURE 10.8 Interruptible Jet Sensor.

Back Pressure Sensor

A back pressure sensor, as the name implies, works on the principle of back pressure. The supply of fluid is provided from supply orifice 'A.' Figure 10.9 shows the working of a back pressure sensor.

When the object whose position is to be located at some considerable distance from the back pressure sensor, air escapes out on the other side following the same straight horizontal path. As a result, there is high output at outlet orifice Y and low output at output orifice X. On the other hand, when the object is brought closer to output port Y of back pressure sensor, in such a way that objects block the passage of fluid to pass from the output orifice. Then the flow of air will be directed to output orifice X. As a result, there will be no output at Y and high output at X. This sensor is used to find the location of objects when the distance between the back pressure sensor and object is not too large.

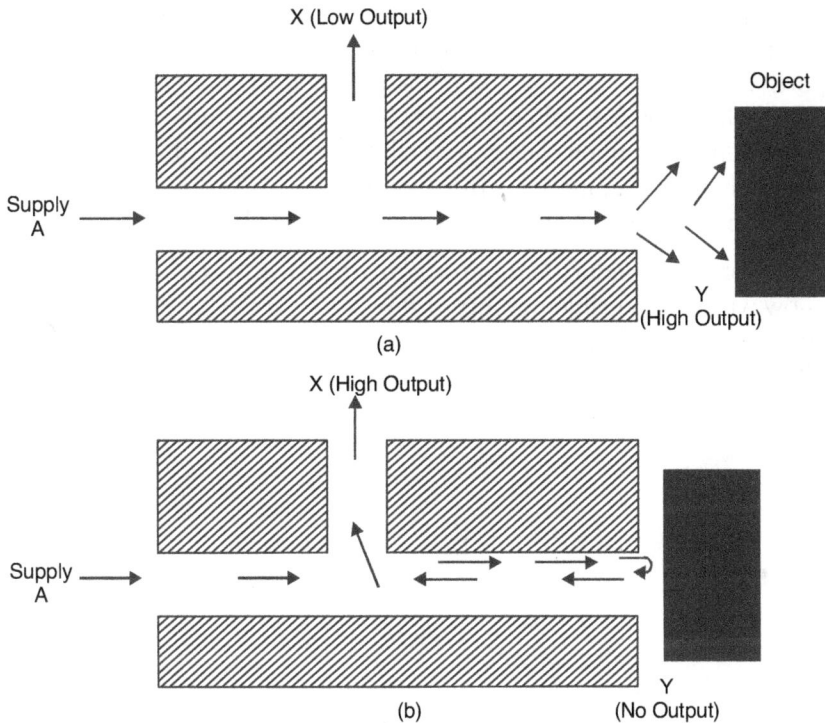

FIGURE 10.9 Back Pressure Sensor.

FLUIDIC AMPLIFIER

Fluidic amplifiers are devices which have no moving parts and use a gas or liquid as the working medium. A high energy stream of fluid is accelerated in a nozzle in the amplifier to create a power jet. Lower energy control jets are supplied transverse to the power jet flow to produce a change in direction of the power jet and a corresponding change of the flow in various output ports in the device. Hence, an amplification of the signal is obtained. This amplification process is similar to that occurring in an electronic tube. A few of them are explained below.

Vortex Amplifier

A vortex amplifier is a fluidic device used to regulate the flow of fluid by utilizing properties of a vortex. A vortex amplifier consists of a cylindrical, disc-like container, which is divided by cylindrical porous elements into two chambers, namely, an outer chamber and a vortex chamber as shown in Figure 10.10.

Supply port A is provided for the inlet of fluid in a cylindrical disc. Fluid enters along the outer periphery of the porous element from the supply port. There is a control jet, which is provided to generate a vortex in the vortex chamber. Several control jets along the circumference can be provided depending upon capacity, which throws streams of fluid in tangential direction and generate the vortex motion in fluid. At the center, there is an output port Y from where the signal is transmitted.

In order to understand the working of the vortex amplifier, one must know what the possible outputs of this device are. This fluidic device provides either an 'amplified signal' or 'no signal' as an output. These two conditions of the vortex amplifier are explained below:

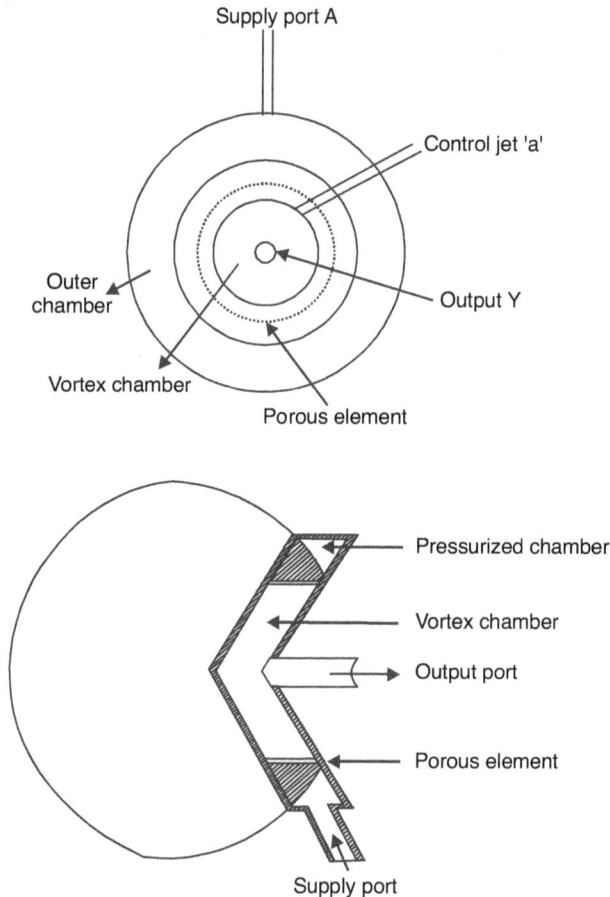

FIGURE 10.10 Vortex Amplifier.

Case 1: Amplified signal

Fluid is entered from the supply port inside the cylindrical disc. Fluid passes from the porous element and moves towards the output port under the effect of gravity. The moment fluid enters the vortex chamber, it comes in contact with control jet 'a.' Due to pressure and direction, these control jets are already moving with the vortex motion. Now, the fluid from supply port 'A' follows this tangential component and approaches the output port. Because of this, some angular velocity is imparted to it as it leaves the vortex chamber. The reason behind the increase in angular velocity is because of the drop in pressure and the increase in velocity of the fluid due to the vortex motion. In order to conserve angular momentum, angular velocity of fluid must increase as it approaches the output port. Similarly, radial velocity of the fluid also increases while flowing upwards to downwards or from the supply port to the output port, respectively. Thus, there are two kinds of signal amplifications in a vortex amplifier, *i.e.*, (1) rise in radial velocity, (2) rise in angular velocity. In the way explained above, the amplified signal is transmitted from the output port.

Case 2: No signal

The second condition of 'no signal' can be attained by regulating the supply at supply port A. With this, centrifugal effects in the vortex chamber becomes prominent due to centrifugal force. A high-pressure region is built up and the supply at output port Y is almost stopped. In fact, due to this pressure rise, flow to supply port A is slightly reversed. However, there may still be some flow at outlet port Y. It is solely due to the flow of control jets. It is important to note here that for certain supply, pressure at least equivalent to a control jet is required. Control pressure should not be less than the supply pressure in any case.

Turbulence Amplifier

A turbulence amplifier is a logic device based on fluid dynamic phenomenon, which was described by Lord Rayleigh in 19th century. According to Rayleigh, a jet of fluid flowing with a Reynolds number less than 1500 in laminar mode becomes turbulent when interrupted by a transverse control jet. Refer to Figure 10.11. Typical turbulence amplifiers that are used in industry consist of a thin supply and output tubes having 15–20mm gap between them. Supply/output tubes are housed in another cylindrical pipe perfectly closed at ends, having a base size equal to more than 25 times the supply/output tubes. Control jets and drain vent are provided in this outer pipe.

The supply of fluid from the supply tube is directed to the output tube, in the absence of control jets. The fluid will be collected by an output tube. However, if any one of the control jets is active at that time, the laminar flow becomes turbulent and there will be no fluid in the outlet pipe. In normal arrangements, 4–6 control jets can be used in one turbulence amplifier. However, it is important to note that any one of the control jets can turn the amplifier 'OFF.'

The power required to turn off the amplifier is much less than the power in output line. It works on low operating pressure. The pressure in the output tube is approximately 10 cm of water, to turn off the signal pressure is approximately 1 cm of water, and the supply pressure is of the order around 25 cm of water. The time needed to turn ON and OFF the turbulence amplifier is 2–3 milliseconds and 5–7 mil-

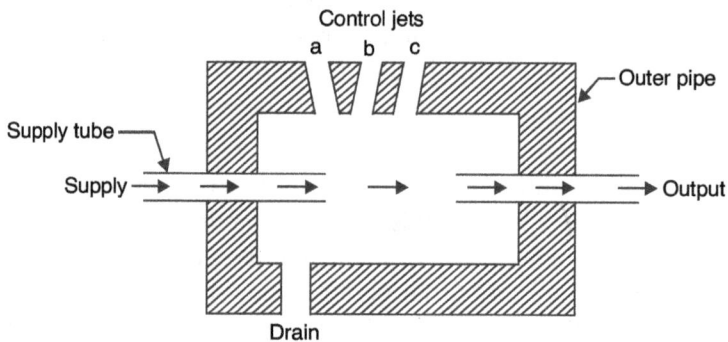

(a) Stable output in absence of control signal.

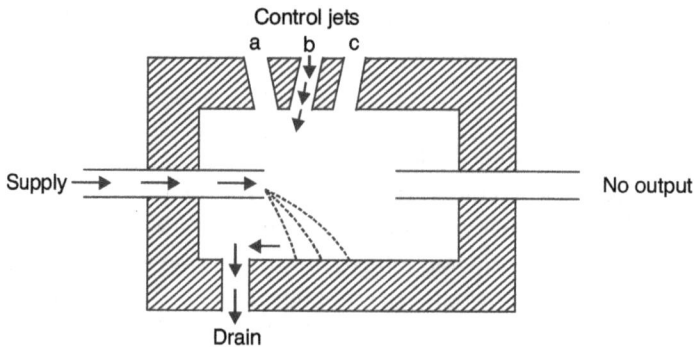

(b) No output when control jet b issues signal.

FIGURE 10.11 Turbulence Amplifier.

liseconds, respectively. Turbulence amplifiers are based on laminar/turbulent flows as well as concept of jet destruction, so they are also called laminar-turbulent flow devices or jet destruction devices.

ADVANTAGES AND DISADVANTAGES OF FLUIDICS

Advantages

1. *Fluidic devices are reliable*: Fluidic devices have no moving parts. These are robust systems, *i.e.*, they have practically no effect of electromagnetic and nuclear radiation. These systems are preferred over electronic devices where there is a question of high temperature service. These systems are however sensitive to dirt of fluid medium. But one can easily cope-using good quality filters for air or liquids prior to their entry.

2. *Fluidic devices are noise free*: Practically, there are no moving parts in fluidic devices, so they do not produce any noise.

3. *Fluidic devices are compact*: Fluidic devices are compact in comparison with pneumatic, hydraulic, electro pneumatic devices. But this statement is not correct in context with microelectronics because electronic gates and other components are still several times more compact than fluidic systems.

4. *Fluidic devices are simple in design*: Fluidic devices are based on few basic operating principles. These principles are easy to understand. Once these are understood it is easy for an individual to design, modify, and repair these fluidic devices. No specialized knowledge is required for this purpose.

5. *Fluidic devices are hazard free*: These devices are preferred in hazardous areas where electronic components fail to perform.

Disadvantages

1. *Limited development of the field*: As discussed earlier, fluidics is based on a few simple principles, which are scientifically approved, but some phenomenon, for example, the Coanda effect is still not clearly understood.

2. *Fluidic devices have slow speeds and low power outputs*: Fluidic devices generally use air as a medium and operate on pressure less than 0.1 bar because the use of such a low pressure, fluidic devices suffer

from relatively slow speeds and low power output. Actuation of power-operated equipment is not directly possible; some mode of amplification is required. Further, these kind of additional components increase the overall expense of a fluidic system.

3. *Fluidic devices are not suitable for incompressible fluids*: At low-pressure operation, incompressible liquids like oils are not economical. Their use is limited in a fluidic system.

4. *Complex fluidic systems are impracticable*: Because of size and cost limitations, a maximum 1,000 logic functions in one system is an extreme limit for a fluidic system. A fluidic microcomputer is neither practicable nor desirable. Another reason behind this statement is speed. Simple fluidic systems are comparatively slower, further when complexity is added to the fluidic system it becomes too slow.

5. *Fluidic devices are not suitable for intermittent operation control systems*: In control systems where operation is intermittent and cycle time is long, fluidic devices are not desirable because of continuous power consumption.

6. *Fluidic devices are inefficient*: The power of controlling medium (*e.g.*, air) is generally lost by traveling long distances in transmission lines. Further, fluidic devices which work on the principle of wall attachment recover about 15% of the power passing through them and a vortex amplifier recovers as much as 40% of supplied power. A fluidic signal over long distances presents problems both in phase shift and signal degradation.

EXERCISES

1. Identify the various laws of Boolean algebra.

2. Identify three logic elements of Boolean algebra. How are they expressed?

3. Explain the construction, working, and performance characteristics of fluidic elements.

4. Explain NAND gate and NOR gate.

5. Using a truth table, show that $\overline{A.B} \quad \overline{A} \quad \overline{B}$.

6. Complete the Boolean expression. $A \quad \overline{A}.B$

7. What is the Coanda effect? Describe with a diagram.

8. What is a truth table?

9. Write the truth table for OR gate and draw the symbol for representing OR gate.

10. Draw the diagram for NOR gate and bi-stable flip-flop. Also, state their working along with constructional details.

11. Define fluidic bi-stable switch.

12. Explain the fluidic NOR gate with a sketch.

13. Describe the working of any fluidic device.

14. Explain the working of a fluidic diode and prepare a truth table for a fluidic bi-stable switch.

15. Sketch any fluidic device and explain its operation. State its applications.

16. What are the advantages and disadvantages of fluidics?

17. How can fluidic devices be classified?

18. What are fluidic sensors? Explain the working of any one sensor with the help of sketch.

19. Is there any similarity between fluidics and electronics?

20. Discuss the construction and working of the following fluidic component: (*i*) OR/NOR. (*ii*) Proximity detector (any one type).

21. What are the major advantages of fluidic systems over traditional systems?

22. Differentiate between analog and digital devices.

23. How are cone jet and backpressure sensors different from interruptible jet sensors?

24. Write the truth table for AND gate and draw the symbol for the same.

11

ELECTRICAL AND ELECTRONIC CONTROLS

INTRODUCTION TO SENSORS AND TRANSDUCERS

A *transducer* can be defined as a device capable of converting energy from one form into another. Transducers can be found both at the input as well as at the output stage of a measuring system. The input transducer is called the *sensor*, because it senses the desired physical quantity and converts it into another energy form. The output transducer is called the *actuator*, because it converts the energy into a form to which another independent system can react, whether it is a biological system or a technical system. So, for a biological system, the actuator can be a numerical display or a loudspeaker to which the visual or aural senses react respectively. For a technical system, the actuator could be a recorder or a laser, producing holes in a ceramic material. Humans can interpret the results.

A *sensor* is a device that produces a signal for purposes of detecting or measuring a property, such as position, force, torque, pressure, temperature, humidity, speed, vibration, etc. Sensor technology has become an important component of manufacturing processes and systems, because they convert one quantity to another. Sensors are also referred as transducers. A *sensor* is a physical device or biological organ that detects, or *senses*, a signal or physical condition and chemical compounds. Sensors are devices that provide an interface between electronic equipment and the physical world. Often the active element of a sensor is referred to as a transducer.

SENSOR TERMINOLOGY

(a) *Sensitivity*: Sensitivity of a sensor is defined as the change in output of the sensor per unit change in the parameter being measured. The factor may be constant over the range of the sensor (linear), or it may vary (nonlinear).

(b) *Range*: Every sensor is designed to work over a specified range. The design ranges are usually fixed, and, if exceeded, result in permanent damage to or destruction of a sensor. Range is the difference between maximum and minimum values of the applied parameter that can be measured.

(c) *Precision*: Precision is the degree of reproducibility of the measurements.

(d) *Resolution*: Resolution is defined as the smallest change that can be detected by a sensor. In other words, it is the response of the measuring instrument for small variations in the input parameter.

(e) *Accuracy*: A very important characteristic of a sensor is accuracy, which really means inaccuracy. Inaccuracy is measured as a ratio of the highest deviation of a value represented by the sensor to the ideal value. It may be represented in terms of measured value.

(f) *Hysteresis*: Hysteresis is the difference in response for increasing and decreasing values of the applied parameter.

(g) *Response time*: The time taken by a sensor to approach its true output when subjected to a step input is sometimes referred to as its response time.

(h) *Offset*: Offset is the sensor output that exists when it should be zero.

(i) *Linearity error*: It is defined as an expression of the extent to which the measured curve departs from the ideal theoretical curve.

(j) *Span*: Span is defined as the range of measured variable for which an instrument is designed to measure with full linearity.

(k) *Calibration*: It is defined as the comparison of specific values of the input and output of an instrument with the corresponding reference standard values.

SELECTION OF A TRANSDUCER

The following factors should be kept in mind while selecting a transducer. The transducer should:

- Recognize and sense the desired input signal and should be sensitive to other signals.
- Have good accuracy.
- Have good precision.
- Have amplitude linearity.
- Have environmental compatibility, *i.e.*, corrosive fluids, pressure, shocks, size, etc.

CLASSIFICATION OF SENSORS

Sensors can be classified according to the type of energy they detect:

1. *Thermal energy.* For measuring temperature, flux, conductivity, and specific heat.

- *Temperature sensors:* thermometers, thermocouples, thermistors
- *Heat sensors:* calorimeter

2. *Electromagnetic sensors.* For measuring voltage, current, charge, magnetic field, flux, and permeability.

- *Electrical resistance sensors:* ohmmeter, multimeter
- *Electrical current sensors:* galvanometer, ammeter
- *Electrical voltage sensors:* voltmeter
- *Electrical power sensors:* watt-hour meters
- *Magnetism sensors:* magnetic compass, magnetometer, Hall effect device

3. *Mechanical sensors.* For measuring quantities such as position, shape, velocity, force, torque, pressure, strain, and mass.

- *Pressure sensors:* barometer, barograph, pressure gauge, air speed indicator.
- *Gas and liquid flow sensors:* flow sensor, flow meter, gas meter, water meter.
- *Strain gauge*

4. *Chemical sensors*

- Ion-selective electrodes, pH glass electrodes.

5. *Optical and radiation sensors*

- Bubble chamber, dosimeter.
- Photocells, photodiodes, phototransistors, photo-electric tubes.

6. *Acoustic sensors*

- *Sound sensors:* microphones, hydrophones, seismometers.

CLASSIFICATION OF TRANSDUCERS

Transducers can be classified in different ways:

1. *Self-generating and non–self-generating transducers.* Self-generating transducers are those which produce their own electrical signal (either current or voltage). For example, thermocouple, thermopile, moving coil generator, piezoelectric pick up, photovoltaic cell, etc.

Non–self-generating type transducers are those which are not capable of generating their own signals. These will not produce an electrical signal of their own but show some variations of resistance, capacitance, and inductance. For example, thermistor, linear variable differential transducer, capacitive pick up, strain gauge, resistance temperature detector, etc.

2. *Input and output transducers.* Input transducers are those that have electronic output and another form of energy as input, *i.e.*, input transducers convert a quantity to an electrical signal (voltage) or to resistance (which can be converted to voltage). *Examples:* Light dependent resistor (LDR) converts brightness (of light) to resistance, thermistor converts temperature to resistance, microphone converts sound to voltage, variable resistor converts position (angle) to resistance, etc.

Output transducers are those that have electronic input and another form of energy for output. *Examples:* lamp converts electricity to light, LED converts electricity to light, loudspeaker converts electricity to sound, motor converts electricity to motion, heater converts electricity to heat, etc.

3. *Analog and digital transducers.* Analog transducer converts input signal into output signal, which is a continuous function of time such as thermistor, strain gauge, thermocouple, LVDT, etc.

A digital transducer converts the input signal into the output signal of the form of pulse that gives discrete output.

4. *Transducers are also classified as:*

▪ Temperature transducers, flow transducers, magnetic transducers, etc.
▪ Force and pressure transducers.
▪ Displacement transducers.

TEMPERATURE SENSORS

The measurement of temperature is important in many industrial applications. These applications require temperature sensors of different physical construction and often-different technology. Several factors must be considered when selecting the type of sensor to be used in a specific application: temperature range, accuracy, response time, stability, linearity, and sensitivity. The commonly used sensors for temperature measurement are:

▪ Resistance Temperature Detector (RTD)
▪ Thermocouple
▪ Thermistor
▪ Fiber Optic Temperature Sensors

Resistance Temperature Detector (RTD)

A resistance temperature detector (RTD) is a temperature sensor that senses temperature by means of changes in the magnitude of current through, or voltage across an element whose electrical resistance varies with temperature. These types of sensors provide a change in resistance proportional to a change in temperature. Resistance temperature detectors have been used for making accurate temperature measurements. They utilize a resistance element whose resistance changes with the ambient temperature in a precise and known manner. The resistance temperature detector may be connected in a bridge circuit, which drives a display, calibrated to show the temperature of the resistance element. Most metals become more resistant to the passage of an electrical current as the metal increases in temperature. The increase in resistance is generally proportional to the rise in temperature. Thus, a constant current passed through a metal of varying resistance produces a variation in voltage that is proportional to the temperature change.

The basic construction of an RTD is quite simple. It consists of a length of fine-coiled wire wrapped around a ceramic or glass core. The element is usually quite fragile, so it is often placed inside a sheathed probe to protect it. (Refer to Figure 11.1.)

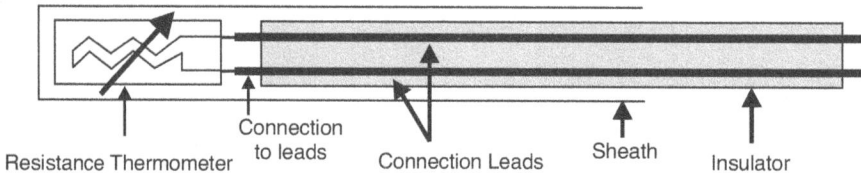

FIGURE 11.1 Resistance Temperature Detector.

Common resistance materials for RTDs are platinum, nickel, and copper. Platinum is the most commonly used metal for RTDs due to its stability and nearly linear temperature. It can measure temperatures up to 800°C. The resistance of the RTD changes as a function of absolute temperature, so it is categorized as one of the absolute temperature devices. (In contrast, the thermocouple cannot measure absolute temperature; it can only measure relative temperature.)

Advantages

- Stable and accurate.
- More linear than thermocouples.

Disadvantages

- More expensive.
- Self heating.
- Requires a current source.
- Response time may not be fast enough for some applications.

Thermocouple

When two dissimilar metal conductors are connected together to form a closed circuit and the two junctions are kept in different temperatures, thermal electromotive force (EMF) is generated in the circuit (Seebeck's effect). Thermocouples make use of this so-called *Peltier-Seebeck effect*. Thus, when one end (cold junction) is kept constant at a certain temperature, normally at 0°C, and the other end (measuring junction) is exposed to an unknown temperature, the temperature at latter end can be determined by measurement of EMF so generated. This combination of two dissimilar metal conductors is called "thermocouple".

Thermocouple is an active transducer, which is used to measure very high temperatures of furnaces in industrial plants. Thermocouple consists of a pair of dissimilar metals/wires joined together to form a junction as shown in Figure 11.2. One end of the junction is the sensing end, which is to be immersed in the medium of temperature. This is called hot junction. The other end of the junction is called cold or reference junction, which is maintained at a constant reference temperature. When the hot junction is being heated by keeping the sensing end of the thermocouple in the medium whose temperature is to be measured, a temperature difference exists between the hot and reference junctions. This produces an EMF, which causes a current in the circuit and can be measured with the help of voltmeter.

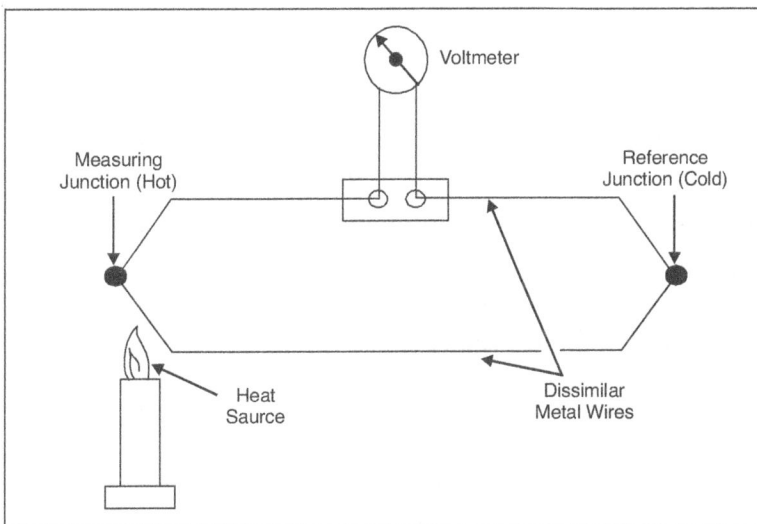

FIGURE 11.2 Thermocouple.

A thermoelectric circuit containing two junctions is illustrated in Figure 11.3. Two wires of metals A and B form junctions at two different temperatures T_1 and T_2, resulting in a potential V that can be measured.

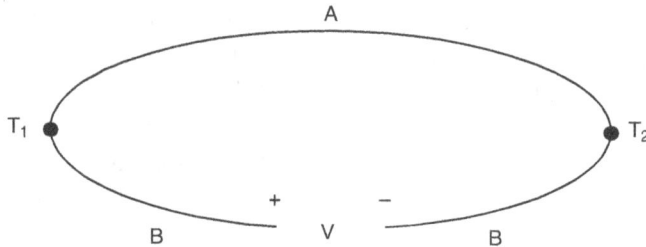

FIGURE 11.3 Thermoelectric Circuit.

The thermocouple voltage is directly proportional to the junction temperature difference:

$$V = \alpha \, (T_1 - T_2)$$

where α is Seeback coefficient.

Advantages

- Self powered requiring no external power supply.
- Simple, rugged, inexpensive, and commonly available.
- Can withstand harsh environments.

Disadvantages

- Nonlinear and require a cold junction compensation for linearization.
- Not as accurate as RTDs or thermistors.
- Low voltage, least stable, and sensitive.

Thermistor

Like RTD, thermistor is also a temperature sensitive resistor. A *thermistor* is an electronic component that exhibits a large change in resistance with a change in body temperature. Thermistors are highly sensitive to temperature variation; hence, they are also called temperature sensitive resistors. Thermistors are manufactured from metal oxide semiconductor material, which is encapsulated in a glass or epoxy bead. Thermistors also have a low thermal mass that results in fast response times, but are limited by a small temperature range. Some of different types of thermistors are shown in Figure 11.4.

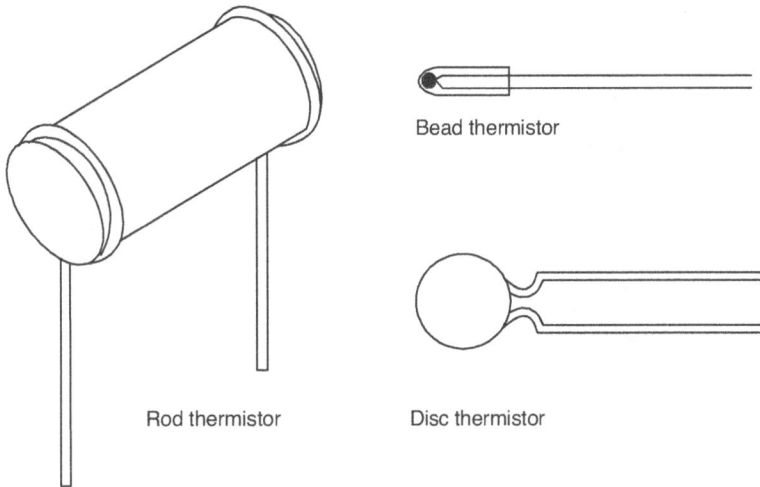

FIGURE 11.4 Thermistors.

Thermistors are divided into negative temperature coefficient (NTC) and positive temperature coefficient (PTC) types. The temperature coefficient of a material can be defined as change in resistance of the material for a unit degree change in temperature. Although positive temperature coefficients units are available, most thermistors have a NTC, *i.e.*, their resistance decreases with increasing temperature. The NTC can be as large as several percent per degree Celsius, allowing the thermistor circuit to detect minute changes in temperature, which could not be observed with an RTD, or thermocouple circuit.

If the relationship between resistance and temperature is assumed to be linear, then:

$$\Delta R = k\, \Delta T$$

where
ΔR = change in resistance

ΔT = change in temperature

k = first-order temperature coefficient of resistance

If k is positive, the resistance increases with increasing temperature, and the device is called a PTC thermistor, *posistor* or *sensistor*. If k is negative, the resistance decreases with increasing temperature, and the device is called a NTC thermistor.

Advantages

- Inexpensive, rugged, and reliable.
- Respond quickly.

Disadvantages

- Smaller temperature range.
- Signal is not linear.
- Self-heating.

Fiber Optic Temperature Sensors

Optical-based temperature sensors provide accurate and stable remote measurement of online temperatures in hazardous environments and in environments having high ambient electromagnetic fields without the need for calibration of individual probes and sensors.

Optical temperature sensor systems measure temperatures from -200°C to 600°C safely and accurately even in extremely hazardous, corrosive, and high electro-magnetic field environments. They are ideal for use in these conditions because their glass-based technology is inherently immune to electrical interference and corrosion. Because there is no need to recalibrate individual sensors, operator and technician safety is greatly enhanced as the need for their repeated exposure to field conditions is eliminated. Probes are made from largely nonconducting and low thermal conductance material, resulting in high stability and low susceptibility to interference, and in increased operator safety. Optical cables also have a much higher information-carrying capacity and are far less subject to interference than electrical conductors.

Applications of Temperature Sensors

These include:

- HVAC — room, duct, and refrigerant equipment
- Motors —overload protection
- Electronic circuits — semiconductor protection
- Electronic assemblies — thermal management, temperature compensation
- Process control — temperature regulation
- Automotive — air and oil temperature
- Appliances — heating and cooling temperature

LIGHT SENSORS

When light strikes special types of materials, a voltage may be generated, a change in electrical resistance may occur, or electrons may be ejected from the material surface. As long as the light is present, the condition continues. It ceases when the light is turned off. Any of the above conditions may be used to change the flow of the current or the voltage in an external circuit and, thus, may be used to monitor the presence of the light and to measure its intensity. Some of the commonly used light sensors are discussed below:

Photoresistors

Photoresistors, as their name suggests, are resistors whose resistance is a function of the amount of light falling on them. Their resistance is very high when no light is present and significantly lower when they are illuminated. These are also often called light-dependent resistors (LDRs) (refer to Figure 11.5). Photoresistors can be used as light sensors, which can enable robot behaviors such as hiding in the dark, moving toward a beacon, etc.

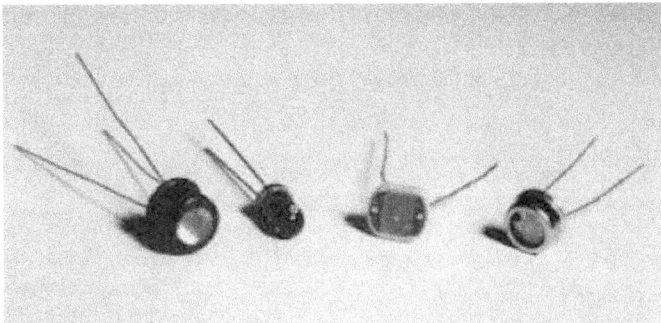

FIGURE 11.5 Typical Photoresistors.

Photodiode

A photodiode is a type of photodetector capable of converting light into either current or voltage, depending upon the mode of operation. Photodiodes are used both to detect the presence of light and to measure light intensity. Most photodiodes consist of semiconductor *pn* junctions housed in a container designed to collect and focus the ambient light close to the junction. They are normally biased in the reverse, or blocking direction; the current, therefore, is quite small in the dark. When they are illuminated, the current is proportional to the amount of light falling on the photodiode.

Phototransistor

A second optoelectronic device that conducts current when exposed to light is the phototransistor. A phototransistor, however, is much more sensitive to light and produces more output current for a given light intensity that does a photodiode.

POSITION SENSORS

A position, or linear displacement sensor, is a device whose output signal represents the distance an object has traveled from a reference point. Types of position/displacement sensors are:

- Inductive sensors
- Capacitive displacement sensors
- Magnetostrictive sensors

Inductive Sensors

These sensors measure inductance variations caused by movement of a flux-concentrating element. They are probably the most versatile of all position sensors, with a wide range of operating characteristics. Inductive sensors are contact-free, inherently robust, and have infinite resolution with high repeatability. They are often used where long-term reliability is important, particularly in harsh and hostile environments. There are two basic types of inductive sensors:

- Linear variable differential transducer (LVDT)
- Rotary variable differential transducer (RVDT)

(a) *Linear Variable Differential Transducer (LVDT)*: Linear variable differential transducer (LVDT) is a common type of electromechanical transducer that can convert the rectilinear motion of an object to which it is coupled mechanically into a corresponding electrical signal. The basic LVDT design shown in Figure 11.6 consists of three elements:

- One primary winding
- Two identical secondary windings
- A movable magnetic armature or "core"

With excitation of the primary coil, induced voltages will appear in the secondary coils. Because of the symmetry of magnetic coupling to the

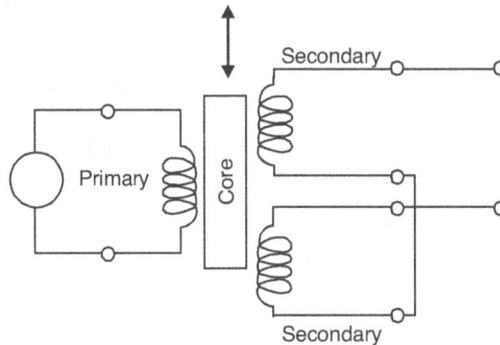

FIGURE 11.6 Design of LVDT.

primary, these secondary induced voltages are *equal* when the core is in the central ("null" or "electric zero") position. When the secondary coils are connected in series opposition, as shown in the figure, the secondary voltages will cancel and (ideally) there will be no net output voltage.

If, however, the core is *displaced* from null position, in either direction, one secondary voltage will increase, while the other decreases. Because the two voltages no longer cancel, a net output voltage will now result. The difference in induced voltages produces an output that is linearly proportional in magnitude to the displacement of the core.

Advantages

- Relative low cost due to its popularity.

- Solid and robust, capable of working in a wide variety of environments.

- No friction resistance, because the iron core does not touch the transformer coils, resulting in an very long service life.

- Short response time, only limited by the inertia of the iron core and the rise time of the amplifiers.

- No permanent damage to the LVDT if measurements exceed the designed range.

Disadvantages

- The core must contact directly or indirectly with the measured surface, which is not always possible or desirable. However, a non-contact thickness gage can be achieved by including a pneumatic servo to maintain the air gap between the nozzle and the work piece.

(b) *Rotary Variable Differential Transducer (RVDT):* The rotational variable differential transducer (RVDT) is used to measure rotational angles and operates under the same principles as the LVDT sensor. Whereas the LVDT uses a cylindrical iron core, the RVDT uses a rotary ferromagnetic core. It is similar to the LVDT except that its core is cam shaped and may be rotated between the windings by means of a shaft. (Refer to Figure 11.7.)

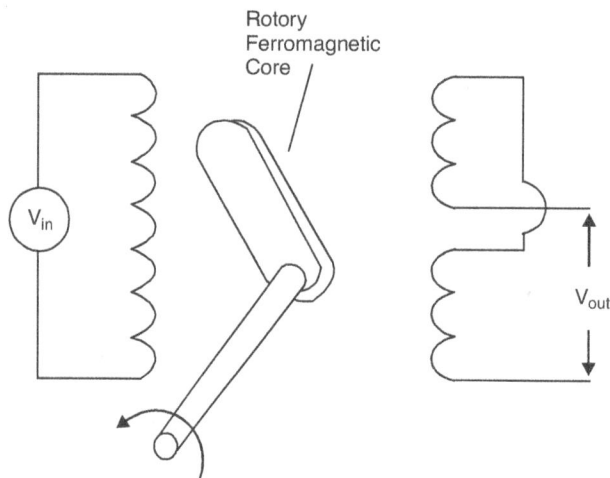

FIGURE 11.7 RVDT.

Capacitive Displacement Sensors

Capacitive sensors detect virtually any material (paper, cardboard, plastic, etc.) at an operating distance of up to 10 mm. They are also suitable for the detection of metallic or fluid objects. They offer high speed and no contact sensing at an extremely long life. Capacitive sensors detect an extremely wide variety of materials, primarily non-metallic materials, at close range.

Capacitive proximity sensors are designed to operate by generating an electrostatic field and detecting changes in this field caused when a target approaches the sensing face. The sensor's internal workings consist of a capacitive probe, an oscillator, a signal rectifier, a filter circuit, and an output circuit as shown in Figure 11.8.

In the absence of a target, the oscillator is inactive. As a target approaches, it raises the capacitance of the probe system. When the capacitance reaches

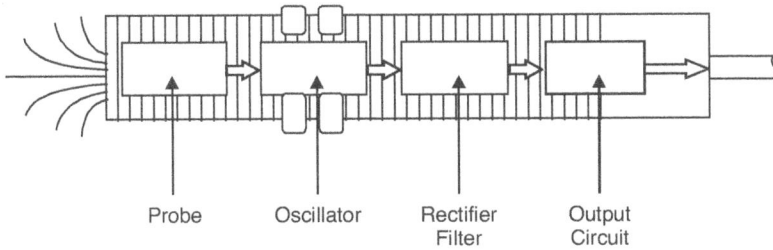

FIGURE 11.8

a specified threshold, the oscillator is activated which triggers the output circuit to change between "on" and "off." The capacitance of the probe system is determined by the target's size, dielectric constant, and distance from the probe. The larger the size and dielectric constant of a target, the more it increases capacitance. The shorter the distance between target and probe, the more the target increases capacitance.

Magnetostrictive Sensor

The magnetostrictive effect is the change of the resistivity of a material due to a magnetic field. Magnetostriction is a property of ferromagnetic materials such as iron, nickel, and cobalt. When placed in a magnetic field, these materials change size and/or shape. Magnetostrictive materials convert magnetic energy to mechanical energy and vice versa. As a magnetostrictive material is magnetized, it strains; that is, it exhibits a change in length per unit length. Magnetostrictive transducers consist of a large number of nickel (or other magnetostrictive material) plates or laminations arranged in parallel with one edge of each laminate attached to the bottom of a process tank or other surface to be vibrated. A coil of wire is placed around the magnetostrictive material. When a flow of electrical current is supplied through the coil of wire, a magnetic field is created. This magnetic field causes the magnetostrictive material to contract or elongate thereby introducing a sound into the surface to be vibrated.

Magnetic Sensors or Hall-Effect Sensors

These produce output voltages proportional to the strength of a nearby magnetic field generated by a moving magnet. They have relatively poor temperature performance, but can be effectively used for short-range position sensing where cost is most important and temperature is not an issue. Hall sensors work best when movements are less than an inch (25 mm).

PIEZOELECTRIC SENSORS

Piezoelectric sensors are considered to be a versatile tool for the measurement of various processes. These sensors are used to measure strain or force by converting them to an electrical signal. They are used for quality assurance, process control, and process development in many different industries. Piezoelectric sensors rely on the piezoelectric effect, which was discovered by the Curie brothers in the late 19th century. While investigating a number of naturally occurring materials such as tourmaline and quartz, Curie brothers realized that these materials had the ability to transform energy of a mechanical input into an electrical output. More specifically, when a pressure is applied to a piezoelectric material, it causes a mechanical deformation and a displacement of charges. Those charges are highly proportional to the applied pressure.

Piezoelectric sensors are used to sense movement or vibrations in many applications. A piezoelectric sensor comprises a piezoelectric crystal, which is typically mechanically coupled to an object that produces a mechanical movement. In piezoelectric materials, an applied electric field results in elongations or contractions of the material. These sensors are able to convert electric energy directly into mechanical energy and offer several advantages, such as high actuating resolution, high actuating power, and very short response times, while their size is small. Piezoelectric sensors are used as transducers because a potential difference is generated when the sensor is subject to a pressure change. The common uses of piezoelectric devices are "buzzers," which produce a buzzing noise when a voltage is applied. The single disadvantage of piezoelectric sensors is that they cannot be used for true static measurements.

PRESSURE SENSORS

A pressure transducer is a transducer that converts pressure into an analog electrical signal. Although there are various types of pressure transducers, one of the most common is the strain-gauge base transducer. The conversion of pressure into an electrical signal is achieved by the physical deformation of *strain gauges*, which are bonded into the diaphragm of the pressure transducer and wired into a wheatstone bridge configuration. Pressure applied to the pressure transducer produces a deflection of the diaphragm, which introduces strain to the gauges. The strain will produce an electrical resistance change proportional to the pressure.

STRAIN GAUGES

There are several methods of measuring strain; the most common is with a strain gauge. The strain gauge has been in use for many years and is the fundamental sensing element for many types of sensors, including pressure sensors, load cells, torque sensors, position sensors, etc. The use of strain gauges is based on the fact that the resistance of a conductor changes when the conductor is subjected to strain. When external forces are applied to a stationary object, stress and strain are the result. Stress is defined as the object's internal resisting forces, and strain is defined as the displacement and deformation that occur. The strain gauge is one of the most important tools of the electrical measurement technique applied to the measurement of mechanical quantities. Strain consists of tensile and compressive strain, distinguished by a positive or negative sign. A strain gauge is a thin piece of conducting material that may look like as shown in Figure 11.9.

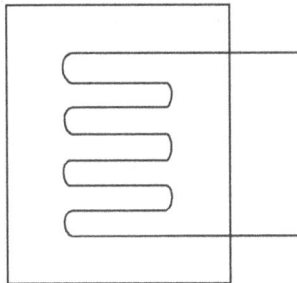

FIGURE 11.9 Strain Gauge.

Theory of Operation

The operation of the resistance strain gauge is based on the principle, that the electrical resistance of a conductor changes when it is subjected to a mechanical deformation, since the resistivity changes with a change in length and area. Figure 11.10 shows a resistance wire in its original state, and after that subjected to a strain. The stretched wire has higher resistance, as it is longer and thinner.

The resistance of a conductor can be expressed as:

$$R = \frac{L}{A}$$

where,

R is the resistance,

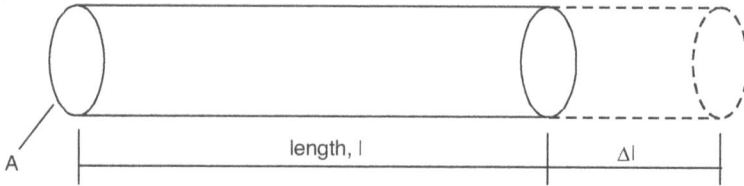

FIGURE 11.10

ρ is the material resistivity,

L is the length of the conductor, and

A is the cross-sectional area of the conductor.

Types of Strain Gauges

Strain gauges can be classified as mechanical, optical, or electrical depending upon the principle of operation and their constructional features. Of these, electrical strain gauges are the most popular. The principle of an electrical strain gauge is based upon the measurement of the changes in resistance, capacitance, or inductance that is proportional to the strain transferred from the object to the basic gauge element. Some of the commonly used strain gauges are discussed below:

1. Resistance Strain Gauges

The resistance of an electrically conductive material changes with dimensional changes, which take place when the conductor is deformed elastically. When such a material is stretched, the conductors become longer and narrower, which causes an increase in resistance. A wheatstone bridge then converts this change in resistance to an absolute voltage. The resulting value is linearly related to strain by a constant called the gauge factor.

2. Capacitance Strain Gauges

Capacitance devices, which depend on geometric features, can be used to measure strain. The capacitance of a simple parallel plate capacitor is proportional to:

$$C \approx \frac{ak}{t}$$

where,　　　　　　　　　C is the capacitance,

a is the plate area,

k is the dielectric constant, and

t is the separation between plants.

The capacitance can be varied by changing the plate area, a, or the gap t. The electrical properties of the materials used to form the capacitor are relatively unimportant, so capacitance strain gauge materials can be chosen to meet the mechanical requirements. This allows the gauges to be more rugged, providing a significant advantage over resistance strain gauges.

3. Photoelectric Strain Gauges

An extensometer (an apparatus with mechanical levers attached to the specimen) is used to amplify the movement of a specimen. A beam of light is passed through a variable slit, actuated by the extensometer, and directed to a photoelectric cell. As the gap opening changes, the amount of light reaching the cell varies, causing a varying intensity in the current generated by the cell.

4. Semiconductor Strain Gauges

In piezoelectric materials, such as crystalline quartz, a change in the electronic charge across the faces of the crystal occurs when the material is mechanically stressed. The piezoresistive effect is defined as the change in resistance of a material due to an applied stress and this term is used commonly in connection with semiconducting materials. The resistivity of a semiconductor is inversely proportional to the product of the electronic charge, the number of charge carriers, and their average mobility. The effect of applied stress is to change both the number and average mobility of the charge carriers.

Features of a Good Strain Gauge

- Small size and mass
- Ease of production over a range of sizes
- Robustness
- Good stability, repeatability, and linearity over large strain range
- Good sensitivity
- Freedom from (or ability to compensate for) temperature effects and other environmental conditions
- Suitability for static and dynamic measurements and remote recording
- Low cost

MICROPROCESSOR

Microprocessors are regarded as one of the most important devices in our everyday machines called computers. A *microprocessor* (abbreviated as μP or uP) is a computer electronic component made from miniaturized transistors on a single semiconductor integrated circuit (IC).

OR

Microprocessor is an electronic circuit that functions as the central processing unit (CPU) of a computer, providing computational control.

The microprocessor communicates and operates in the binary numbers 0 and 1, called bits. Each microprocessor has a fixed set of instructions in the form of binary patterns called a machine language. A microprocessor is a single-integrated circuit. The integrated circuit is a complex collection of very small electronic components organized into a circuit that controls the 'on' and 'off' switches of the computer. The circuit is referred to as integrated because all of the components that need to work together are etched into a single silicon chip. The microprocessor processes instructions and communicates with outside devices, controlling most of the operation of the computer. The microprocessor usually has a large heat sink attached to it. Some microprocessors come in a package with a heat sink and a fan included as a part of the package. Microprocessors are also used in other advanced electronic systems, such as computer printers, automobiles, and jet airliners.

Microprocessors are classified by the semiconductor technology of their design (TTL, transistor-transistor logic; CMOS, complementarymetal-oxide semiconductor; or ECL, emitter-coupled logic), by the width of the data format (4-bit, 8-bit, 16-bit, 32-bit, or 64-bit) they process; and by their instruction set (CISC, complex-instruction-set computer, or RISC, reduced-instruction-set computer). TTL technology is most commonly used, while CMOS is favored for portable computers and other battery-powered devices because of its low power consumption. ECL is used where the need for its greater speed offsets the fact that it consumes the most power. Figure 11.11 shows the Pentium 4 microprocessor.

History of Microprocessors

The first digital computers were built in the 1940s using bulky relay and vacuum-tube switches. Relays had mechanical speed limitations.

FIGURE 11.11 Pentium-4 Microprocessor.

Vacuum tubes required considerable power, dissipated a significant amount of heat, and suffered high failure rates. In 1947, Bell Laboratories invented the transistor, which rapidly replaced the vacuum tube as a computer switch for several reasons, including smaller size, faster switching speeds, lower power consumption and dissipation, and higher reliability. In the 1960s, Texas Instruments invented the integrated circuit, allowing a single silicon chip to contain several transistors as well as their interconnections.

The first microprocessor was the Intel 4004, produced in 1971. Originally developed for a calculator, and revolutionary for its time, it contained 2,300 transistors on a 4-bit microprocessor that could perform only 60,000 operations per second. The first 8-bit microprocessor was the Intel 8008, developed in 1972 to run computer terminals. The Intel 8008 contained 3,300 transistors. The first truly general-purpose microprocessor, developed in 1974, was the 8-bit Intel 8080, which contained 4,500 transistors and could execute 200,000 instructions per second. By 1989, 32-bit microprocessors containing 1.2 million transistors and capable of executing 20 million instructions per second had been introduced.

Developed during the 1970s, the microprocessor became most visible as the central processor of the personal computer.

Layout of a Microprocessor System

The microprocessor system consists of three main components as shown in Figure 11.12.

- Central processing unit (CPU)
- Memory
- Input/Output

Central Processing Unit

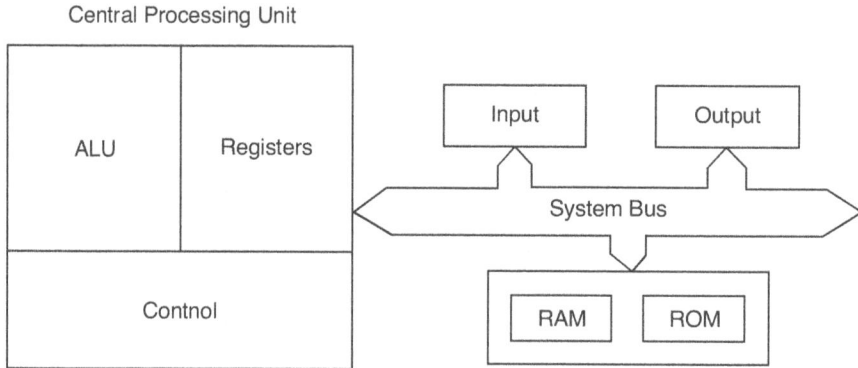

FIGURE 11.12 Layout of Microprocessor.

These three components will work together or interact with each other to perform a given task. Different buses are used to interconnect these components.

1. Central Processing Unit (CPU)

The CPU is called the brain of the microprocessors. The CPU consists of:

- Arithmetic logic unit
- Control unit
- Registers

(a) *Arithmetic Logic Unit*: The ALU performs basic arithmetical calculations (addition, subtraction, multiplication, and division) and logic functions (AND, OR, EXCLUSIVE OR, etc.). The ALU is a fundamental building block of the CPU of a computer. The ALU carries out these operations in the following manner:

- Stores data fetched from memory or I/O in the registers.
- Send this data either to its arithmetic circuitry or logical circuitry, where the necessary arithmetic or logical operations are carried out.
- Send results of its arithmetic or logical operation to relevant accumulator, to the memory, or to the I/O interfaces.

(b) *Control Unit*: The control unit is the part of the microcomputer that controls its basic operations. It is made up of the control signal generating circuitry (clock) and the command (instruction) decoder. The control section fetches pre-programmed instructions from memory as needed and temporarily stores them in the command register (also known as instruction register [IR]). These instructions are then decoded by the operation decoder, which sends control signals to the relevant parts of the microcomputer system (via the system busses) to cause them to carry out the required operation. The clock determines the timing with which these control signals are generated.

(c) *Registers*: When the processor executes instructions, data is temporarily stored in registers. Depending on the type of processor, the overall number of registers can vary from about ten to many hundred. These registers are used to store data temporarily, either 8-bit data or 16-bit data according to their size. Registers are given names, normally an alphabet, such as A, B, C, D, E, H, L and each capable of storing an 8-bit data. These registers can also work as a pair, such as BC, DE, and HL and capable of storing 16-bit data.

For all 8-bit operations, *register A* is used as the accumulator, where the result after an ALU operation will be stored automatically here. For 16-bit operations, register pair HL will be used to store the result. Register F is an 8-bit register used to store the status of the CPU, such as carry, zero, parity, overflow, etc. The others are 16-bit registers, which is used to store memory addresses. A very simple microprocessor may have the following registers:

- *Accumulator register*, which stores the results of arithmetic and logical operations.
- *Program counter*, which determines in which sequence the program instructions, are to be executed.
- *Instruction register*, which hold the last instruction, fetched from memory.
- *Memory data register*, which holds the data that was last read from or written to memory.
- *Memory address register*, which holds the address of the data or the instruction currently being accessed.

2. Memory

All microprocessor systems have some memory. Memory is the term for various storage devices, which are used to store the programs and data for the microprocessor. There are 2 types of memory:

- Read only memory (ROM)
- Random access memory (RAM)

(a) *Read only memory (ROM)*: It is used to store programs and data that don't need to be altered, *i.e.*, permanent storage. The CPU can only read programs and data stored in ROMs. The monitor program is normally stored in the ROM.

(b) *Random access memory (RAM)*: It is used to store user programs and data, and can be altered at any time, *i.e.*, temporary storage. The information stored in RAM can be easily read and altered by the CPU. The contents (data or programs) stored is lost if power supply to this chip is turned off. There are different kinds of random-access memory. *Static RAM (SRAM)* holds information as long as power is turned on and is usually used as cache memory because it operates very quickly. Another type of memory, *dynamic RAM (DRAM)*, is slower than SRAM and must be periodically refreshed with electricity or the information it holds is lost. DRAM is more economical than SRAM and serves as the main memory element in most computers.

3. Input/Output

The input/output unit allows the microprocessor to communicate with the outside world, either to receive or to send data. Most of the time, the input/output unit will also act as an interface for the microprocessor, *i.e.*, to convert the data into a suitable format for the microprocessor. *Input devices* are devices that input data or send data to the computer. Input devices are such as keyboards, punched card readers, sensors, switches, etc. *Output devices* are devices that output data or perform various operations under the control of the CPU. Output devices are LEDs, display unit, speaker, CRT, printer, etc.

4. Buses

The interconnections are known as buses because they contain a large number of parallel connecting wires. The buses are of three types:

- Data bus
- Address bus
- Control bus

The data bus can send data to memory or receive data from memory. The address bus carries memory addresses. The control bus is used to make sure everything works in the correct sequence by sending and receiving timing signals.

Classification of Microprocessors

The different types of microprocessors used most frequently are as follows:

Microprocessors

A microprocessor is a single integrated circuit (IC) which comprises all of the functions of CPU. Before microprocessors, computers would require multiple circuit boards with many ICs. Early Intel microprocessors were the 4000 series (the **4004** was developed in 1972 by Busicom of Japan) which had a four-chip set and the 8000 series (1972-1979) which increased computing power by a factor of 20. The first modern microprocessor was the **80286** in 1982, which came with 16 MB of addressable memory and 1 GB of virtual memory, this 16-bit chip is referred to as the first "modern" microprocessor.

Modern personal computers use multi-core microprocessors which while still a single IC but with multiple processing units operating independently but interconnectedly.

Applications of Microprocessors

Microprocessors are utilized in computer systems ranging from personal computers, to smartphones and tablets to supercomputer-class workstations. Programs ranges from simple word processing, email, Internet browsing, spreadsheets, animation, graphics, and database processing. Owing to their low cost and flexibility, microprocessors appear in many everyday household appliances. All modern cars incorporate microprocessor controlled ignition and emission systems to improve engine operation, increasing fuel economy while reducing pollution.

MICROCONTROLLER

Microcontroller (or *MCU*) is a computer on a chip. Micro suggests that the device is small, and controller tells that the device might be used to control objects, processes, or events. Another term to describe a microcontroller is embedded controller, because the microcontroller and its support circuits are often built into, or embedded in, the devices they control. It is a type of microprocessor emphasizing self-sufficiency and cost-effectiveness, in contrast to a microprocessor. In addition to all arithmetic and logic elements of a microprocessor, the microcontroller usually also integrates additional elements such as read-only and read-write memory, and input/output interfaces.

Microcontrollers are frequently used in automatically controlled products and devices, such as automobile engine control systems, office machines, appliances, power tools, and toys. By reducing the size, cost, and power consumption compared to a design using a separate microprocessor, memory, and input/output devices, microcontrollers make it economical to electronically control many more processes.

Microcontroller differs from a microprocessor in many ways. First and the most important is its functionality. In order for a microprocessor to be used, other components such as memory, or components for receiving and sending data must be added to it. On the other hand, microcontroller is designed to be all of that in one. No other external components are needed for its application because all necessary peripherals are already built into it. Thus, the time and space needed to construct devices is saved. A microcontroller is a specialized form of microprocessor that is designed to be self-sufficient and cost-effective, whereas a microprocessor is typically designed to be general purpose (the kind used in a PC).

Features of a Microcontroller

- Microcontrollers are embedded inside some other device (often a consumer product) so that they can control the features or actions of the product. Microcontroller are also called embedded controller.
- Microcontrollers are dedicated to one task and run one specific program. The program is stored in ROM (read-only memory) and generally does not change.
- Microcontrollers are often low-power devices.
- A microcontroller has a dedicated input device and often (but not always) has a small LED or LCD display for output. A microcontroller also takes input from the device it is controlling and controls the device by sending signals to different components in the device.

Applications of Microcontroller

Microcontrollers are found in all kinds of things these days. Any device that measures, stores, controls, calculates, or displays information is a candidate for putting a microcontroller inside. Some of the common applications are:

- **In automobiles:** Just about every car manufactured today includes at least one microcontroller for engine control, and often more to control additional systems in the car.

- **In desktop computers:** Microcontrollers are found inside keyboards, modems, printers, and other peripherals.

- **In test equipment:** Microcontrollers make it easy to add features such as the ability to store measurements, to create and store user routines, and to display messages and waveforms.

PROGRAMMABLE LOGIC CONTROLLER (PLC)

In digital electronic systems, there are three basic kinds of devices: *memory, microprocessors,* and *logic. Memory* devices store random information such as the contents of a spreadsheet or database. *Microprocessors* execute software instructions to perform a wide variety of tasks such as running a word processing program or video game. *Logic devices* provide specific functions, including device-to-device interfacing, data communication, signal processing, data display, timing and control operations, and almost every other function a system must perform.

The advent of the PLC began in the 1970s, and has become the most common choice for manufacturing controls. A programmable logic controller, also called a *PLC or programmable controller,* is a computer-type device used to control equipment in an industrial facility. The kinds of equipment that PLCs can control include conveyor systems, food processing machinery, auto assembly lines etc. PLCs are often defined as miniature industrial computers that contain hardware and software used to perform control functions. Unlike general-purpose computers, the PLC is designed for multiple inputs and output arrangements, extended temperature ranges, immunity to electrical noise, and resistance to vibration and impact.

National Electrical Manufacturers Association has defined PLC as "a digitally operating electronic apparatus which uses a programmable memory for the internal storage of instructions for implementing specific functions such as logic, sequencing, timing, counting and arithmetic to control, through digital or analog input/output modules, various types of machines or processes."

In a traditional industrial control system, all control devices are wired directly to each other according to how the system is supposed to operate. In a PLC system, however, the PLC replaces the wiring between the devices.

This is shown in *figure 11.13*. Thus, instead of being wired directly to each other, all equipment is wired to the PLC. Then, the control program inside the PLC provides the "wiring" connection between the devices. The *control program* is the computer program stored in the PLC's memory that tells the PLC what's supposed to be going on in the system. The use of a PLC to provide the wiring connections between system devices is called *soft wiring*.

For example, lets assume that a push button is supposed to control the operation of a motor. In a traditional control system, the push button would be wired directly to the motor. In a PLC system, however, both the push button and the motor would be wired to the PLC instead. Then, the PLC's control program would complete the electrical circuit between the two, allowing the button to control the motor.

Components of PLC

A PLC consists of two basic sections: the central processing unit (CPU) and the input/output interface system. The CPU, which controls all PLC activity, can further be broken down into the processor and memory system. The input/output system is physically connected to field devices (*e.g.*, switches, sensors, etc.) and provides the interface between the CPU and the information providers (inputs) and controllable devices (outputs). To operate, the CPU "reads" input data from connected field devices through the use of its input interfaces, and then "executes" or performs the control program that has been stored in its memory system. Programs are typically created in ladder logic, a language that closely resembles a relay-based wiring schematic, and are entered into the CPU's memory prior to operation. Finally, based on the program, the PLC "writes" or updates output devices via the output interfaces. This process, also known as scanning, typically continues in the same sequence without interruption, and changes only when a change is made to the control program.

The schematic diagram of PLC is shown in Figure 11.14. The basic components of PLC are:

- Input module
- Output module
- Processor
- Memory
- Power supply
- Programming device

1. The Input/Output Module

The *input/output (I/O)* module is the connections to the industrial processes that are to be controlled. If the CPU can be thought of as the brain of a PLC, then the I/O system can be thought of as the arms and legs. The I/O system is what actually physically carries out the control commands from the program stored in the PLC's memory. The I/O system consists of two main parts:

- The rack
- I/O modules

The *rack* is an enclosure with slots in it that is connected to the CPU.

Input/Output units are the interfaces between the internal PLC systems and the external processes to be monitored and controlled. *I/O modules* are devices with connection terminals to which the field devices are wired.

Together, the rack and the I/O modules form the interface between the field devices and the PLC. When set up properly, each I/O module is both securely wired to its corresponding field devices and securely installed in a slot in the rack. This creates the physical connection between the field equipment and the PLC.

(In a traditional system, all control devices are wired directly to each other)

(In a PLC system, all control devices are wired to the PLC)

FIGURE 11.13

All of the field devices connected to a PLC can be classified in one of two categories:

▪ Inputs

▪ Outputs

Inputs are devices that supply a signal/data to a PLC. Typical examples of inputs are push buttons, limit sensors, switches, and measurement devices.

Outputs are devices that await a signal/data from the PLC to perform their control functions.

Lights, horns, motors, and valves are all good examples of output devices.

For example, a bulb and its corresponding wall switch are good examples of everyday inputs and outputs. The wall switch is an input, which provides a signal for the light to turn on. The bulb is an output, which waits until the switch sends a signal before it turns on. Let's assume a bulb/switch circuit that contains a PLC. In this situation, both the switch and the bulb will be wired to the PLC instead of to each other. Thus, when the switch is turn on, the switch will send its "turn on" signal to the PLC instead of to the bulb. The PLC will then relay this signal to the bulb, which will then turn on.

2. *The Central Processing Unit*

The *central processing unit (CPU)* is the part of a programmable controller that retrieves, decodes, stores, and processes information. It also executes the control program stored in the PLC's memory. In essence, the CPU is the "brains" of a programmable controller. It functions much the

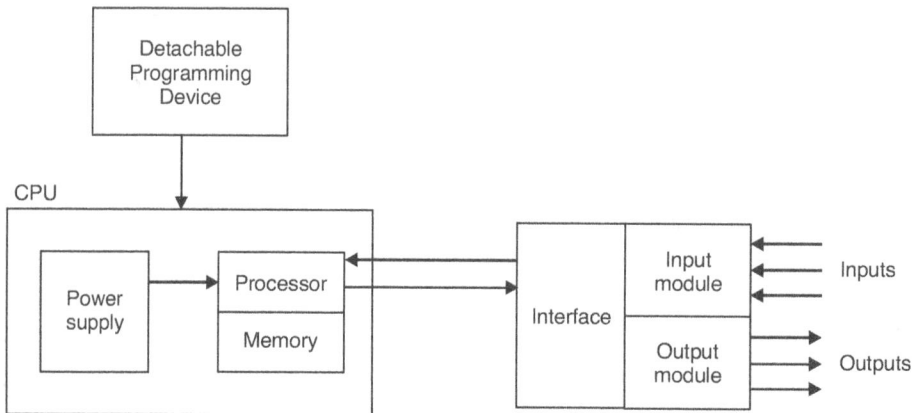

FIGURE 11.14 Block Diagram showing Components of Programmable Logic Controller.

same way the CPU of a regular computer does, except that it uses special instructions and coding to perform its functions. The CPU has three parts:

- The processor
- The memory system
- The power supply

The *processor* is the section of the CPU that codes, decodes, and computes data.

The *memory system* is the section of the CPU that stores both the control program and data from the equipment connected to the PLC. Memory in a PLC system is divided into the program memory, which is usually stored in EPROM/ROM, and the operating memory. The RAM memory is necessary for the operation of the program and the temporary storage of input and output data.

The *power supply* is the section that provides the PLC with the voltage and current it needs to operate.

3. Programming Device

The PLC is programmed by means of a programming device. The programming device is usually detachable from the PLC cabinet so that it can be shared between different controllers.

Working of PLC

A PLC works by continually scanning a program. It consists of three steps as shown in Figure 11.15.

- Check input status
- Execute program
- Update output status

1. Check Input Status

First the PLC takes look at each input to determine if it is on or off. In other words, is the sensor connected to the first input on? How about the second input? How about the third. It records this data into it memory to be used during next step.

2. Execute Program

Next, the PLC executes program, *i.e.*, one instruction at a time. Maybe the program says that if first the input is on then it should turn on the first output. Because it already knows which inputs are on/off from the previous step, it will be able to decide whether the first output should be turned on based on the state of the first input. It will store the execution results to use later during the next step.

3. Update Output Status

Finally the PLC updates the status of the outputs. It updates the outputs based on which inputs were on during the first step and the results of executing your program during the second step. Based on the example in the step 2, it would now turn on the first output.

PLC Programming

PLC programming is done with the help of special programming languages. The function of all programming languages is to allow the user to communicate with the programmable controller (PC) via a programming device. They all convey to the system, by means of instructions, a basic control plan.

Ladder diagrams, function blocks, and the sequential function chart are the most common types of languages encountered in programmable controller system design. Ladder diagrams form the basic PC languages, while function blocks and the sequential function charting are categorized as high-level languages. The basic programmable controller languages consist of a set of instructions that will perform the most common type of control functions like relay replacement, timing, counting, sequencing, and logic.

Ladder Logic

Ladder logic is the main programming method used for PLCs. The ladder logic diagram has been found to be very convenient for shop personnel who are familiar with circuit diagrams because it does not require them to learn an entirely new programming language. Modern PLCs can be programmed in ladder logic or in more traditional programming languages such as C.

The ladder diagram language is a symbolic instruction set that is used to create a programmable controller program. The aim of ladder diagram program is to control outputs based on input conditions. Ladder rung is used for the control. Control rung, in general, consists of a set of input conditions represented by relay contact-type instructions and an output instruction at the end of the rung represented by the coil symbol.

An example of ladder logic can be seen in Figure 11.16. To interpret this diagram, imagine that the power is on the vertical line on the left=hand side, we call this the hot rail. On the right hand side is the neutral rail. In

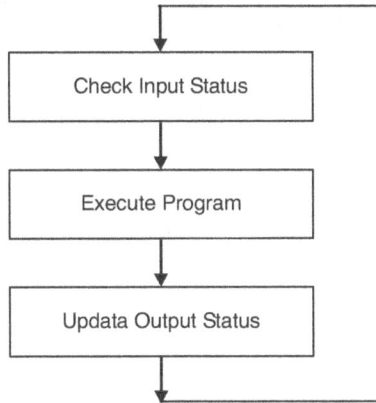

FIGURE 11.15

the figure, there are two rungs, and on each rung there are combinations of inputs (two vertical lines) and outputs (circles). If the inputs are opened or closed in the right combination the power can flow from the hot rail, through the inputs, to power the outputs, and finally to the neutral rail. An input can come from a sensor, switch, or any other type of sensor. An output will be some device outside the PLC that is switched on or off, such as lights or motors. In the top rung, the contacts are normally open and normally closed. Which means if input A is on and input B is off, then power will flow through the output and activate it. Any other combination of input values will result in the output X being off.

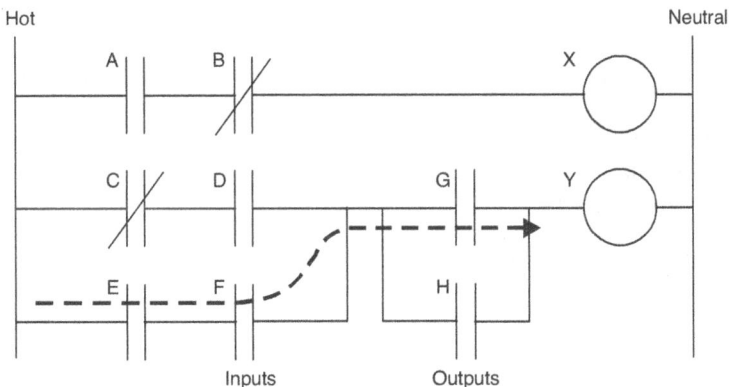

Note: Power needs to flow through some combination of the inputs (A, B, C, D, E, F, G, H) to turn on outputs (X, Y).

FIGURE 11.16 A Simple Ladder Logic Diagram.

Advantages and Disadvantages of PLC

Advantages

There are significant advantages in using a programmable logic controller rather than conventional relays, timers, counters, and other hardware elements. These advantages include:

- Programming the PLC is easier than wiring the relay control panel.
- High reliability.
- The PLC can be reprogrammed. Conventional controls must be rewired and are often scrapped instead.
- PLCs take less floor space then relay control panels.
- Little maintenance due to no moving parts.
- No special programming skills required by maintenance personnel.
- Computing capabilities.
- Reduced costs.
- Ability to withstand harsh environments.
- Expandability.
- High speed of operation.

Disadvantages

Although the PLC systems have many advantages, there are also disadvantages. These include:

- Fault finding, as PLC systems are often much more complex than the hardwired relay systems.
- Failure of the PLC may completely stop the controlled process, whereas a fault in a conventional control system would only disrupt the process.
- External electrical interference may disrupt the PLC memory.

Applications of PLC

PLCs have now become a very convenient tool for flexible automation. Applications of PLC include:

- Control of electrical motors in industrial drives.
- CNC machines.
- Robot control.
- Home and medical equipment.
- Operation of lifts in buildings.

▪ Control of traffic signals.

▪ Safety control of presses.

EXERCISES

1. Write short notes on "transducer types."

2. Distinguish between a transducer and a sensor.

3. Discuss the various types of sensors used for position or displacement measurement.

4. Name any four advantages of PLC over conventional control systems.

5. Describe the construction and working principle of a microprocessor with the help of a schematic diagram.

6. Identify the various types of integrating devices used to integrate mechanical systems with computer systems. Describe any two of such devices in detail giving their applications.

7. Identify the desirable features of a good transducer. Describe the construction and working principle of LVDT with the help of a sketch.

8. Discuss the construction of PLC and microprocessors and describe their use in industrial applications with the help of suitable examples.

9. Define a programmable logic controller (PLC).

10. What is a sensor? Identify desirable features of transducers.

11. By means of a sketch, describe the working of any piezoelectric sensor.

12. What are microprocessors?

13. Identify the various components of a PLC. Describe a PLC with the help of a schematic diagram.

14. Where are PLC employed?

15. Name any three parts required for integration of mechanical system with electrical system.

16. Discuss the advantages and limitations of microprocessor based controllers.

17. Mention two significant differences between a microprocessor and a programmable logic controller.

18. What are the factors to be considered in selection of a transducer?

19. What are the main functions of the A register in a microprocessor?

20. What are ports in a microprocessor system? Explain the difference between accessing ports and memory.

21. Write a short paragraph on thermocouple.

22. What are the main registers and their functions in a microprocessor?

23. What are programmable logic controllers? Discuss the applications for which these are used. Discuss three significant advantages and disadvantages.

24. What is a microcontroller?

25. Explain the architecture of a programmable logic controller with the help of a sketch.

26. Distinguish between LVDT and RVDT.

27. What are the application of microprocessors and PLC?

28. Why were ladder diagrams used for programming PLC systems?

29. How can transducers be classified?

30. List the various electric and electronic control elements used in automation.

31. Discuss the advantages and limitations of microprocessor-based controllers.

32. Explain the constructional features of a PLC.

33. List the various electric and electronic control elements used in automation.

34. Explain the constructional features of a micro controller.

CHAPTER 12

TRANSFER DEVICES AND FEEDERS

INTRODUCTION

Since the beginning of the nineteenth century, the increasing need for finished goods in large quantities has led engineers to search for and to develop new methods of manufacturing or production. As a result of development in the various manufacturing processes, it is now possible to mass-produce high-quality durable goods at low cost. One of the most important manufacturing processes is the assembly process that is required when two or more components parts are to be secured together. The history of assembly process development is closely related to the history of the development of mass-production methods. The pioneers of mass production are also the pioneers of modern assembly techniques. Their ideas and concepts have brought significant improvement in the assembly methods employed in high volume production.

However, many aspects of manufacturing engineering, especially the parts fabrication process, have been revolutionized by the application of automation, the technology of the basic assembly process has failed to keep pace. Although, during the last few decades, efforts have been made to reduce assembly costs by the application of high speed automation and more recently by the use of assembly robots, success has been quite limited and many assembly workers are still using the same basic tools as those employed at the time of the industrial revolution. So, in these days, it is necessary that manufacturing engineers and designers must learn about automatic assembly. This in turn provides means to improve design, productivity, and competitiveness.

FUNDAMENTALS OF PRODUCTION LINES

A production line consists of a series of workstations arranged so that the product moves from one station to the next, and at each location, a portion of the total work is performed on it. Production lines are associated with mass production. If quantities of the product are very high and the work can be divided into separate tasks that can be assigned to individual workstations, then a production line is the most appropriate manufacturing system. In terms of the capacity of a production line to cope with model variations, three types of line can be distinguished:

Single-model line produces only one model and there is no variation in the model. The tasks performed at each station are the same on all product units.

Batch-model line produces each model in batches. The workstations are set up to produce the desired quantity of the first model, and then the stations are reconfigured to produce the desired quantity of the next model, and so on.

Mixed-model line also produces multiple models; however, the models are intermixed on the same line rather than being produced in batches. While a particular model is being worked on at one station, a different model is being processed at the next station.

TYPES OF ASSEMBLY LINES

There are two types of assembly lines:

- Manual assembly lines
- Automated assembly lines

Manual Assembly Lines

A manual assembly line consists of multiple workstations arranged sequentially, at which assembly operations are performed by human workers. Because all of the operations are in the control of the worker, so the tools required are simple and less expensive than those used in automated assembly. Manual assembly systems are best suited for low volume products that have high product variety. Processes accomplished on manual assembly lines include mechanical fastening operations, spot welding, hand soldering, and adhesive joining.

Automated Assembly Lines

The term automated assembly refers to the use of mechanized and automated devices to perform the various assembly tasks in an assembly line. Most automated assembly systems are designed to perform a fixed sequence of assembly steps on a specific product. An automated assembly line consists of automated workstations connected by a parts transfers system whose actuation is coordinated with the stations. In an ideal line, no human workers are on the line, except to perform auxiliary functions such as tool changing, loading and unloading parts, and repair and maintenance activities. Modern automated lines are integrated systems operating under computer control. Automated production lines are applied in processing operations as well as assembly.

An automated assembly system performs a sequence of automated assembly operations to combine multiple components into a single entity. The single entity can be a final product or a subassembly in a larger product. In many cases, the assembled entity consists of a base part to which other components are attached. The components are joined one at a time (usually), so the assembly is completed progressively. A typical automated assembly system consists of the following subsystems:

- One or more workstations at which the assembly steps are accomplished.
- Parts feeding devices that deliver the individual components to the workstations.
- A work handling system for the assembled entity.

REASONS FOR USING AUTOMATED ASSEMBLY LINES

Automated assembly technology should be considered when the following conditions exist:

- *High product demand*: Automated assembly systems should be considered for products made in large quantity (in millions).

- *Stable product design*: In general, any change in the product design means a change in workstation tooling and possibly the sequence of assembly operations. Such changes can be very costly.

- The assembly consists of no more than a limited number of components.

  Automated production lines can be divided into two basics categories:

  (a) *Transfer lines*, which consist of a sequence of workstations that performs processing operations, with the automatic transfer of work units between stations. Transfer lines are usually expensive pieces of equipment; they are designed for a job requiring high quantities of parts.

  (b) *Dial indexing machine,* is a device used to convey parts for assembly, machining, packaging, finishing, or other manufacturing operations. In a dial indexing machine, the workstations are arranged around a circular worktable called a dial. The worktable is actuated by a mechanism that provides partial rotation of the table on each work cycle.

TRANSFER SYSTEMS IN ASSEMBLY LINES

The four main types of transfer systems are:

- Continuous transfer system
- Intermittent/synchronous transfer system
- Asynchronous transfer system
- Stationary transfer system

In *continuous transfer systems*, the work carriers are moving at a constant speed while the work head index moves back and forth. Assembly operations are carried out during the period when the work heads are moving forward, keeping pace with the work carriers. Examples of continuous transfer systems can be seen in bottling operations, manual assembly where the worker can move with the moving line.

In *intermittent transfer*, the work carriers are transferred intermittently and the work heads remain stationary. All the parts are moved at same time, hence, the term *synchronous transfer system*, which is also used to describe this method of workpiece transport. For example, intermittent transfer systems are used in machining operations, press-working operations, etc. This system is stressful to human workers, but good in automated operations.

The *asynchronous transfer system* allows each work part to advance to the next station when processing at the current station has been completed. Each part moves independent of the other parts, increasing flexibility. This type of system is good for both manual and automated operations.

In the *stationary system*, the part is placed in a fixed location where it remains during the entire assembly process. This system is used when the assembled product is bulky or difficult to handle, *e.g.*, airplanes, ships, etc.

AUTOMATIC MACHINES

Automatic machines are those machines in which both the workpiece handling and the metal cutting operations are performed automatically. These machines have played an important role in increasing the production rate and have been in use for long time. In automatic machines, operations from feeding of stock to clamping, machining, and even inspection of the workpiece are carried out automatically.

TRANSFER DEVICES/MACHINES

A transfer machine is an automatic machine capable of performing many operations. It consists of many machine tools properly linked together. These are special purpose machines where the components are automatically transferred from one machining head to another. Operations are performed sequentially. Each machining head carries out one operation until the component reaches the end of the line and all the necessary operations have been performed. Transfer of the workpiece and its fixture from station to station is done automatically. Each station could be considered as simple work head with its own motor mounted on a base. It could also be defined as a combined material processing and material-handling machine.

Transfer machines perform a variety of machining, inspecting, and quality control functions. They drill, mill, grind, as well as, control and inspect the operations. Transfer machines range from comparatively small units having only two to three workstations to long straight-line machines with more than 100 workstations. These machines are used primarily in automobile industry. A wide variety of transfer machines with different models, designs, and sizes are available in the market with salient features like automatic reset timer, sliding headstock, increased platen clamping speed, pressure adjustment system, and many more. Some of these machines are flexible, precise, fast, and some are used for high production jobs. They are used in many industrial sectors. There are different types of transfer machines available in the market like the CNC rotary transfer machine, economy precision transfer machine, rotary transfer machines for brass goods, automatic transfer line machine, transfer machines for production of valve bodies, six-station dial rotary transfer machines, and many more suiting different industrial purposes.

SELECTION OF TRANSFER DEVICES

Selection of a particular transfer device is based upon the following factors:

- Accuracy required in components.
- Various forces acting at stations on the components.
- Physical size of parts and production rate.
- Number of operations to be performed.
- Type of drive required, *i.e.*, pneumatic, hydraulic, electric, or a combination of these.

TRANSFER MECHANISM IN TRANSFER DEVICES

Transfer mechanisms are used to move the workpiece from one station to another in the machine or from one machine to another to enable various operations to be performed on the part. Sensors and other devices usually control transfers of parts from station to station. Tools on transfer machines can be changed easily using tool holders with quick-change features. These machines may be equipped with various automatic gauging and inspection systems. These systems are utilized between operations to ensure that the dimensions of a part produced in one station are within acceptable tolerances before that part is transferred to the next station. The two transfer mechanisms used the are:

- Linear transfer mechanism
- Rotary transfer mechanism

The goal is the same for both—put a blank, bar, casting, or forging in the first station and get a completely machined part at the other end.

LINEAR TRANSFER MECHANISM

Linear transfer systems provide a linear motion for workpart transfer in automated production systems. Linear transfer mechanisms are used for inline machines. Some of the commonly used linear transfer mechanisms are:

Walking Beam System

In this type of linear transfer mechanism (refer to Figure 12.1), the workpants are lifted up from their workstation locations by a transfer bar and moved one position ahead to the next station. The transfer bar then lowers the parts into nests that position them for processing at their stations. The beam then retracts to make ready for the next transfer cycle. A walking beam transfer system is used in transferring cylindrical parts, motor shafts, camshafts, crankshafts, tubing, and piping of all types. The advantages of walking beam machines include a lower machine chassis cost and significant reductions in fixturing costs. This system lends itself well to very high-speed machines.

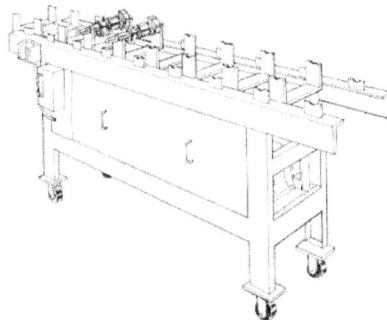

FIGURE 12.1 Walking Beam System.

Powered Roller Conveyor System

This type of system (refer to Figure 12.2) is used in automated flow lines. The conveyor can be used to move parts or pallets. Roller conveyors are flexible, robust, and highly efficient.

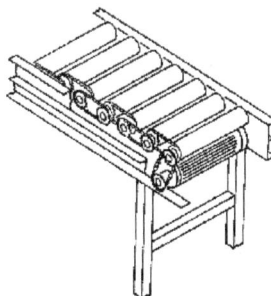

FIGURE 12.2 Powered Roller Conveyor System.

Chain Drive Conveyor System

A chain or steel belt is used to transport the work carriers. The chain drive conveyor system shown in Figure 12.3 can be used for continuous, intermittent, or non-synchronous movement of work parts.

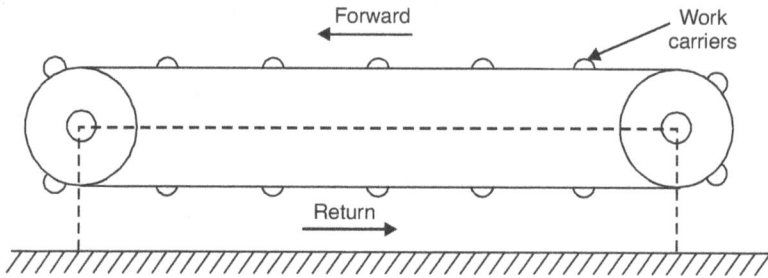

FIGURE 12.3 Chain Drive Conveyor System.

ROTARY TRANSFER MECHANISM

In rotary transfer system, the workpieces are held in fixtures on a continuous rotating table. There are various methods used to index a circular table or dial at various equal angular positions corresponding to workstation locations. Some of the methods are described below:

Rack and Pinion

The rack and pinion mechanism is not suited for high-speed operation often associated with indexing machines. The device uses a piston to drive the rack, which causes the pinion gear and attached indexing table to rotate.

Ratchet and Pawl

A ratchet mechanism is based on a wheel that has teeth cut out of it and a pawl that follows as the wheel turns. Looking at Figure 12.4, as the ratchet wheel turns, the pawl falls into the dip between the teeth. The ratchet wheel can only turn in one direction (in this case anticlockwise). Its operation is simple but somewhat reliable, owing to wear and sticking of several of the components.

Geneva Mechanism

The Geneva-type mechanism has more general application in assembly machines but its cost is higher than the mechanisms described earlier. A

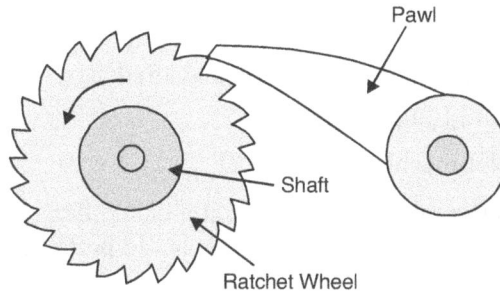

FIGURE 12.4 Ratchet and Pawl System.

Geneva mechanism as shown in Figure 12.5 is used to provide an intermittent rotational motion of the driven part while the driver wheel rotates continuously. If the driven member has eight slots for an eight-station dial indexing machine, each turn of the driver will cause the table to advance one-eighth of a turn. The driver only causes movement of the table through a portion of its rotation.

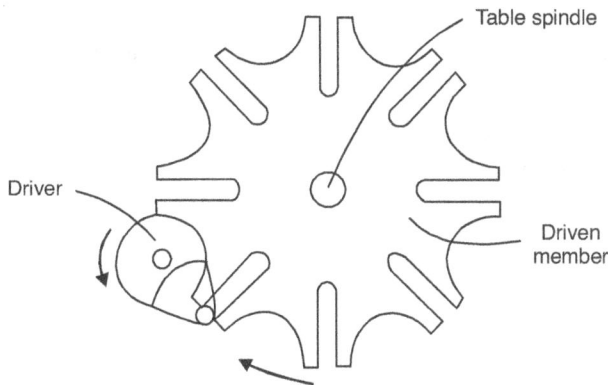

FIGURE 12.5 Geneva Mechanism.

Because the driven wheel in a Geneva motion is always under full control of the driver, there is no problem with overrunning. Impact is still a problem unless the slots of the driven wheel are accurately made and the driving pin enters these slots at the proper angle. The main characteristic of the Geneva mechanism is its restriction on the number of stops per revolution. The smaller the number of stops, the greater the adverse mechanical advantage between the driver and the driven members. This results in a high indexing velocity at the center of the indexing movement.

CLASSIFICATION OF TRANSFER DEVICES

Transfer devices/machines can be classified as:

- Inline transfer machines
- Rotary transfer machines

The goal is the same for all, *i.e.*, put a blank, bar, casting, or forging in the first station and get a completely machined part at the other end.

Inline Transfer Machines

The inline assembly machine consists of a series of automatic workstations located along an inline transfer system (refer to Figure 12.6). Inline transfer machines have a load station at one end and an unload station at the other end. The parts feeders, workstations, and inspection stations are arranged along the workflow. The components are transferred automatically from one machining station to another either by pulling supporting rails by means of an endless chain conveyor or by pushing along the continuous rails by air or hydraulic pressure. The components are loaded manually or automatically onto the central bed. All types of machining operations are carried out at various stations and the chips produced are removed so that these do not foul the working parts.

These machines are generally suitable for operations involving large workpieces and for those in which large workstations are required. Inline

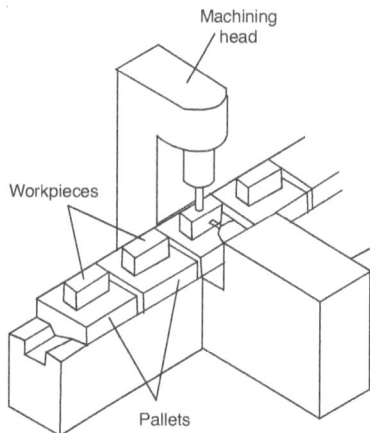

FIGURE 12.6 Inline Transfer Machine.

machines have the advantages of an unlimited number of workstations and efficient operator loading. In comparison with rotary transfer systems, the manufacturing costs of linear transfer systems are generally 10 to 20% higher according to type. Space requirements are greater as compared to the rotary transfer system.

Inline transfer machines can be broadly classified as:

- Pallet type transfer machines
- Plain type transfer machines

In pallet type transfer machines, the workpieces are transferred from stations by holding them in fixtures called pallets. In plain type transfer machines, the parts move in an unclamped position. Pallet type transfer machines are used for producing components requiring a high degree of accuracy.

Rotary Transfer Machines

In rotary transfer machines, the workpieces are located and clamped in pallet type fixtures that are indexed in a circular path as shown in Figure 12.7. The table rotates about a vertical axis and its movement could be continuous or intermittent. Rotary transfer machines automatically feed multiple workstations from a rotating turret. This combines an automated part feed with multiple simultaneous operations, streamlining the machining process significantly. The rotary transfer technology indexes a workpiece from station to station via a rotary table, with operations performed at each station. The number of stations allow for easier balance between long and short operation cycles. Because the tools rather than the workpiece are rotating, it is possible to insert machine stock of virtually any shape. This type of machining method permits the workpiece to be loaded and unloaded at a single location without interrupting the machining.

Machining operations are typically hole-making operations (drilling, cross drilling, tapping, boring, counter boring, etc.) but can also include milling, turning, cutoff, broaching, crimping, threading, tapping, broachving, and other secondary machining and assembly operations. Although the rotary transfer machine is best suited for multi-million part runs, its flexibility also makes it effective for family of parts production. These machines are used when few machining stations need to be employed. Rotary arrangement saves floor space and presents a more compact arrangement.

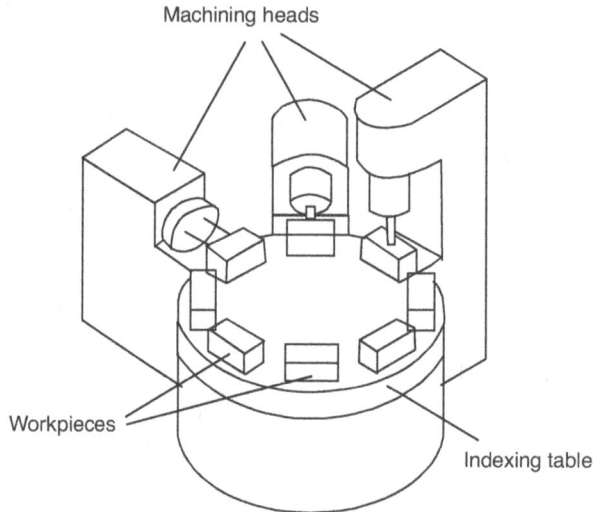

FIGURE 12.7 Rotary Transfer Machine.

Rotary Index Table

Rotary indexing tables are used to index/transfer parts and components in defined, angular increments so that they can be machined, worked, or assembled in multiple operations. Tables consist of a circular steel plate, one or more spindles, a drive system, and pins that hold parts and components in place. A rotary indexing table as shown in Figure 12.8 has either fixed or adjustable indexing angles. During each revolution, the table stops for a specified period of time so that an operation can be performed at each station. Rotary indexing tables are powered by pneumatic and electric motors, hydraulic drives, and manual actuation. Drive mechanisms can be located above, below, behind, or to the side of the table surface. Pneumatic rotary indexing tables are suitable for small and medium loads. Electrically powered tables are generally faster than pneumatic devices and can handle heavier loads. Tables that are powered by hydraulic drives use a pressurized fluid that transfers rotational kinetic energy. Manually actuated rotary indexing tables often include a hand crank or are loosened, turned, and adjusted by hand.

Besides boosting output, indexing tables improve safety. Hazardous operations can be placed on the opposite side of the table from the operator. A light curtain or safety glass installed above the table, along its diameter, will ensure that the hazardous operation and the operator remain separated. Applications for rotary indexing tables include assembly and equipment positioning as well as various automation, inspection, and machining applications.

FIGURE 12.8 Rotary Indexing Table.

ADVANTAGES AND DISADVANTAGES OF TRANSFER MACHINES

Advantages

Transfer machines offer a number of advantages:

- Complex shaped components can be easily machined.
- Optimum utilization of floor space with these machines.
- Increased productivity.
- Components produced are of high quality, accurate, and cheap.
- Elimination of worker's fatigue.
- Loading and unloading time is eliminated.

Disadvantages

- High initial investment.
- These machines are limited to high production industries.
- A breakdown of one machine means stoppage of whole production line.
- Overhauling and maintenance costs are high.

CONVEYOR SYSTEMS USED IN TRANSFER DEVICES

A conveyor system is a common piece of material handling equipment that moves materials from one location to another. Conveyors work by two methods: through manual operation or through a power source. Conveyors are especially useful in applications involving the transportation of heavy or bulky materials. Conveyor systems allow quick and efficient transportation for a wide variety of materials, which makes them very popular in the material handling and packaging industries. Many kinds of conveying systems are available, and are used according to the various needs of different industries. By laying the conveyors along the line of production machines, the integration of material movement with the production processes is facilitated. The various types of conveyors used are:

- Roller conveyor
- Wheel conveyor
- Chute conveyor
- Belt conveyor
- Chain conveyor
- Magnetic belt conveyor
- Bucket conveyor

(a) *Roller Conveyor*: Roller conveyors form an important means of conveying unit material in the industry. Usually, the load should possess a flat base, or should be mounted on a flat-bottomed pallet/container, so that it can roll on top of straight cylindrical rollers. These may be powered (or live) or non-powered (or gravity) roller conveyors. Gravity conveyors do not require a motor, but use wheels, rollers, and the pull of gravity to move materials along a conveyor. Power conveyors, unlike gravity conveyors, require a pneumatic or an electrical power source. Power is transmitted from the drive system to a drive pulley, which is fastened to the drive shaft. The drive pulley then transmits the power to the conveyor belt, which moves the conveyor bed, upon which the materials rest. Roller conveyors are shown in Figure 12.9.

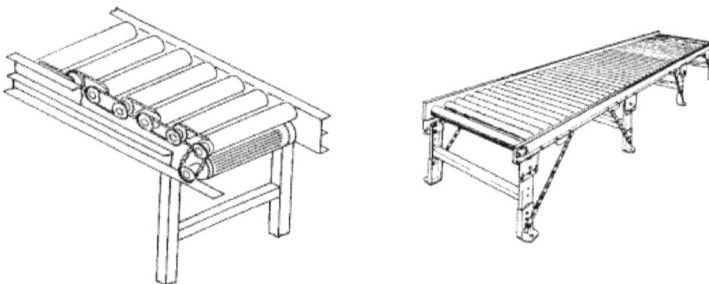

FIGURE 12.9 Roller Conveyor.

(b) *Wheel Conveyor*: The wheel conveyor as shown in Figure 12.10 uses a series of skate wheels mounted on a shaft (or axle), where the spacing of the wheels is dependent on the load being transported. Slope for gravity movement depends on load weight. These are more economical than the roller conveyor.

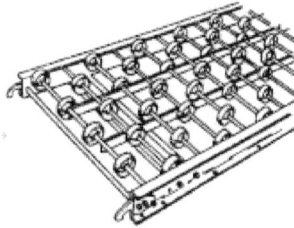

FIGURE 12.10 Wheel Conveyor.

(c) *Chute Conveyor*: An inexpensive chute conveyor is used to link two handling devices. These are used to provide accumulation in shipping areas and to convey items between floors. These are inexpensive but have the limitation of difficulty to control position of the items. (Refer to Figure 12.11.)

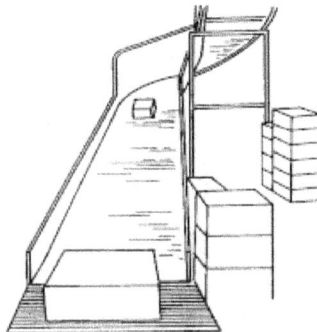

FIGURE 12.11 Chute Conveyor.

(d) *Belt Conveyor*: Belt conveyors as shown in Figure 12.12 are used for the controlled movement of a large variety of both regular and irregular shaped products. They can move light, fragile to heavy, rugged unit loads on a horizontal, inclined, or declined path within the limits of product stability and the conveyor component capacities. The items being conveyed are carried by the top surface of the belt.

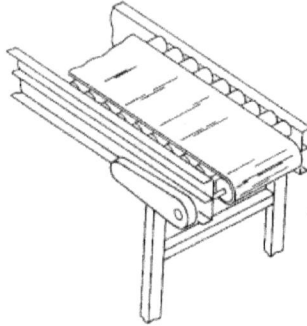

FIGURE 12.12 Belt Conveyor.

(e) *Chain Conveyor*: These conveyors offer high flexibility of layout and conditions of work. (Refer to Figure 12.13.)

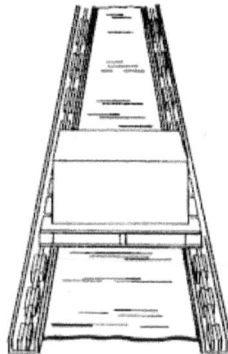

FIGURE 12.13 Chain Conveyor.

(f) *Magnetic Belt Conveyor*: A magnetic belt conveyor is used to transport ferrous materials vertically, upside down, and around corners. (Refer to Figure 12.14.)

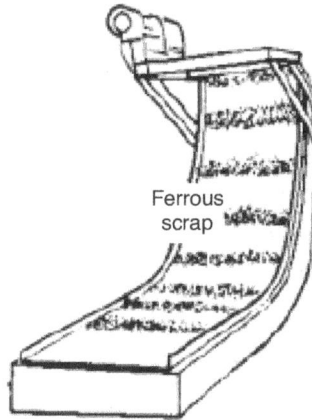

FIGURE 12.14 Magnetic Conveyor.

(g) *Bucket Conveyor*: Bucket conveyors are used to move bulk materials in a vertical or inclined path. Buckets are attached to a cable, chain, or belt. Buckets are automatically unloaded at the end of the conveyor run. (Refer to Figure 12.15.)

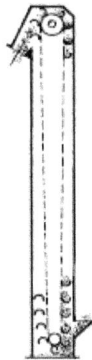

FIGURE 12.15 Bucket Conveyor.

FEEDERS

A feeder is an extremely important element in a bulk material handling system, because it is the means by which the rate of solids flow from a hopper or bin is controlled. When a feeder stops, solids flow should cease. When a feeder is turned on, there should be a close correlation between its speed of operation and the rate of discharge of the bulk solid. Feeders feed material to conveyors, processors, and other equipment at a controlled rate to maximize efficiency and production. In essence, they are short conveyors that come in multiple shapes and sizes. Feeders are also used in recycling applications.

Feeders differ from conveyors in that the latter are only capable of transporting material, not modulating the rate of flow. Dischargers are not feeders. Such devices are sometimes used to encourage material to flow from a bin, but they cannot control the rate at which material flows. This requires a feeder.

CLASSIFICATION OF FEEDERS

There are two basic types of feeders used in industrial plants:

- Volumetric feeders
- Gravimetric feeders

As the name implies, a *volumetric feeder* modulates and controls the volumetric rate of discharge from a bin (*e.g.*, cu.ft./hr.). The four most common types of such feeders are screw, belt, rotary valve, and vibrating pan.

A *gravimetric feeder*, on the other hand, modulates the mass flow rate. This can be done either on a continuous basis (the feeder modulates the mass per unit time of material discharge) or on a batch basis (a certain mass of material is discharged and then the feeder shuts off). The two most common types of gravimetric feeders are loss-in-weight and weight belt. Gravimetric feeders should be used whenever there is a requirement for close control of material discharge. A gravimetric feeder should also be used when the bulk density of the material varies.

CRITERIA FOR FEEDER SELECTION

The feeder used (whether volumetric or gravimetric) should provide the following:

- Reliable and uninterrupted flow of material from some upstream device (typically a bin or hopper).
- The desired degree of control of discharge rate over the necessary range.
- Uniform withdrawal of material through the outlet of the upstream device.

▨ Interface with the upstream device such that load acting on the feeder from the upstream device is minimal. This minimizes the power required to operate the feeder, particle attrition, and abrasive wear of the feeder components.

PARTS FEEDING DEVICES

Parts feeding devices are the devices that deliver the individual components to the workstations. Some of these parts feeding devices are explained as follows:

▨ Hopper

▨ Chute

▨ Magazines

▨ Separator

▨ Parts feeder

▨ Feed track

▨ Escapement and placement device

▨ Selectors and orientor

▨ Ejectors and pushers

▨ Reel

▨ Cradle

▨ Bowls

▨ Pallets

(a) *Hopper*: Hoppers are receptacles for the temporary storage of material. They are designed so that stored material can be dumped easily. This means that the parts are initially randomly oriented in the hopper. Hoppers can be made out of aluminum or steel for heavy-duty use, and plastic for light-duty applications. Hoppers are generally one of two types, bottom or tilt. Bottom hoppers are designed so that stored material can be dumped from the bottom of the hopper. Live bottom hoppers have hydraulically or mechanically driven screws to aid in discharging the material. They are typically used to aid in the discharge of sluggish or viscous materials. Some hoppers have a base that allows them to be lifted by forklifts for dumping.

Important parameters to consider when specifying hoppers are the *weight capacity, volume capacity, height,* and *weight* of the hopper when empty. The weight capacity is the maximum weight the hopper is designed to hold. The volume capacity is the maximum volume of material the hopper is designed to hold. The height of the hopper is necessary to consider for applications where overhead space is limited.

Common features for hoppers include *agitation, perforations, self-dumping, side dumping, stackable*, and *wheels*. A hopper will use agitation to dislodge material while dumping. Hoppers with agitation may also be called vibratory hoppers. Perforations in the hopper allow for the drainage of water or other liquids. Self-dumping hoppers have a mechanism to allow them to self-dump the material. Side dumping hoppers dump their contents from the side rather than the front. Stackable hoppers allow for greater economy of space, most are only to be stacked when empty, not when loaded. Some hoppers will come equipped with wheels for easy moving.

(b) *Chute*: Chutes are smooth surfaced, inclined troughs that allow materials to move down the chute under the force of gravity. Chutes can be made of steel, plastic, or wood. Chutes may be straight or spiral. Chutes are designed to move materials without damage and convey objects from high-to-low levels under the force of gravity.

(c) *Magazines*: Magazines are loading devices adopted to hold a pile of work of various shapes and sizes, for further loading into the machine. Magazines can be subdivided into flat (palettes) and chute magazines. Three different types of magazines are shown in Figure 12.16. Figure 12.16 A shows a flat magazine. Figure 12.16 B shows a chute magazine in zigzag form for the magazining of symmetrical cylindrical parts. Figure 12.16 C shows a chute magazine for plate parts.

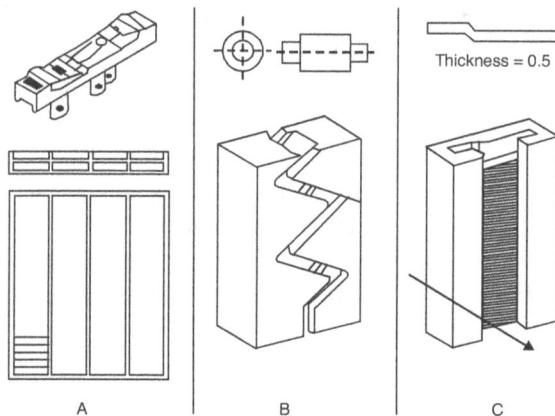

FIGURE 12.16 Types of Magazines.

(d) *Separator*: Many times workpieces are to be fed into the machine at set intervals of time. So workpieces are separated by interrupting the flow of parts to machine. This can be done with the help of separators.

(e) *Parts Feeders*: It is very costly to maintain part order throughout the manufacture cycle. For example, instead of keeping parts in pallets, they are often delivered in bags or boxes, where they must be picked out and sorted. A parts feeder is a machine that orients such parts before they are fed to an assembly station (refer to Figure 12.17). The most common type of parts feeder is the *vibratory bowl feeder*, where parts in a bowl are vibrated using a rotary motion, so that they climb a helical track. As they climb, a sequence of baffles and cut-outs in the track create a mechanical "filter" that causes parts in all but one orientation to fall back into the bowl for another attempt at

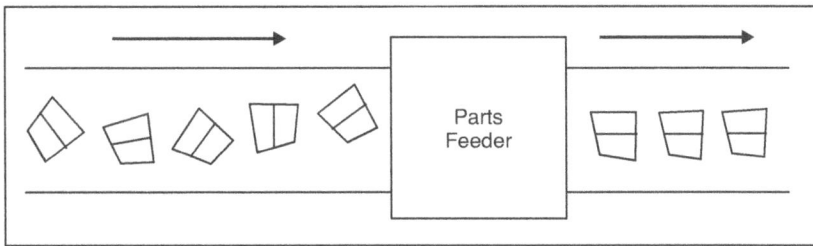

FIGURE 12.17 A Parts Feeder Orients Parts as They Arrive on the Left-hand Conveyor Belt.

running the gauntlet. Other common devices use centrifugal forces, reciprocating forks, or belts to push parts through filters. These devices have the disadvantage that design and setup for new parts requires manual trial and error, which is slow and error-prone. Parts feeders provide a cost-effective alternative to manual labor, saving manufacturers valuable time and labor costs. One operator can over-see a number of automated machines, as opposed to one worker hand loading one machine. Hand selecting and inspecting is also time consuming and labor intensive. The tediousness of the process can subject the workers to repetitive motion injuries. Using vibratory feeders typically results in a better product, as well. When selecting a parts feeder, several factors must be taken into account, including the industry, application, material properties, and product volume.

(f) *Feed Track*: A feed track is used to transfer the components from the hopper and part feeder to the exact location of the assembly workhead, maintaining proper orientation of the parts during the transfer. Feed tracks can be powered or gravity type.

(g) *Escapement and Placement Devices*: The purpose of an escapement device is to remove components from the feed track at timed intervals that are consistent with the cycle time of the assembly workhead. The placement device physically places the component in the correct location at the workstations for the assembly operation by the worker. (Refer to Figure 12.18.)

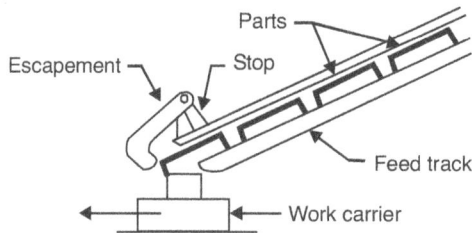

FIGURE 12.18 Escapement Device.

(h) *Selector and Orientor*: The purpose of the selector and/or orientor is to establish the proper orientation of the components for the assembly workhead. A selector is a device that acts as a filter, which allows only those parts to pass through which are in correct orientation. Improperly oriented components are rejected back into the hopper. An orientor is a device that allows properly oriented parts to pass through but provides a reorientation of components that are not properly oriented initially. (Refer to Figure 12.19.)

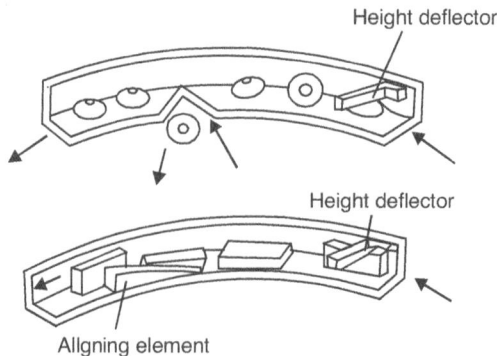

FIGURE 12.19 Selector and Orientor.

TYPES OF FEEDERS

- Apron feeders
- Belt feeders
- Vibratory feeders
- Rotary feeders
- Reciprocating feeders
- Disc feeders
- Screw feeders
- Centrifugal feeders
- Flexible feeders

Some of the commonly used feeders are described below:

APRON FEEDERS

Apron feeders are useful for feeding large tonnages of bulk solids being particularly relevant to heavy, abrasive, ore-type bulk solids and materials requiring feeding at elevated temperatures. They are also able to sustain extreme impact loading. Apron feeders consist of a continuous steel belt that is made up of overlapping flights or pans that are connected to and supported by steel chains or bars. The underside of these flights is reinforced and designed to withstand impact and pressure. An endless conveyor travelling over rollers is created. Some feeders are made of extra-heavy structural steel and are particularly suited to handling coarse, abrasive material. Apron feeders provide a positive material flow and with variable speed drives can provide close control of the feed rate to the crusher. These feeders have the limitation of high cost and limited length. (Refer to Figure 12.20.)

RECIPROCATING FEEDERS (PLATE FEEDERS)

Plate feeders, also known as reciprocating feeders (refer to Figure 12.21), are in widespread use, usually at the tail end of a conveyor or elevator to relieve pressure and drag. Designed for feeding at a fixed rate, a plate is driven reciprocally under a head of bulk material. Size varies and the feed rate can be controlled easily. These feeders use rollers to support the belt and are engineered for maximum uptime. They often are used in wet applications, and products ranging from sand and gravel to crushed stone that pass over belt feeders.

FIGURE 12.20 Apron Feeders.

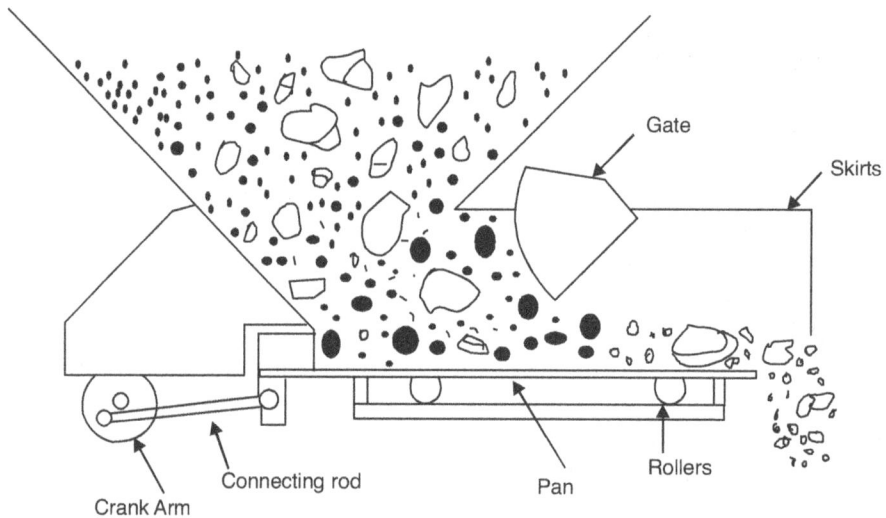

FIGURE 12.21 Reciprocating Feeders.

Advantages

- Low cost.
- Ability to handle a wide range of miscellaneous materials.

Disadvantages

- Not self-cleaning.
- Not recommended for highly abrasive materials.

RECIPROCATING-TUBE HOPPER FEEDER

A reciprocating-tube hopper feeder (refer to Figure 12.22) consists of a conical hopper with a hole in the center through which the delivery tube passes. Relative vertical motion between the hopper and the tube is achieved by reciprocating either the tube or the hopper. During the period when the top of the tube is below the level of parts, some parts will fall into the delivery tube.

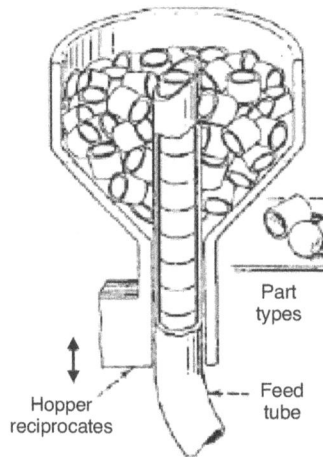

FIGURE 12.22 Reciprocating Tube Hopper Feeder.

RECIPROCATING PLATE FEEDER

A reciprocating plate feeder consists of a plate mounted on four wheels and forms the bottom of the hopper. (Refer to Figure 12.23.) When the plate is moved forward, it carries the material with it; when moved back, the plate is withdrawn from under the material, allowing it to fall into the chute. The plate is moved by connecting rods from cranks or eccentrics. The length and number of strokes, width of plate, and location of the adjustable gate determine the capacity of this feeder.

VIBRATORY BOWL FEEDER

Vibratory parts feeding is a technology used to orient (proper position), singulate (proper quantity), and differentiate (separate/sort) and move parts to a desired location. Vibratory bowl feeders are used for feeding in oriented

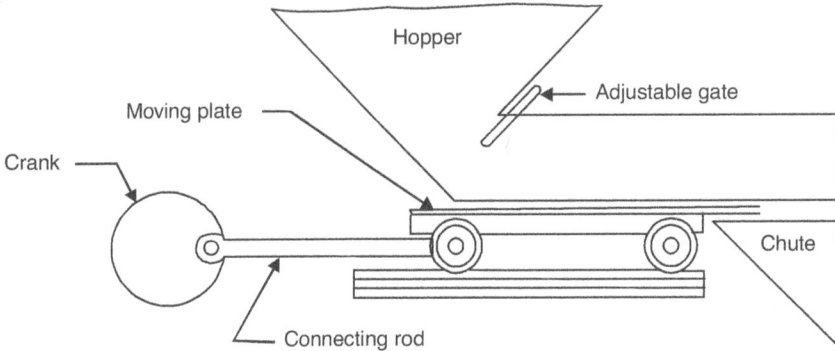

FIGURE 12.23 Reciprocating Plate Feeder.

form, a wide range of components such as steel balls, nuts, belts, washers, rivets, nails, caps, plugs, spoons droppers, rings, and various other components having very odd shapes. The vibratory bowl feeder is the oldest and still most common approach to the automated feeding (orienting) of industrial parts. These feeders are very useful for automatic feeding of components to various machines and on automatic assembly lines such as presses, grinders, threading machines, wadding machines, knurling machines, etc. The size of the feeders varies from 200 mm–1000 mm diameters depending upon size, shape, and weight of components. There are no moving parts in the equipment and thus no wear and tear and needs no maintenance.

Vibratory Feeder Terminology

(a) *Hopper (Storage Hopper):* The storage hopper is the storage area provided to backlog bulk parts prior to entering the feeder bowl. This hopper eliminates overloading or insufficient loads of parts, causing the bowl not to function as required. Feed rate from the hopper to a bowl is metered by a level control switch.

(b) *Basic Bowl:* Basic bowls are not off-the-shelf standard items. They are individually designed and can be supplied for any profile of part up to approximately 5" long.

(c) *Feeder Bowl:* The feeder bowl is the actual orienting and feeding device that orients the part according to machine requirements and is the heart of the system. The feeder bowl is always custom tooled to a specific part configuration. Feeder bowls are almost always round and constructed of stainless steel for long life.

(d) *Rate (or Feed Rate):* Feed rate is the number of parts per minute, or per hour, which meets the production requirements.

(e) *Selector*: An area of the system designed and custom fit to profile only the properly positioned part. Parts entering a selector, which are not in the proper position, are diverted out of the feed line.

(f) *Confinement*: The fabrication installed to ensure 100% control of the part after a selector ejects improperly positioned parts.

(g) *Return Pan*: An extra pan-like area welded to the outside of the bowl, which catches excess and rejected parts falling from the track. The pan guides these parts back into the interior of the bowl for recirculation.

(h) *Discharge of the Feeder Bowl*: This is the last section of the bowl. In most cases, it is a straight exit that confines the parts after they are oriented.

(i) *Track Switch*: A switch of any type, which turns the feeder off whenever, the tracking to the machine becomes full. The track switch also helps to eliminate wear, noise, and jams within the feeder. Air, photoelectric, proximity, and electro-mechanical track switches can be used.

Construction and Working of Vibratory Bowl Feeders

Vibratory bowl feeders are commonly used for aligning and feeding small parts. A typical vibratory bowl feeder consists of a bowl mounted on a base by three or four inclined leaf springs. The springs constrain the bowl so that, as it travels vertically, it also twists about a vertical axis. As parts move up an inclined track along the edge of the bowl, tooling in the bowl orients parts into the proper orientation or rejects misaligned parts into the center of the bowl where they begin their travel up the track again. An electromagnet, mounted between the base and the bowl, generates the force to drive the bowl feeder. The feeder base rests on rubber feet, which serve to isolate the vibration of the feeder from the surrounding environment.

The drive unit, equipped with a variable-amplitude controller, vibrates the bowl, forcing the parts to move up a circular, inclined track. The track is designed to sort and orient the parts in consistent, repeatable positions, according to certain requirements. The length, width, and depth of the bed of the vibratory feeder can be adjusted, and special bed liners can be installed if the material to be handled is abrasive. Dust-proof outlet covers can be attached to the inlet and discharged to reduce dusting caused by dusty material.

A part feeder takes in a stream of identical parts in arbitrary orientations and outputs them in a uniform orientation. It consists of a bowl filled with parts surrounded by a helical metal track. The bowl and track undergo an asymmetric helical vibration that causes parts to move up the track, where they encounter a sequence of mechanical devices such as wiper blades, grooves, and traps. Most of these devices are filters, that serve to reject (force back to the bottom of the bowl) parts in all orientations except for the desired one. Thus, a stream of oriented parts emerges at the top after successfully running the gauntlet.

A picture of a section of the vibratory bowl feeder track is given in Figure 12.24. The parts move from the right to the left on the feeder track. Parts in undesired orientations fall back into the bowl, other orientations remain supported.

Figure 12.25 shows a typical vibratory bowl feeder used to sort small parts for an automated assembly system. Large numbers of unoriented parts are placed into the device and move up the spiral track on the bowl's interior by means of vibratory motion. Parts reach the top of the track in

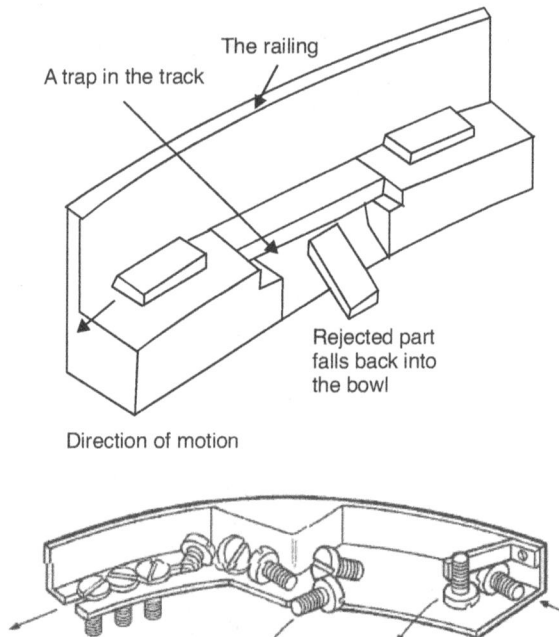

FIGURE 12.24 Vibratory Bowl Feeder Track.

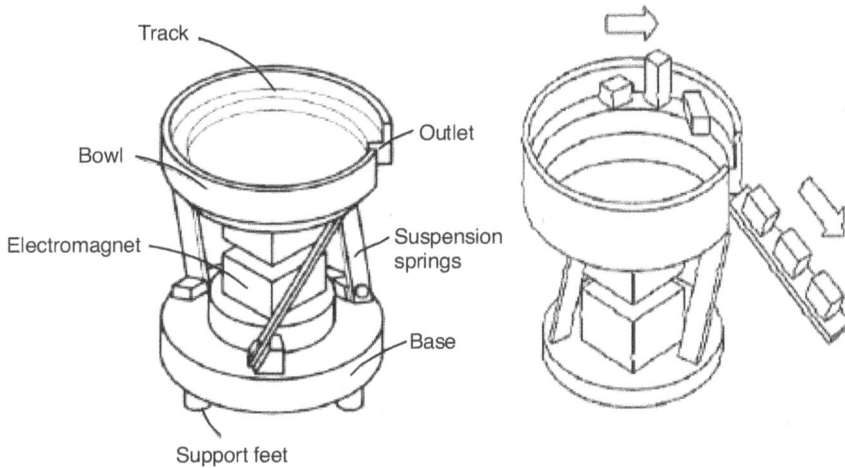

FIGURE 12.25 Vibratory Bowl Feeder.

single file in one of a finite number of stable orientations where they interact with a series of features built into the track and bowl wall.

Controls for Vibratory Feeders

There are a variety of controls available for the different types of vibratory feeders. Pneumatic feeder controls include a quick-acting valve, an airline filter, a pressure regulator gauge, a lubricator, and a long air hose. A transformer-type device, available for electromagnetic vibratory feeders, adjusts the intensity of vibration by varying the applied voltage. Electromechanical feeders have a wall-mounted control box with an on/off button or switch and overload protection. Special controls for remote operation include two speed, maximum-to-minimum material flow controls for batch weighing, and panel board controls for multiple feeder installation. An accelerometer can be attached to the drive unit to monitor the amplitude and to apply a correction to the feeder, which tends to vibrate more quickly as the material level drops.

Applications of Vibratory Feeders

Vibratory feeders are utilized in the pharmaceutical, automotive, chemical, and mining industries. Other industries include steel, glass, foundry, concrete, recycling, bakery, railroad unloading, and plastics. Chemical plants typically use vibratory feeders to control the flow of ingredients to the mixing tanks. Foundries use them to add binders and carbons to sand reprocessing systems. The pulp and paper industry uses vibratory feeders for chemical additive feeding in the bleaching process, while the metal working industry uses them for feeding metal parts to beat treating furnaces. Water and sewage treatment plants also use vibratory feeders in chemical additive handling. Other materials that are separated by vibratory feeders include powder, plastic pellets, dry chemicals, coal, metals, ore, minerals, aluminum, mining and aggregates, grains, seed, dry detergents, ceramics, textiles, rubber, fibers, wood chips, salt, sugar, and many more.

SCREW FEEDERS

Screw feeders are among the most widely used types of bulk-solids-handling equipment in the chemical industry. These are well suited for use with bins having elongated outlets. A screw feeder is shown in Figure 12.26. Screw feeders can be used to deliver bulk solids at tremendous range of feed rates. Bulk solid materials are usually metered into production processes from storage containers, such as hoppers and bins. These feeders have an advantage over belt feeders that in these feeders there is no return element to spill solids. Because a screw is totally enclosed, it is excellent for use with fine, dusty materials. Screw feeders require less maintenance than belt feeders. All screw feeders are volumetric devices. This means that they are intended to deliver a certain volume of bulk material per revolution of the screw. The volume delivered depends on the screw outside and inside (shaft) diameter and the pitch of the screw. Commercial feeders often use

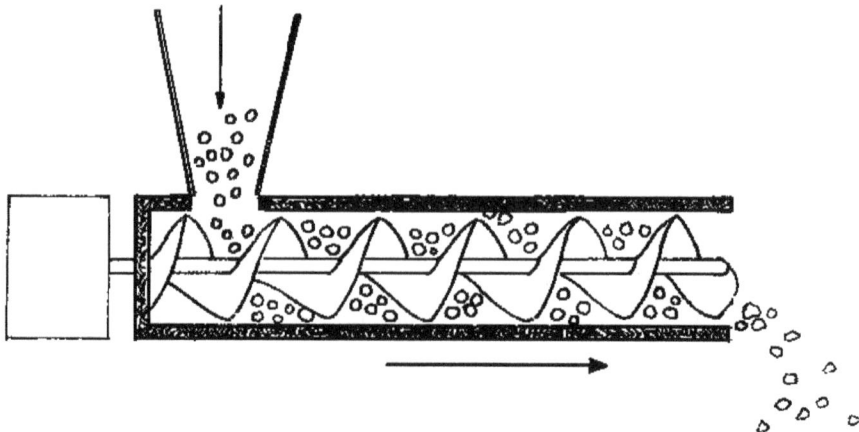

FIGURE 12.26 Screw Feeder.

a variety of agitation devices to ensure that the solids enter the screw in a highly flowable state. These devices may stir, massage, or vibrate the solids in order to fill the screws as completely, or at least as uniformly, as possible.

BELT FEEDERS

Belt feeders are used to provide a controlled volumetric flow of bulk solids from storage bins and bunkers (refer to Figure 12.27). Like screw feeders, belt feeders can be an excellent choice when there is a need to feed material from an elongated hopper outlet. Belt feeders generally can handle a higher flow rate than screw feeders. They generally consist of a flat belt supported by closely spaced idlers and driven by end pulleys. These types of feeders have more moving parts and, therefore, generally require more attention and maintenance than a well-designed screw or vibrating bowl feeder. They usually also require more installation space than screw feeders.

Some particular *features* of belt feeders include:

▪ Suitable for withdrawal of material along slotted hopper outlets when correctly designed.

▪ Can sustain high-impact loads from large particles.

▪ Flat belt surfaces can be cleaned quite readily allowing the feeding of cohesive materials.

FIGURE 12.27 Belt Feeder.

- Suitable for abrasive bulk solids.

- Capable of providing a low initial cost feeder.

ROTARY PLOW FEEDERS

Rotary plow feeders are often used under large stockpiles because of lower capital and operating costs, as well as greater ease of maintenance. This system can be used to move minerals ranging from coal to iron ore stored at mine sites, processing facilities, and power plants. The mechanism by which a rotary plow moves material is as follows:

When a rotary plow begins operating, it loosens material in a narrow vertical channel above it. If twin plows are used and both are operating, two channels will form independent of each other. The pressures exerted on the material adjacent to the channels are generally low and proportional to the size of the flow channel. If a plow does not traverse under the stockpile, and if the material has sufficient cohesive strength, the channel eventually empties out, forming a rat hole. However, as a plow traverses, the narrow plow channel lengthens. Whether material on either side of it remains stationary or slides depends on the wall friction angle along the sloping wall, the wall angle, and the head of the material.

If the material slides, it does so for only a short distance, because, as the material in the flow channel is compressed, the pressure it exerts on the adjacent material increases, resulting in a stable mass. If the side material does not slide, the level of material in the flow channel drops and material sloughs off the top surface.

ROTARY TABLE FEEDERS

The rotary table feeder can be considered as an inverse of the plough feeder. It consists of a power-driven circular plate rotating directly below the bin opening, combined with an adjustable feed collar which determines the volume of bulk material to be delivered. The aim is to permit equal quantities of bulk material to flow from the complete bin outlet and spread out evenly over the table as it revolves. The material is then ploughed off in a steady stream into a discharge chute. This feeder is suitable for handling cohesive materials, which require large hopper outlets, at flow rates between 5 and 125 metric tons per hour. Feed rates to some extent are dependent on the degree to which the material will spread out over the table. This is influenced by the angle of repose of the material, which varies with moisture content, size distribution, and consolidation. These variations prevent high feed accuracy from being obtained. Rotary table feeders are suitable for bin outlets up to 2.5 m in diameter; the table diameter is usually 50 to 60% larger than the hopper outlet diameter. With some materials, a significant dead region can build up at the center of the table. This can sometimes be kept from becoming excessive by incorporating a scraping bar across the hopper outlet. It is important to ensure that the bulk material does not skid on the surface of the plate, severely curtailing or preventing removal of the bulk material.

CENTRIFUGAL HOPPER FEEDER

A centrifugal hopper feeder shown in Figure 12.28 is a feeder system, with a central flat or conical turntable, which drives the working materials

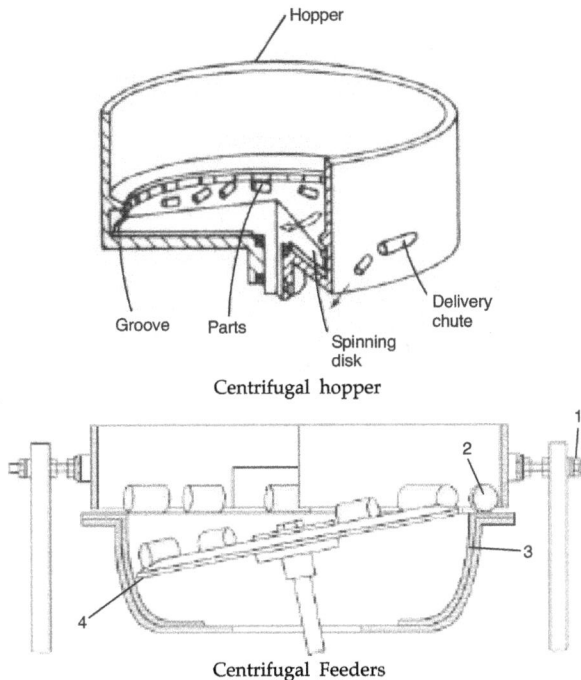

Centrifugal hopper

Centrifugal Feeders

1. Adjustable hopper suspension mounting 2. Workpiece 3. Rotating delivery ring with drum lining
4. Continuously rotating turntable

FIGURE 12.28 Centrifugal Hopper Feeder.

via a rotary action. The resulting centrifugal force causes workpieces to separate out of the heap and move towards the edge of the drum. Here, they meet the delivery ring and slide onto the ramp. The speeds of the turntable and the delivery ring can be adjusted separately. Separated out workpieces can be aligned by orienting devices and, thus, proceed to the pick-up point correctly orientated. Incorrectly orientated workpieces pass back into the hopper. Complex workpieces can be fed by means of a conveyor belt. Any excess conveyed workpieces fall back into the heap in the hopper.

CENTERBOARD HOPPER FEEDER

In this feeder, a blade made of hardened steel, with a shaped top is oscillated up and down (by a crank mechanism) through a mass of parts. Properly oriented parts are picked by the blade and discharged by gravity into the track. This type of feeder is suitable for parts having simple shape like balls, cylinders, nuts and bolts, rivets, etc. These are very robust and have a long working life. The capacity of a centerboard hopper is large. The disadvantage of these is that they cannot be used for fragile components and the degree to which they can orient is rather limited.

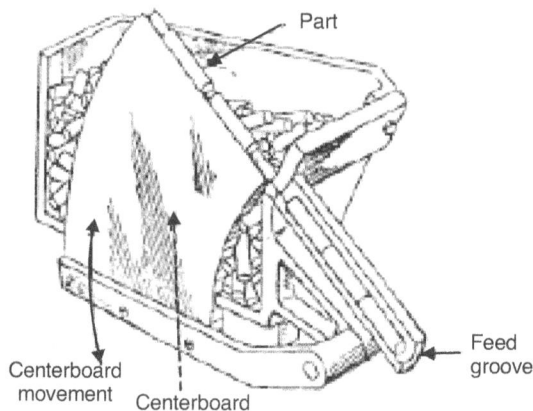

FIGURE 12.29 Centerboard Hopper Feeder.

FLEXIBLE FEEDERS

A crucial component of any assembly system is the parts feeders. Unfortunately, parts feeders are also one of the most specialized components of such systems. While a modular approach could be employed which would allow specialized feeders to be quickly brought into the work cell, there are several difficulties, which present themselves using this approach. First, a feeding system tends to be large and bulky in comparison to other specialized components, such as a robot's gripper. Size alone could make storing all the specialized feeders difficult. Second, a feeding system is generally more expensive than other specialized components. It might be difficult to economically justify building several new feeders for each new product being assembled. Third, the lead-time to build and adjust most current feeding systems is rather long. This diminishes the ability of the work cell to be rapidly adapted to new products. Therefore, simply placing current feeding technology behind a generic, modular interface is not a feasible solution.

A typical robotic mechanical assembly work cell may have several feeders that are tooled specifically for particular parts. Any change to the design of a part requires that the feeder be either re-tooled or completely replaced. With today's product life cycles as short as a few months for many consumer products, this is no longer acceptable. The need for greater flexibility, lower cost of automation, and faster product change over time has brought about a new approach to parts feeding, termed *flexible feeding*. Flexible feeding is an emerging alternative to traditional part feeding methods. This alternative greatly enhances the versatility of a manufacturing workcell by using a robot manipulator and sophisticated sensing devices such as machine vision; thereby significantly reducing both cost and set up time.

Flexible feeding combines several different technologies into a subsystem including servo controlled conveyors that can be reprogrammed for different speed and motion profiles, mechanical transfer devices such as hoppers and buckets used to re-circulate parts, machine vision and lighting, and a robot manipulator.

While vibratory bowl feeders are "hard tooled" to manipulate the fed part into a single orientation, flexible feeders take advantage of the vision system's capability to determine part position and orientation. The flexible part feeder need only de-bulk, disentangle, and re-circulate parts, which is a much simpler task. Conversely, because the bowl feeder is more complicated, it may become jammed, thereby causing expensive downtime. In cases where robot and machine vision systems are already required for the assembly task itself, it becomes very cost effective to use the same equipment in the upstream parts feeding process. In some cases, it is possible to feed multiple parts in the same feeder and use the vision system to determine the part type, thus reducing the number of feeders, and saving money and floor space.

This system shown in Figure 12.30 is composed of three conveyors working together.

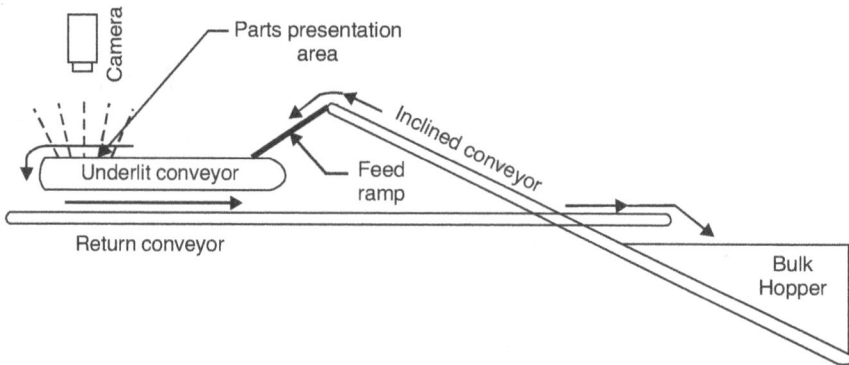

FIGURE 12.30

The *first conveyor*, under servo control, is mounted at an inclined angle and is used to lift parts from a bulk hopper. Parts slide down a ramp at the end of the *inclined conveyor* and onto the horizontal conveyor. Overhead cameras are used to locate parts on the horizontal conveyors. An array of compact fluorescent lights is installed within each of the horizontal conveyors. These lights together with a translucent conveyor belt provide an under lit area in which parts can be presented to the vision system. Using binary vision tools parts on the feeder belts are examined. First, the vision system looks to see if a part is graspable (*i.e.*, the part is in a recognized, stable, and enough clearance exists between the part and its neighbors to grasp it with a gripper). Second, the pose of the part in the robot's world coordinates is determined. This pose, and the motion associated with acquiring the part, is checked to make sure that they are within the work envelope of the robot. Parts, which are not in useful orientations or are overlapping, are dropped from the end of the horizontal conveyor onto the return conveyor. The *return conveyor* transports the parts back to the bulk hopper for re-feeding.

Flexible Feeder Sub-system

Flexible feeders are divided into three distinct sub-systems, which work together to feed parts. They are the parts presentation sub-system, the part pose determination sub-system, and the part retrieval sub-system, discussed below.

1. Parts Presentation

The first major component of any flexible feeding system is the parts presentation sub-system. This is the component of the feeder, which is responsible for removing parts from the bulk supply and presenting them in a quasi-singulated fashion to the rest of the system. There are two major modes of operation, which may be identified for this sub-system: incremental and continuous. In the incremental mode, a higher-level object requests that the sub-system move more parts into the range of the sensing element. When the sub-system has accomplished this task, it signals the calling object that it has finished. The second mode of operation is continuous, in which parts are continuously moved past the sensor. In the continuous mode, a higher-level object first requests that the sub-system begin operation, and then waits for a signal that an additional batch of parts has moved into range. As long as the sub-system is operating, additional new parts-in-range signals are generated.

2. Part Pose Determination

The second major component of any flexible feeder is the part pose determination sub-system. This is the component of the feeder responsible for examining the parts, which are within sensing range, and determining which of those parts are advantageous to retrieval and assembly. This sub-system is most often constructed using an industrial vision system. In general, a vision system places the least amount of restriction on the types of parts that can be sensed. The part pose determination sub-system can also operate in two modes. The first is to find a single retrievable part in the sensing area and the second is to find all retrievable parts in the sensing area.

3. Part Retrieval

The final component of any flexible feeder is the part retrieval sub-system. This is the component, which is responsible for actually retrieving the parts, which have been identified by the part pose sub-system. While this component of the feeder is most often an industrial robot, the fundamental requirement is a mechanism capable of reaching any location and orientation within a certain space. Industrial robots are generally used because they are well understood, well accepted, and often come with a programmable controller, which can save substantial time in system development.

Advantages of Flexible Feeding

Some of the advantages of the flexible feeding concept of automation are:

- *Tooling is soft*: The tooling is the vision and robot software. No long lead times to fabricate tooling.

- *Tooling is flexible*: Simultaneous product and tooling development can occur with flexible feeding. Flexible feeding easily accommodates design changes that occur in product development.

- The ability to quickly produce *new products at production volumes* is a competitive advantage.

- *Avoids the expense* of building prototype and manual assembly tooling.

- Can automate with lower production volumes through combining of products on *same flex feeding cell*.

- Simultaneous *quality vision inspection* using the vision system that guides robot.

- Majority of tooling capital expense is *reusable* if new product is not successful in the marketplace.

EXERCISES

1. How are transfer devices classified? Describe the construction, working principle, and important applications of any two types of transfer devices.

2. What is a transfer machine?

3. What are the advantages and disadvantages of transfer machines?

4. Define automated material handling.

5. What are automated assembly lines? What are the reasons for using automated assembly lines?

6. What is a material transport system? List the various types of material transport equipment.

7. Briefly explain the various types of transfer systems used in assembly lines.

8. Name three linear and three rotary transfer systems.

9. Name some important applications of transfer devices and feeders in the industry.

10. State the role of transfer devices in assembly operations.

11. List various transfer devices available for movement of components in industry.

12. Differentiate between inline and rotary transfer of work parts.

13. Identify the various types of transfer devices. Explain the working of any one type of transfer device with a sketch.

14. Describe any one type of job rotating device.

15. Identify the main elements of a parts feeding system at an assembly workstation.

16. What are the various devices used for alignment of jobs for assembly?

17. List the factors to be considered in the selection of transfer devices.

18. What is the function of feeders in material handling?

19. Describe the construction and working principle of any two types of job orienting devices used in automatic feeding systems with the help of sketches.

20. Classify the various feeders.

21. Compare transfer devices with feeders.

22. Discuss the various types of conveyors used in automated material handling.

23. List the various parts feeding devices.

24. Differentiate between manual and automated assembly lines.

25. Write a short note on Geneva mechanism.

26. List the advantages and disadvantages of transfer machines.

27. What is a gravity conveyor?

28. Differentiate between volumetric and gravimetric type feeders.

29. Explain the construction and working of a vibratory bowl feeder with the help of a sketch.

30. What are the advantages of flexible parts feeding?

31. Differentiate between flexible and vibratory bowl feeders.

32. What is the use of an escapement and placement device in a part feeding system?

33. With the help of a sketch, explain the construction and working of reciprocating tube hopper feeder.

34. What are the various devices used for alignment of jobs for assembly?

35. Describe any one type of job rotating devices.

CHAPTER 13

ROBOTICS

INTRODUCTION

The subject of robotics covers many different areas. Robots alone are hardly ever useful. They are used together with other devices, peripherals, and manufacturing machines. They are generally integrated into a system, which as a whole is designed to perform a task or to do an operation. Robots are very powerful elements of today's industry. They are capable of performing many different tasks and operations precisely and do not require common safety and comfort elements that humans needs. However, it takes much effort and many resources to make a robot function properly. As with humans, robots can do certain things, but not other things. As long as they are designed properly for the intended purpose, they are very useful and will continue to be used. *Robotics* is the art, knowledge base, and the know-how of designing, applying, and using robots in human endeavors. Robotic systems consist of not just robots, but also other devices and systems that are used together with the robots to perform the necessary tasks.

HISTORY OF ROBOTS

The word robot always refers to an automated multifunctional manipulator that works by energy, to perform a variety of tasks. The word *robot* was introduced in 1920 in a play by Karel Capek called R.U.R.,or Rossum's Universal Robots. Robot comes from the Czech word *robota*, meaning forced labor or drudgery. In the play, human-like mechanical creatures produced in Rossum's factory are docile slaves. Because they are just machines, the robots are badly treated by humans. One day a misguided scientist gives them emotions, and the robots revolt, kill nearly all the humans, and take over the world. However, because they are unable to reproduce themselves, the robots are doomed to die. However, the sole surviving human creates a male and a female robot to perpetuate their species.

One of the first types of robots was a feedback (self-correcting) control mechanism. It was a watering trough that used a float to sense the water level. When the water gets to low, the float drops, opens a valve, and more water dumps into the trough. As the water rises, so does the float. Once it reaches a certain height, the valve is closed and the water is shut off. In 1954, the American inventor George Devol, Jr. developed a primitive arm that could be programmed to perform specific tasks. In 1975, the American mechanical engineer Victor Scheinman developed a truly flexible multipurpose manipulator known as the Programmable Universal Manipulation Arm (PUMA). PUMA was capable of moving an object and placing it with any orientation in a desired location. Before the 1960s, robot usually meant a manlike mechanical device (*mechanical man* or *humanoid*) capable of performing human tasks or behaving in a human manner. Today robots come in all shapes and sizes, including small robots and larger wheeled robots that play football with a full-size ball. In 1995, there were about 700,000 robots operating in the world. A major user of robots is the automobile industry. General Motors uses approximately 16,000 robots for tasks such as spot welding, painting, machine loading, parts transfer, and assembly.

DEFINITION OF A ROBOT

A robot is a computer-controlled machine that is programmed to move, manipulate objects, and accomplish work while interacting with its environment. According to the Robot Institute of America (1979), a robot is defined as "a reprogrammable, multifunctional manipulator designed to move material, parts, tools, or specialized devices through various programmed motions for the performance of a variety of tasks."

Or

Industrial robot is defined as "a number of rigid links connected by joints of different types that are controlled and monitored by computer."

One feature of a robot is the ability to operate automatically, on its own. This means that there must be in-built intelligence, or a programmable memory, or simply an arrangement of adjustable mechanisms that command manipulation.

INDUSTRIAL ROBOT

Industrial robots are advanced automation systems, mainly controlled by a computer. Today computers form an important part of industrial automation. They supervise production lines and control manufacturing systems (*e.g.*, machine tools, welders, laser cutting devices, etc.). The new generation of robots executes various tasks in industrial systems and they participate in the full automation of factories. Japanese defined industrial robots in four levels:

- *Manual manipulators*: perform fixed or preset task sequences.
- *Playbacks*: repeat pre-programmed fixed instructions.
- *NC robot*: carry out tasks through numerically loaded information.
- *Intelligent robots*: perform through their own recognition capabilities.

A robot and a crane are very similar in the way they operate and in the way they are designed. Both possess a number of links attached serially to each other with joints, where some type of actuator can move each joint. In both systems, the hand of the manipulator can be moved in space and be placed in any desired location within the workspace of the system, each one can carry a certain amount of load and each one is controlled by a central controller which controls the actuators. However, one is called a robot and the other manipulator (or, in this case, a crane). The fundamental difference between the two is that in the case of a crane, a human operates and controls the actuators, whereas a computer that runs a program controls the robot manipulator. This difference between the two determines whether a device is a simple manipulator or a robot. In general, robots are designed and meant to be controlled by a computer or similar device. The motions of the robot are controlled through a controller that is under the supervision of the computer, which itself is running some type of a program. Thus, if the program is changed, the actions of the robot is changed accordingly. The robot is designed to be able to perform any task that can be programmed (within limit, of course) simply by changing the program. The simple manipulator (or crane) cannot do this without an operator running it all the time.

LAWS OF ROBOTICS

Isaac Asimov proposed three laws of robotics and he later added a "zeroth law."

Law One: A robot may not injure a human being, or, through inaction, allow a human being to come to harm, unless this would violate a higher order law.

Law Two: A robot must obey orders given to it by human beings, except where such orders would conflict with a higher order law.

Law Three: A robot must protect its own existence as long as such protection does not conflict with a higher order law.

Law Zero: A robot may not injure humanity, or, through inaction, allow humanity to come to harm.

MOTIVATING FACTORS

For robotics systems to be introduced to the industrial world, they must have positive factors that would make a difference in using them. The motivating factors can be categorized as:

1. Technical factors
2. Economic factors
3. Social factors

Technical Factors

Robots can do incomparable tasks that humans can't do. It is generally considered that humans can't match the speed, quality, reliability, and the endurance of a robotic system. In that they offer:

(a) High flexibility of product type and variation.

(b) Lower preparation time than hard automation.

(c) Better quality of products.

(d) Fewer rejects and less waste than labor intensive production.

Economic Factors

(a) The needs to increase production rates to remain competitive.

(b) Pressure from the market place to improve quality.

(c) Increasing costs.

(d) Shortage of skilled labor.

Social Factors

Some people think that the use of robotized systems increases the unemployment of workers and prevents many people from a main income. But the usage of robots causes a reduction of workload on workers and prevents dangerous working conditions as robots can be used in hazardous environments.

ADVANTAGES AND DISADVANTAGES OF ROBOTS

Robots offer specific benefits to workers, industries, and countries. If introduced correctly, industrial robots can improve the quality of life by freeing workers from dirty, boring, dangerous, and heavy labor. It can be said, therefore, that robots give the possibility to humans to occupy with jobs, that they can execute better. It is true that robots can cause unemployment by replacing human workers but robots also create jobs such as robot technicians, salesmen, engineers, programmers, and supervisors.

The *advantages* of robots are:

- Increase in productivity, safety, efficiency, quality, and consistency of products with the use of robots.

- Robots can work in hazardous environments without the need for life support, comfort, or concern about safety.

- Robots need no environmental comfort such as lighting, air conditioning, ventilation, and noise protection.

- Robots can work continuously without experiencing fatigue or boredom, do not have hangovers, and need no medical insurance or vacation.

▪ Robots have repeatable precision at all times, unless something happens to them or unless they wear out.

▪ Robots can be much more accurate than humans.

The *disadvantages* of robots are:

▪ Robots replace human workers creating economic problems, such as lost salaries, and social problems such as dissatisfaction and resentment among workers.

▪ Robots lack capability to respond in emergencies, unless the situation is predicted and the response is included in the system. Safety measures are needed to ensure that they do not injure operators and machines working with them.

▪ Robots are costly due to initial cost of equipment, installation costs, need for training, and need for programming.

CHARACTERISTICS OF AN INDUSTRIAL ROBOT

An industrial robot has *a hand, wrist, arm, base, lifting power, repeatability, manual control, automatic control, memory, programs, safety interlock, speed of operation, computer interface, reliability, and easy maintenance.*

The *hand* of a robot is known as a gripper or end effector or end-of-arm tooling. It is the driven mechanical device(s) attached to the end of the manipulator.

The *wrist* of the robot is used to aim the hand at any part of the work piece. The wrist may have three motions: pitch (up-and-down motion), yaw (side-to-side motion), and roll (rotating motion).

The *arm* is used to move the hand within reach of a part or work piece. It can pivot at its elbow and at its shoulder joint.

The *waist*, or *base*, of the robot supports the arm and is called the shoulder. The arm can rotate about the shoulder.

Repeatability is the replication of motion within some specified precision or tolerance.

An *RCC* or *remote center compliance* device helps pull the hand or tool into the required position by acting as a multi-axis float.

A *manual control* device is used to teach the robot how to do a new task.

An *automatic control* system is used to carry out the instructions stored in the robot's memory.

The robot's *memory* holds a library of programs to use in executing different tasks.

Safety interlocks prevent the robot from inserting a hand into a machine and causing damage to the robot and machine.

The robot's *speed of operation* in performing a task should be at least equal to that of the human worker it is replacing.

The robot's *computer interface* enables the robot to use the computer's larger memory to hold more task programs and to synchronize its actions with a complete production line of robots and other machines.

COMPONENTS OF AN INDUSTRIAL ROBOT

Industrial robot systems consist of four major subsystems as shown in Figure 13.1.

- Mechanical unit
- Drive
- Control system
- Tooling

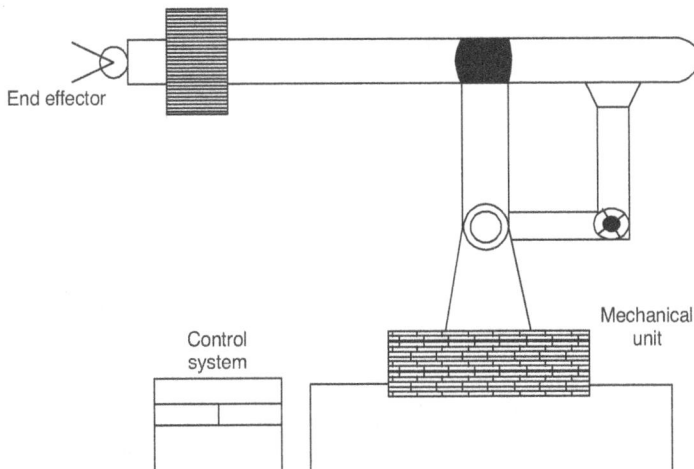

FIGURE 13.1 Components of an Industrial Robot.

A short description of each is given below:

Mechanical Unit. The mechanical unit refers to the robot's manipulative arm and its base. Tooling such as end effectors, tool changers, and grippers are attached to the wrist-tooling interface. The mechanical unit consists of a fabricated structural frame with provisions for supporting mechanical linkage and joints, guides, actuators, control valves, limiting devices, and sensors. The physical dimensions, design, and loading capability of the robot depends upon the application requirements.

Drive. An important component of the robot is the drive system. The drive system supplies the power, which enables the robot to move. Drive for a robot may be hydraulic, pneumatic, or electric. Hydraulic drives have been used for heavier lift systems. Pneumatic drives have been used for high speed, non-servo robots and are often used for powering tooling such as grippers. Electric drive systems can provide both lift and/or precision, depending on the motor and servo system selection and design. An AC or DC powered motor may be used depending on the system design and applications.

Control System. Controller is the brain of the robot. Controller is a communication and information-processing device that initiates, terminates, and coordinates the motions and sequences of a robot. Most industrial robots incorporate computer or microprocessor based controllers. These perform computational functions and interface with sensors, grippers, tooling, and other peripheral equipment.

Controller programming may be done on-line or from off-line control stations. Programs may be on cassettes, floppy disks, internal drives, or in memory; and may be loaded or downloaded by cassettes, disks, or telephone modem. Some robot controllers have sufficient computational ability, memory capacity, and input/output capability to serve as system controllers for other equipment and processes.

Tooling. Tooling is manipulated by the robot to perform the functions required for the application. Depending on the application, the robot may have one functional capability, such as making spot welds or spray-painting. These capabilities may be integrated with the robot's mechanical system or may be attached at the robot's wrist-end effector interface. Alternatively, the robot may use multiple tools that may be changed manually (as part of set-up for a new program) or automatically during a work cycle.

Tooling and objects that may be carried by a robot's gripper can significantly increase the envelope in which objects or humans may be struck. Tooling manipulated by the industrial robot and carried objects can cause more significant hazards than motion of the bare robotic system. The hazards added by the tooling should be addressed as part of the risk assessment.

Sensors. Sensors are used to collect information about the internal state of the robot or to communicate with the outside environment. As in humans, the robot controller needs to know where each link of the robot is, in order to know the robot's configuration. The state of the human body is determined because feedback sensors in human's central nervous system embedded in their muscle tendons send information to the brain. The brain uses this information to determine the length of their muscles, and thus, the state of their arms, legs, etc. The same is true for robots; sensors integrated into the robot send information about each joint or link to the controller, which determines the configuration of the robot. Robots are often equipped with external sensory devices such as a vision system, touch and tactile sensors, speech synthesizers, etc., which enable the robot to communicate with the outside world.

Figure 13.2 given below compares the physical parts of the human being and those of the industrial robot.

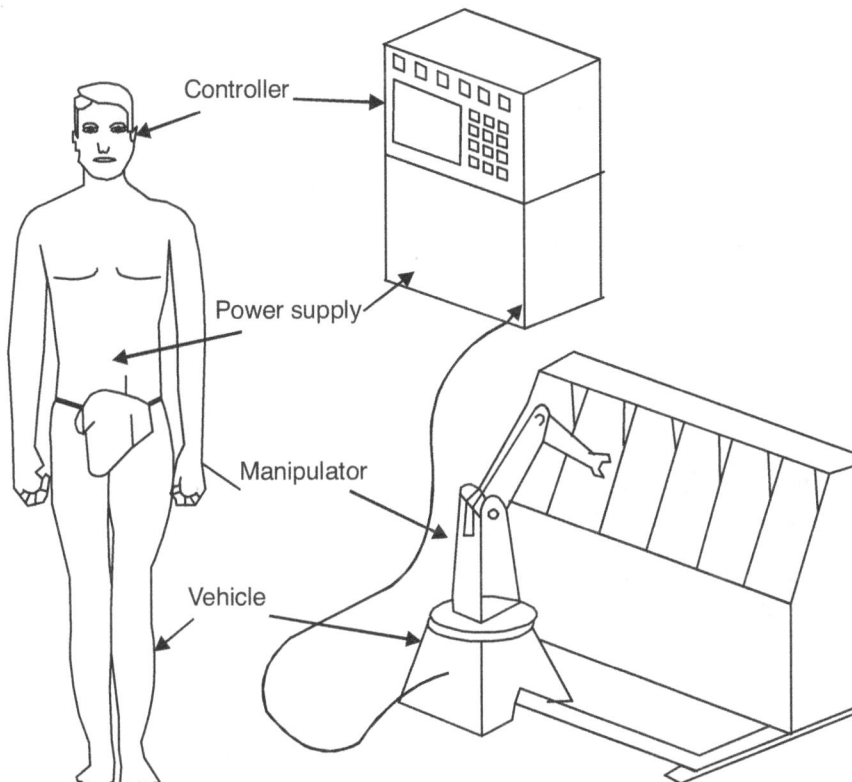

FIGURE 13.2 Comparison of Parts of Robot and Human Being.

COMPARISON OF THE HUMAN AND ROBOT MANIPULATOR

The parts of a robot's manipulator are named after similar parts in a human's chief manipulators, *i.e.*, the arm and the hand. (Refer to Figure 13.3.)

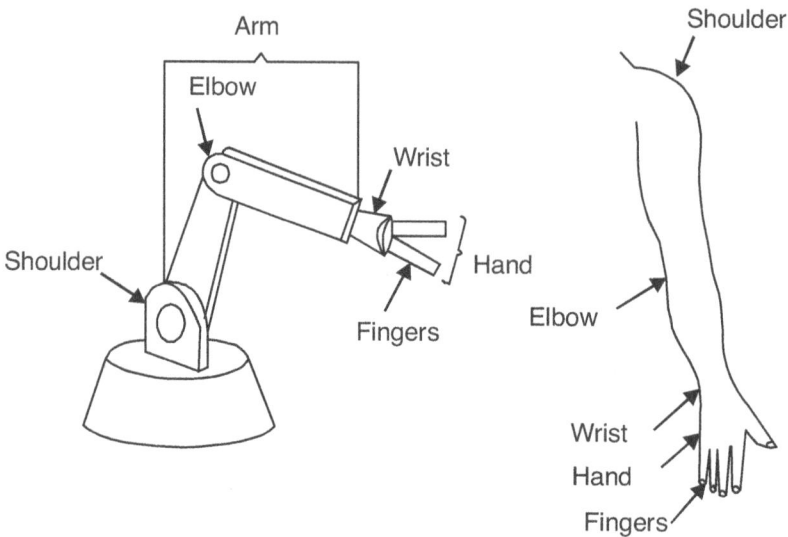

FIGURE 13.3 Comparison of Human and Robot Manipulator.

ROBOT WRIST AND END OF ARM TOOLS

Six axis co-ordinates are required by a robot in order to completely specify the location and orientation of an object. Three co-ordinates can locate the center of gravity of an object (e.g., x, y, and z co-ordinates-in a rectangular co-ordinate system). Three other co-ordinates are normally achieved by adding wrist and hand movements with the end of arm tooling. There are three basic types of wrist motions:

■ Pitch - Rotational *or* bending movement in a vertical plane.

■ Yaw - Rotational *or* twisting movement in a horizontal plane.

■ Roll - Rotational *or* swivel movement.

Additionally, the wrist (refer to Figure 13.4) serves as the mounting point for a variety of devices. These end effectors are either hand, or gripping devices, or job specific tools.

Figure 13.5 shows the various arms and wrist motions of a robot.

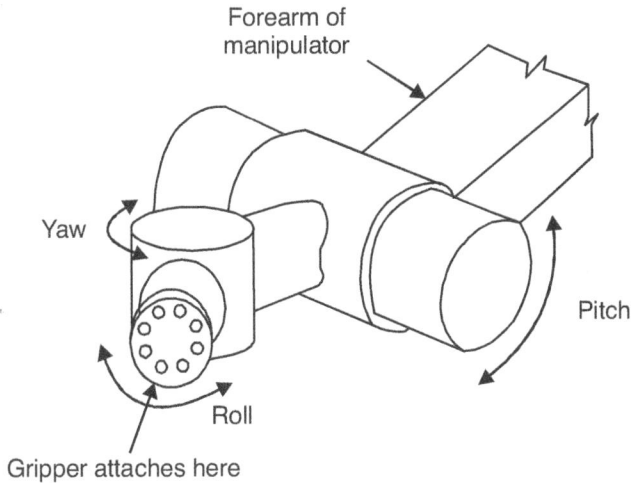

Forearm of
manipulator

Yaw

Pitch

Roll

Gripper attaches here

FIGURE 13.4 Robotic Wrist.

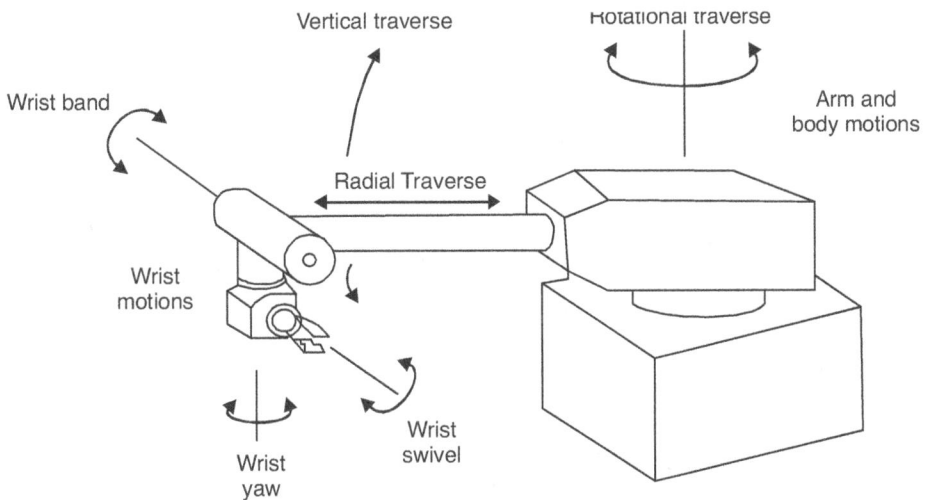

Vertical traverse

Rotational traverse

Wrist band

Arm and
body motions

Radial Traverse

Wrist
motions

Wrist
yaw

Wrist
swivel

FIGURE 13.5 Arm and Wrist Motions of a Robot.

End Effector. The end effector is the device at the end of the robot arm. The end effector attached to the robot wrist is shown in Figure 13.6. There are two main types of end effectors: *grippers* and *tools*.

(*i*) *Grippers*: Grippers are devices, which can be used for holding or gripping an object. These include mechanical hands and anything like hooks, magnets, and suction devices, which can be used for holding or gripping.

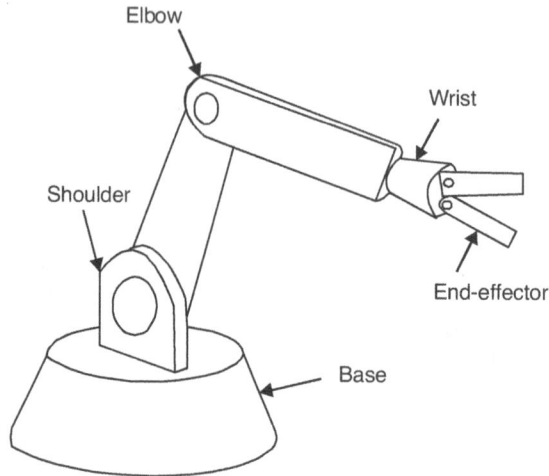

FIGURE 13.6 End Effector Attached to Robot Wrist.

(ii) *Tools*: Tools are devices, which robots use to perform operations on an object, *e.g.*, drills, paint sprays, grinders, welding torches, and any other tool which get a specific job done.

ROBOT TERMINOLOGY

There is a set of basic terminology and concepts common to all robots. These terms follow with brief explanations of each.

(i) *Links and Joints*: Links are the solid structural members of a robot, and joints are the movable couplings between them.

(ii) *Degree of Freedom (dof)*: Degree of freedom is the number of independent movements a robot can realize with respect to its base. The number of axes is normally the same as the number of degrees of freedom of the robot. Each joint on the robot introduces a degree of freedom. Each degree of freedom can be a slider, rotary, or other type of actuator. Robots typically have five or six degrees of freedom as shown in Figure 13.7. Three of the degrees of freedom allow positioning in 3D space, while the other two or three are used for orientation of the end effector. Six degrees of freedom are enough to allow the robot to reach all positions and orientations in 3D space. Six degrees of freedom are commonly available with articulated arm and gantry robots. Four degrees of freedom are typical with the selective compliance assembly robot arm (SCARA) configuration. Seven or more axes are used for some special applications.

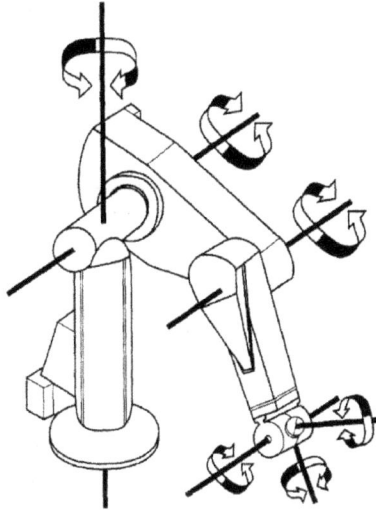

FIGURE 13.7 Robot with Six Degrees of Freedom.

Robot with Seven Degrees of Freedom

Figure 13.8 shows an industrial robot with three basic degrees of freedom plus three degrees of freedom in the wrist and a seventh in its ability to move back and forth along the floor.

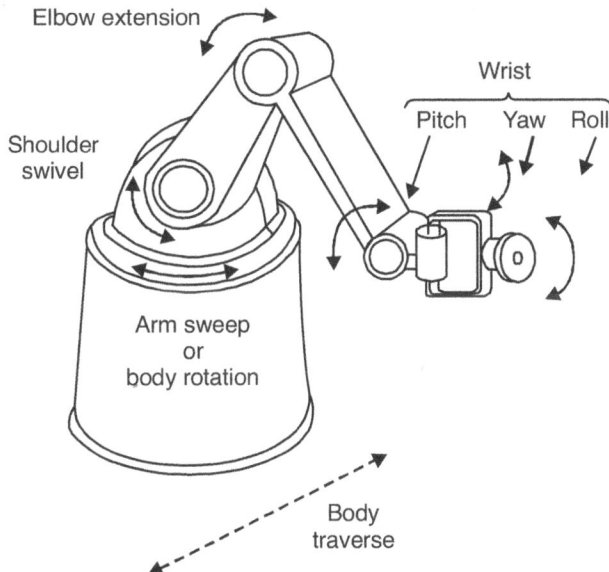

FIGURE 13.8 Robot with Seven Degrees of Freedom.

(iii) Orientation Axis: Basically, if the tool is held at a fixed position, the orientation determines which direction it can be pointed in. Roll, pitch, and yaw are the common orientation axes used. (Refer to Figure 13.9.)

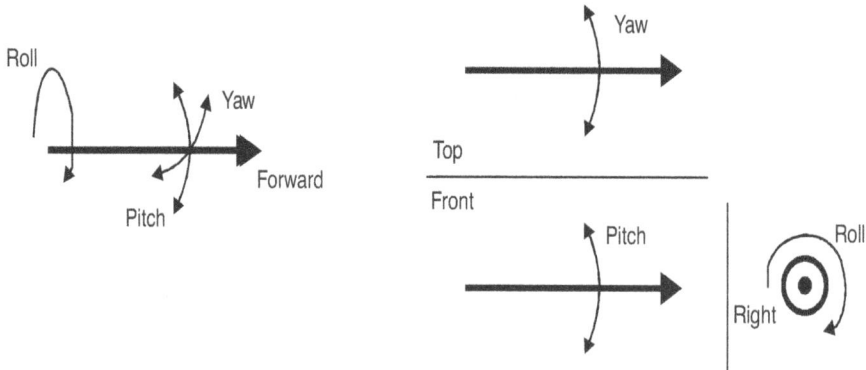

FIGURE. 13.9 Orientations.

(iv) Position Axis: The tool, regardless of orientation, can be moved to a number of positions in space.

(v) Tool Center Point (TCP): The tool center point is located either on the robot, or the tool as shown in Figure 13.10. Typically, the TCP is used when referring to the robots position, as well as the focal point of the tool (*e.g.*, the TCP could be at the tip of a welding torch). The TCP can be specified in Cartesian, cylindrical, spherical, etc., co-ordinates depending on the robot. As tools are changed, one often reprograms the robot for the TCP.

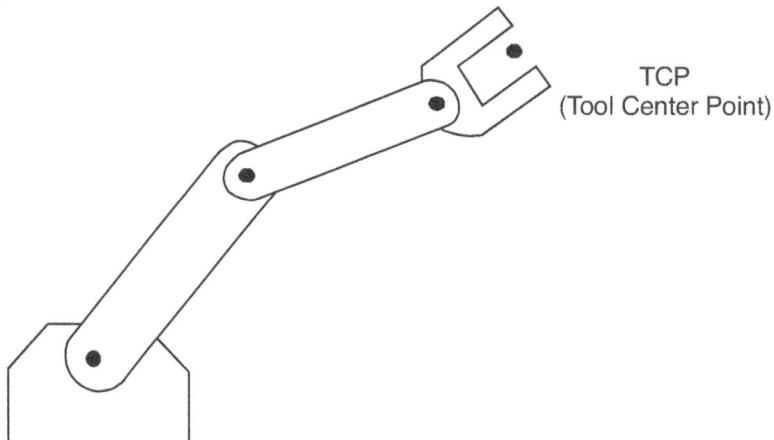

FIGURE 13.10 Tool Center Point.

(vi) *Accuracy*: Accuracy specification describes how close the arm will be when it moves to the desired point.

(vii) Precision (validity): Precision is defined as how accurately a specified point can be reached. This is a function of the resolution of the actuators, as well as its feedback devices.

(viii) *Repeatability (variability)*: Repeatability is how accurately the same position can be reached if the motion is repeated many times (Refer to Figure 13.11). Suppose that a robot is driven to the same point 100 times. Because many factors may affect the accuracy of the position, the robot may not reach the same point every time, but will be within a certain radius from the desired point. The radius of a circle that is formed by this repeated motion is called repeatability. Repeatability is much more important than precision. If a robot is not precise, it will generally show a consistent error, which can be predicted and thus, corrected through programming.

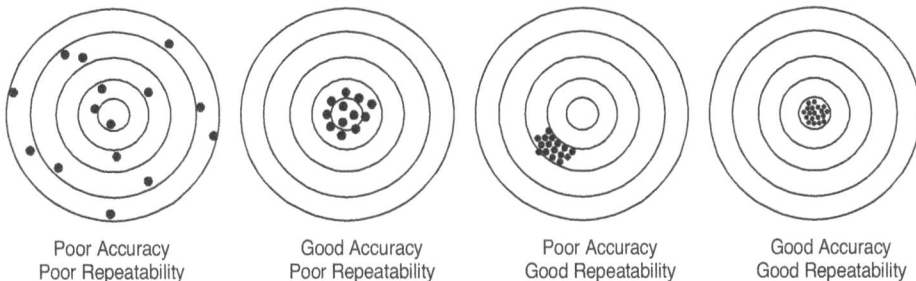

| Poor Accuracy | Good Accuracy | Poor Accuracy | Good Accuracy |
| Poor Repeatability | Poor Repeatability | Good Repeatability | Good Repeatability |

FIGURE 13.11 Accuracy versus Repeatability.

(ix) *Work envelope/Workspace*: A robot can only work in the area in which it can move. This area is called the work envelope. The work envelope is determined by how far the robot's arm can reach and how flexible the robot is. The more reach and flexibility a robot has, the larger the work envelope will be. It is one of the most important characteristics to be considered in selecting a suitable robot. Various robot configurations have different work envelopes. For a Cartesian configuration, the reach is a rectangular-type space. For a cylindrical configuration, the reach is a hollow cylindrical space. For a polar configuration, the reach is part of a hollow spherical shape. Robot reach for a jointed-arm configuration does not have a specific shape. Figure 13.12 shows the work envelope of a robot.

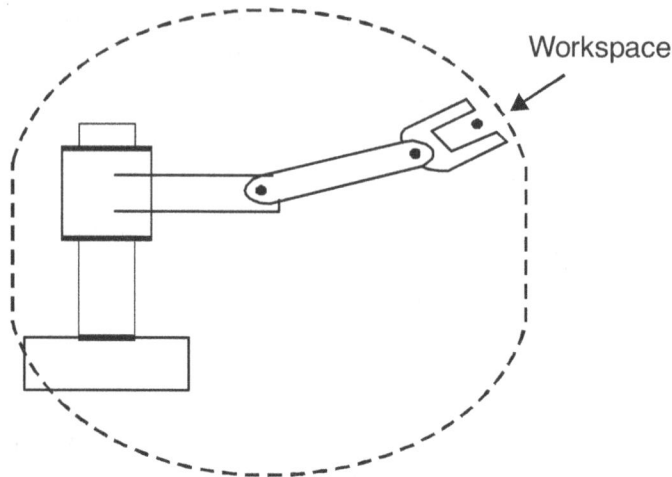

FIGURE 13.12 Work Envelope of a Robot.

(x) *Stability*. Stability refers to robot motion with the least amount of oscillation. A good robot is one that is fast enough but at the same time has good stability.

(xi) *Speed*. Speed refers either to the maximum velocity that is achievable by the Tool Center Point (TCP), or by individual joints. This number is not accurate in most robots, and will vary over the workspace as the geometry of the robot changes (and hence the dynamic effects). The number will often reflect the maximum safest speed possible. Some robots allow the maximum rated speed (100%) to be passed, but it should be done with great care.

(xii) *Payload*. Payload is the weight a robot can carry and still remain within its specifications. For example, a robot's maximum load capacity may be much larger than its specified payload, but at the maximum level, it may become less accurate, may not follow its intended path accurately, or may have excessive deflections. The payload of robots compared with their own weight is usually very small.

(xiii) *Reach*. Reach is the maximum distance a robot can reach within its work envelope.

(xiv) *Settling Time*. During a movement, the robot moves fast, but as the robot approaches the final position it slows down, and slowly approaches at final position. The settling time is the time required for the robot to be within a given distance from the final position.

ROBOTIC JOINTS

A robot joint is a mechanism that permits relative movement between parts of a robot arm. The joints of a robot are designed to enable the robot to move its end-effector along a path from one position to another as desired. The basic movements required for a desired motion of most industrial robots are:

- *Rotational movement*: This enables the robot to place its arm in any direction on a horizontal plane.

- *Radial movement*: This enables the robot to move its end-effector radially to reach distant points.

- *Vertical movement*: This enables the robot to take its end-effector to different heights.

Degrees of freedom, independently or in combination with others, define the complete motion of the end-effector. These motions are accomplished by movements of individual joints of the robot arm. The joint movements are basically the same as relative motion of adjoining links. Depending on the nature of this relative motion, the joints are classified as

1. Prismatic joints
2. Revolute joints

   1. Prismatic joints are also known as sliding as well as linear joints. They are called prismatic because the cross section of the joint is considered as a generalized prism. They permit links to make a linear displacement along a fixed axis. In other words, one link slides on the other along a straight line. These joints are used in gantry, cylindrical, or similar joint configurations.

   2. Revolute joints, the second type of joint is a revolute joint where a pair of links rotates about a fixed axis.

   Figure 13.13 shows the prismatic and revolute joint.

The variations of revolute joints shown in Figure 13.14 include:

- Rotational joint (R)
- Twisting joint (T)
- Revolving joint (V)

A *rotational joint (R)* is identified by its motion, rotation about an axis perpendicular to the adjoining links. Here, the lengths of adjoining links do

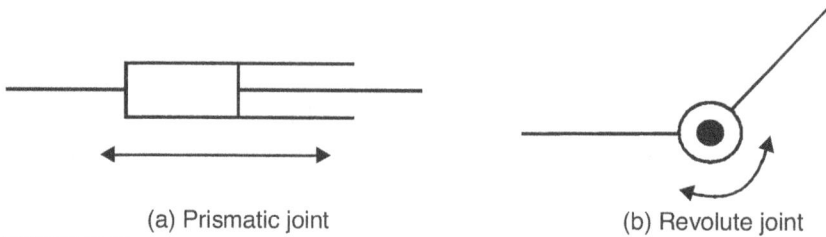

(a) Prismatic joint   (b) Revolute joint

FIGURE 13.13

Rotational joint (Type R)	
Twisting joint (Type T)	
Revolving joint (Type V)	

FIGURE 13.14 Revolute Joints.

not change but the relative position of the links with respect to one another changes as the rotation takes place.

A *twisting joint (T)* is also a rotational joint, where the rotation takes place about an axis that is parallel to both adjoining links.

A *revolving joint (V)* is another rotational joint, where the rotation takes place about an axis that is parallel to one of the adjoining links. Usually, the links are aligned perpendicular to one another at this kind of joint. The rotation involves revolution of one link about another.

CLASSIFICATION OF ROBOTS

Robots can be classified by four fundamental elements of operation:

- Coordinate systems
- Power source
- Method of control
- Programming method

ROBOT CLASSIFICATION ON THE BASIS OF CO-ORDINATE SYSTEMS

Structurally, robots are classified according to the wrist's co-ordinate system as follows:

1. *Cartesian/Rectilinear Robot*: The axis or dimensions of these robots are 3 intersecting straight lines (x-y-z) as shown in Figure 13.15. The Cartesian co-ordinate robot is one that consists of a column and an arm. It is sometimes called an x-y-z robot, indicating the axis of motion. The x-axis is lateral motion, the y-axis is longitudinal motion, and the z-axis is vertical motion. Thus, the arm can move up and down on the z-axis; the arm can slide along its base on the x-axis; and then it can telescope to move to and from the work area on the y-axis. The features of a Cartesian robot (electronic equipment, control program) are same as those of CNC machine tools. Cartesian robots are not preferred in the industry because they do not have mechanical flexibility (*i.e.,* they cannot reach objects that lie on the floor or that are not visible from their base). Also, speed of operation on a horizontal plane is usually less than the corresponding speed of robots with a revolute base.

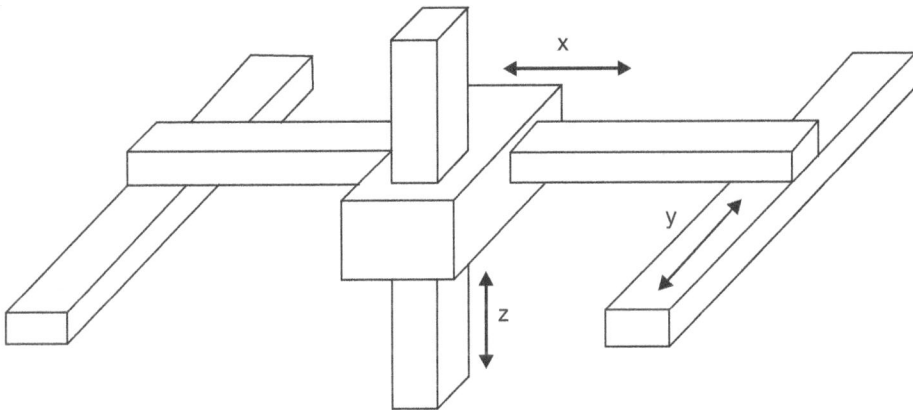

FIGURE 13.15 Cartesian Robot.

Applications

These types of robots are used for:

- Pick and place work
- Application of sealant

- Assembly operations
- Handling machine tools
- Arc welding

   Advantages

- Ability to do straight line insertions into furnaces
- Easy computation and programming
- Most rigid structure for given length

   Disadvantages

- Requires large operating volume
- Exposed guiding surfaces require covering in corrosive or dusty environments
- Can only reach front of itself
- Axis hard to seal

   2. *Cylindrical Robot*: Cylindrical robots have one angular dimension and 2 linear dimensions as shown in Figure 13.16. The rigid structure of this system offers them the capability to lift heavy loads through a large working envelope. The main body of such a robot consists of a horizontal arm mounted on a vertical column. The column is mounted on a rotating base. The horizontal arm moves forward and backward on the direction of the longitudinal axis and it also moves up and down on the column. Column and arm are rotating on the base around the vertical axis. Resolution of a cylindrical robot is not constant, but depends on the distance between the column and the tool along the horizontal arm.

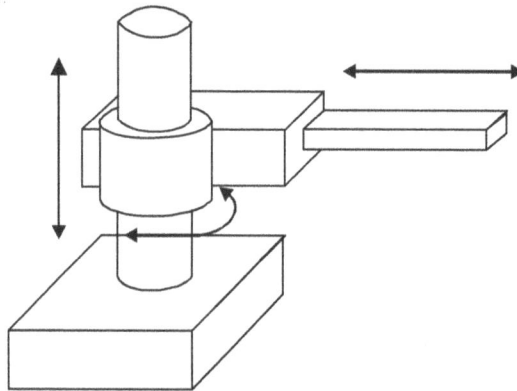

FIGURE 13.16 Cylindrical Robot.

Applications

These types of robots are used for:

- Assembly operations
- Handling machine tools
- Spot-welding
- Handling die-casting machines

Advantages

- Can reach all around itself
- Rotational axis easy to seal
- Relatively easy programming
- Rigid enough to handle heavy loads through large working space
- Good access into cavities and machine openings

Disadvantages

- Can't reach above itself
- Won't reach around obstacles
- Exposed drives are difficult to cover from dust and liquids
- Linear axes is hard to seal

3. *Spherical (Polar) Robot*: Robots of this type consist of a rotating base, a lifting part, and a telescopic arm, which moves inwards and outwards (refer to Figure 13.17). The 2 dimensions of spherical robots are angles and the third is a linear distance from the point of origin. These robots operate according to spherical co-ordinates and offer greater flexibility. This design is used where a small number of vertical actions are adequate.

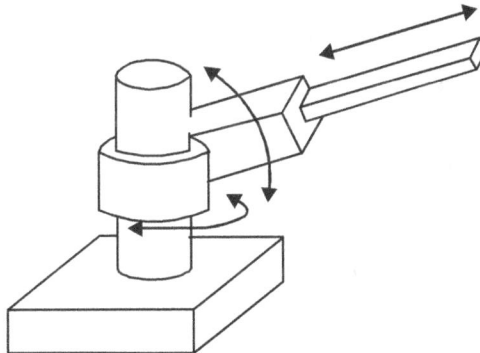

FIGURE 13.17 Spherical Robot.

Applications

These types of robots are used for:

- Handlings at die casting or fettling machines
- Handling machine tools
- Arc/spot welding

Advantages

- Large working envelope
- Two rotary drives are easily sealed against liquids/dust

Disadvantages

- Complex co-ordinates more difficult to visualize, control, and program
- Exposed linear drive
- Low accuracy

4. *Articulated Robot*: Articulated robots consist of three constant parts (links) that are joined with revolute joints and are placed on a rotating base as shown in Figure 13.18. The kinematics layout is similar to a human arm. The tool (gripper) is similar to a palm and is adjusted to the lower part of the arm through the wrist. The elbow connects the lower and upper part of the arm and the shoulder connects the upper part of the arm with the base. Many times the shoulder joint has a rotational motion in the horizontal plane. The articulated robot has all three axes revolute, so the position resolution is completely dependent on the arm's position. The total accuracy of an articulated robot is small, because joint errors are accumulated at the end of the arm, which is at the wrist position.

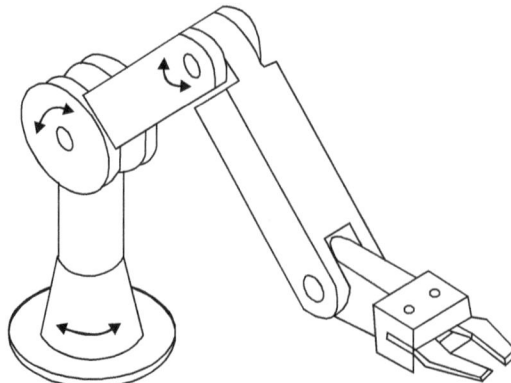

FIGURE 13.18 Articulated Robot.

Applications

These types of robots are used for:

- Assembly operations
- Die casting
- Fettling machines
- Gas welding
- Arc welding
- Spray-painting

Advantages

- High mechanical flexibility
- They can move with high speed at three degrees of freedom
- All joints can be sealed from the environment

Disadvantages

- Extremely difficult to visualize, control, and program
- Restricted volume coverage
- Low accuracy

5. *SCARA Robot*: One style of robot that has recently become quite popular is a combination of the articulated arm and the cylindrical robot. This robot has more than three axes and is called a SCARA robot (refer to Figure 13.19). It is used widely in electronic assembly. The rotary axes are mounted vertically rather

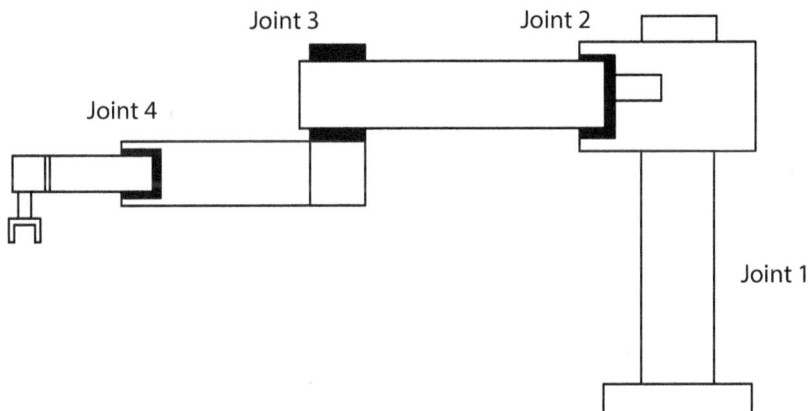

FIGURE 13.19 SCARA Robot.

than horizontally. This configuration minimizes the robot's deflection when it carries an object while moving at a programmed speed. The acronym SCARA stands for *Selective Compliance Assembly Robot Arm*, a particular design developed in the late 1970s.

Applications

SCARA robots are commonly used for:

- Pick and place work
- Assembly operations
- Application of sealant
- Handling machine tools

Advantages

- High speed
- Excellent repeatability
- Good payload capacity
- Large work area for floor space
- Moderately easy to program

Disadvantages

- Limited applications
- Two ways to reach point
- Difficult to program off-line
- Highly complex arm

ROBOT CLASSIFICATION ON THE BASIS OF POWER SOURCE

Actuators drive the mechanical linkages and joints of a manipulator, which can be various types of motors and valves. The energy for these actuators is provided by some power source such as hydraulic, pneumatic, or electric. There are three major types of drive systems for industrial robots:

- Hydraulic drive system
- Pneumatic drive system
- Electric drive system

1. Hydraulic Drive System

The most popular form of the drive system is the *hydraulic drive system* because hydraulic cylinders and motors are compact and allow for high levels of force and power, together with accurate control. These systems are driven by a fluid that is pumped through motors, cylinders, or other hydraulic actuator mechanisms. A hydraulic actuator converts forces from high pressure hydraulic fluid into the mechanical shaft rotation or linear motion. Hydraulic robots are preferred in environments in which the use of electric drive robots may cause fire hazards, for example, in spray painting.

Advantages

- A hydraulic device can produce an enormous range of forces without the need for gears, simply by controlling the flow of fluid
- Preferred for moving heavy parts
- Preferred to be used in explosive environments
- Self-lubrication and self-cooling
- Smooth operation at low speeds
- There is need for return line

Disadvantages

- Occupy large space area
- There is a danger of oil leak to the shop floor

2. Pneumatic Drive System

Pneumatic drive systems are found in approximately 30 percent of today's robots. These systems use compressed air to power the robots. Because machine shops typically have compressed air lines in their working areas, the pneumatically driven robots are very popular. These robots generally have fewer axis of movement and can carry out simple pick-and-place material-handling operations, such as picking up an object at one location and placing it at another location. These operations are generally simple and have short cycle times. The pneumatic power can be used for sliding or rotational joints.

Advantages

- Less expensive than electric or hydraulic robots
- Suitable for relatively less degrees of freedom design
- Do not pollute work area with oils

- No return line required
- Pneumatic devices are faster to respond as compared to a hydraulic system as air is lighter than fluid

  Disadvantages

- Compressibility of air limits control and accuracy aspects
- Noise pollution from exhausts
- Leakage of air can be of concern

3. *Electric Drive System*

Electrical drive systems are used in about 20 percent of today's robots. These systems are servomotors, stepping motors, and pulse motors. These motors convert electrical energy into mechanical energy to power the robot. Compared with a hydraulic system, an electric system provides a robot with less speed and strength. Electric drive systems are adopted for smaller robots. Electrically driven robots are the most commonly available and used industrial robots. There are three major types of electric drive that have been used for robots:

(a) *Stepper Motors:* These are used mainly for simple pick and place mechanisms where cost is more important than power or controllability.

(b) *DC Servos:* For the early electric robots, the DC servo drive was used extensively. It gave good power output with a high degree of control of both speed and position.

(c) *AC Servos:* In recent years, the AC servo has taken over from the DC servo as the standard drive. These modern motors give higher power output and are almost silent in operation. As they have no brushes, they are very reliable and require almost no maintenance in operation.

  Advantages

- Good for small and medium size robots
- Better positioning accuracy and repeatability
- Less maintenance and reliability problems

  Disadvantages

- Provides less speed and strength than hydraulic robots
- Not all electric motors are suited for use as actuators in robots
- Require more sophisticated electronic controls and can fail in high temperature, wet, or dusty environments

ROBOT CLASSIFICATION ON THE BASIS OF METHOD OF CONTROL

The motions of a robot are controlled by a combination of software and hardware that is programmed by the user. Robots are classified by control method into *servo and non-servo robots*.

1. Non-Servo Controlled Robots

Non-servo control is a purely mechanical system of stops and limit switches, which are pre-programmed for specific repetitive movements. This can provide accurate control for simple motions at low cost. The motions of non-servo controlled robots are controlled only at their end-points, not throughout their paths. *Non-servo robots* are often referred as "endpoint," "pick and place," or "limited sequence" robots. These robots are used primarily for materials transfer.

Characteristics of non-servo robots include:

- Relatively high speed possible due to smaller size of the manipulator
- These robots are low cost and simple to operate, program, and maintain
- These robots have limited flexibility in terms of program capacity and positioning capability

2. Servo Controlled Robots

Servo control system is capable of controlling the velocity, acceleration, and path of motion, from the beginning to the end of the path. It uses complex control programs. These systems use programmable logic controllers (PLCs) and sensors to control the motions of robots. They are more flexible than non-servo systems, and they can control complicated motions smoothly. Sensors are used in servo-control systems to track the position of each of the axis of motion of the manipulator. Servo controlled robots are classified according to the method that the controller uses to guide the end-effector. These are:

- Point-to-point (PTP) control robot
- Continuous-path (CP) control robot
- Controlled-path robot

(a) Point-to-Point Control Robot (PTP)

A point-to-point robot is capable of moving from one discrete point to another within its working envelope. During point-to-point operation the robot moves to a position, which is numerically defined, and it stops there. The end effector performs the desired task, while the robot is halted. When task is completed, the robot moves to the next point and the cycle is repeated. Such robots are usually taught a series of points with a teach pendant. The points are then stored and played back.

Applications: Point-to-point robots are severely limited in their range of applications. Common applications include:

- Component insertion
- Spot-welding
- Hole drilling
- Machine loading and unloading
- Assembly operations

An example of a typical point-to-point system is found in a spot welding robot. In this case, the robot moves until the point to be welded is positioned between the two electrodes of the welding pistol, and welding is performed. Then, the robot moves to another point, which is again welded. This procedure is repeated until all necessary points are welded.

(b) Continuous-path (CP) Control Robot

In a *continuous-path robot*, the tool performs its task, while the robot (its axes) is in motion, like in the case of arc welding, where the welding pistol is driven along the programmed path. All axes of continuous path robots move simultaneously, each with a different speed. The computer co-ordinates the speeds so that the required path is followed. The robot's path is controlled by storing a large number of spatial points in the robot's memory during the teach sequence. During teaching, and while the robot is being moved, the co-ordinate points in space of each axis are continually monitored and placed into the control system's computer memory. These are the most advanced robots and require the most sophisticated computer controllers and software development.

Applications: Typical applications include:

- Spray painting
- Finishing
- Gluing
- Arc welding operations
- Cleaning of metal articles
- Complex assembly processes

(c) Controlled-path Robot

In controlled-path robots, the robot is moved along a computer-generated, predictable path as the robot travels from point to point. The computer-generated path may be a straight line with end-effector orientation or it may involve curved paths through successive points and/or gradual orientation changes. Good accuracy can be obtained at any point along the specified path. Only the start and finish points and the path definition function must be stored in the robot's control memory. The robot movements are more precise than with point-to-point programming and are less likely to present a hazard to personnel and equipment.

ROBOT CLASSIFICATION ON THE BASIS OF PROGRAMMING METHOD

Robots can be classified according to the programming method such as:

- Manual programming
- Lead-through programming
- Walk-through programming

These methods are discussed in detail in chapter of Robot Programming.

ROBOT SELECTION

Once the application is selected, a suitable robot should be chosen from the many commercial robots available in the market. The characteristics of robots generally considered in a selection process include:

- Size of class
- Degrees of freedom
- Velocity
- Drive type
- Control mode
- Repeatability
- Lift capacity
- Right-left, Up-down, and In-out traverse
- Yaw, pitch, and roll
- Weight of the robot

    1. *Size of class*: The size of the robot is given by the maximum dimension (x) of the robot work envelope.

- Micro (x < 1 m)
- Small (1 m < x < 2 m)
- Medium (2 m < x < 5 m)
- Large (x > 5 m)

2. *Degrees of freedom*: The cost of the robot increases with the number of degrees of freedom. Six degrees of freedom is suitable for most works.

3. *Velocity*: Velocity consideration is effected by the robot's arm structure.

- Rectangular
- Cylindrical
- Spherical
- Articulated

4. *Drive type*:

- Hydraulic
- Electric
- Pneumatic

5. *Control mode*:

- Point-to-point control (PTP)
- Continuous path control (CP)
- Controlled path control

6. *Lift capacity*:

- 0–5 kg
- 5–20 kg
- 20–40 kg and so forth

ROBOT WORKCELL

Industrial applications of robots almost always involve other pieces of equipment, such as machine tools, conveyors, sensors, and fixtures, in addition to the robot itself. Each piece of equipment performs some function in the cell, and for the cell to perform properly, all of these functions must be sequenced and co-ordinated with the actions of the robot. The function of workcell control is to provide this sequencing and co-ordination of the cell components. Robot workcell is shown in Figure 13.20.

FIGURE 13.20 Robot Workcell.

A control unit called the workcell controller accomplishes workcell control. In most applications, the robot's controller serves as the workcell controller. A number of input and output ports are included on the back panel of the robot controller to permit electrical connections to be established with the other cell components. Control commands, interlocks, sensor data, and other similar signals enter and leave the controller through these input/output ports. The sequencing and co-ordination of the signals is done by means of the robot program. In some robot applications, the robot controller is inadequate as the workcell control unit. These applications include situations where the robot lacks sufficient input/output capacity to deal with the number of input/output signals that must be coordinated, or where there are multiple robots in the cell (*e.g.*, a robot spot welding line for car bodies). In these cases, a programmable controller or small computer is used as the workcell controller. These control units download commands to the robots (to activate portions of the robot programs) and other components in the cell in order to accomplish the workcell control function.

The workcell controller performs several important functions in the robot installation. The functions can be divided into three categories:

1. Sequence control
2. Operator interface
3. Safety monitoring

   1. Sequence Control: Sequence control is the basic function of the workcell controller. It is concerned with regulating the sequence of activities in the cell. The sequence is determined not only by controlling the activities as a function of time; it is also determined by using interlocks to ensure that certain elements of the work cycle are completed before other elements are started. For example, consider a robot machine loading and unloading application. Input/output interlocks in sequence control are used for purposes such as the following:

- Making sure that the part is at the pickup location before the robot attempts to grasp it.

- Ensuring that the part is properly loaded into the machine before the processing cycle begins.

- Indicating to the robot that the machine cycle is completed and the part is read for unloading.

  2. *Operator Interface*: A means for the operator to interact with the robot cell must be provided. Reasons for establishing the operator interface include the following:

- Programming the robot.

- Participation in the work cycle by a human operator. The human being and the robot each perform a portion of the work in the cell. The human being typically accomplishes tasks that require judgment or sensory capabilities that the robot does not possess. Certain assembly operations fit this category.

- Data entry by the human operator. The data might simply be the part identification so that the robot can use the correct work cycle. In other cases, alphanumeric data (*e.g.*, part dimensions) must be provided.

- Emergency stopping of the cell activities.

There are a number of ways to provide the operator interface: teach pendant, control panel of the robot controller, alphanumeric keyboard and CRT monitor, and emergency stop buttons located in the cell. Alternatives that might be considered include automatic bar code readers and voice input of data. Another important reason for operator interaction with the robot cell is for stopping the work cycle due to emergency conditions. The emergency might result from a malfunction of the robot (or other equipment in the cell), or from a human worker inadvertently intruding into the cell space. Under these circumstances, it would be desirable to stop the action in the cell to prevent harm to either equipment or people. An emergency stop button in the workcell is provided for this purpose.

3. *Safety Control*: Emergency stopping of the robot cycle requires an alert operator to notice the emergency and take positive action to interrupt the cycle. Safety emergencies are not always so convenient as to occur when an alert operator is present. A more automatic and reliable means of protecting the cell equipment and people who might wander into the work zone is called safety monitoring. Safety monitoring (the term *hazard monitoring* is also sometimes used) is a workcell control function in which sensors are used to monitor the status and activities of the cell and to detect the unsafe or potentially unsafe conditions. Various sensors can be used to implement a safety monitoring system in a robot cell. These sensors include simple limit switches that detect whether the movement of a particular component has occurred correctly, temperature sensors, pressure sensitive floor mats, light beams combined with photosensitive sensors, and machine vision systems.

The safety monitoring system is programmed to respond to the various hazard conditions in different ways. These responses might include one or more of the following: complete stoppage of the cell activity, slowing down the robot speed to a safe level (when human beings are present), warning buzzers to alert maintenance personnel of a safety hazard in the cell, and specially programmed subroutines to permit the robot to recover from a particular unsafe event. This last response is an example of programming for automated systems that is called error detection and recovery.

MACHINE VISION

The use of machine vision technology is growing very rapidly, spurred by the need of manufacturers for increasingly fine control over the quality of manufactured parts. Machine vision (MV) is the application of computer vision to industry and manufacturing. Whereas computer vision is mainly focused on machine-based image processing, machine vision most often requires also digital input/output devices and computer networks to control other manufacturing equipment such as robotic arms. Machine vision technology uses an imaging system and a computer to analyze an image and to make decisions based on that analysis. There are two basic types of machine vision applications—inspection and control. In inspection applications, the machine vision optics and imaging system enable the processor to "see" objects precisely and, thus, make valid decisions about which parts pass and which parts must be scrapped. In control applications, sophisticated optics and software are used to direct the manufacturing process. Machine vision-guided assembly can eliminate any operator error that might result from doing difficult, tedious, or boring tasks; can allow process equipment to be utilized 24 hours a day; and can improve the overall level of quality.

Machine vision systems are programmed to perform narrowly defined tasks such as counting objects on a conveyor, reading serial numbers, and searching for surface defects. Manufacturers favor machine vision systems for visual inspections that require high-speed, high-magnification, 24-hour operation, and repeatability of measurements. Machine vision systems typically are fast enough to inspect 100% of the product being processed without slowing up the manufacturing line. Machine vision systems are more adaptable than traditional optical or mechanical sensors. They offer the versatility and flexibility of a robot. When alterations to the manufacturing process are required, these systems are easily reconfigured, often with only minor software changes.

The following process steps are common to all machine vision applications:

1. *Image acquisition*: An optical system gathers an image, which is then converted to a digital format and stored into computer memory.

2. *Image processing*: A computer processor uses various algorithms to enhance elements of the image that are of specific importance to the process.

3. *Feature extraction*: The processor identifies and quantifies critical features in the image (*e.g.*, the position of holes on a printed circuit board, the number of pins in a connector, the orientation of a component on a conveyor) and sends the data to a control program.

4. *Decision and control*: The processor's control program makes decisions based upon the data. Are the holes within specification? Is a pin missing? How must a robot move to pick up the component? Machine vision technology is used extensively in the automotive, agricultural, consumer product, semiconductor, pharmaceutical, and packaging industries, to name but a few. Some of the hundreds of applications

include vision-guided circuit board assembly, and gauging of components, razor blades, bottles and cans, and pharmaceuticals.

Components of a Machine Vision System

A simple machine vision system will consist of the following:

- An optical sensor
- A black-and-white camera
- Lighting
- Camera interface card for computer, known as "framegrabber"
- Computer software to process images
- Digital signal hardware or a network connection to report results.

The *optical sensor* determines when a part moving on a conveyor is in position to be inspected. The optical sensor triggers the *camera* to take a picture of the part as it passes beneath the camera and lighting. The *lighting* used to illuminate the part is designed to highlight features of interest and obscure or minimize the appearance of features that are not of interest.

The camera's image is captured by the *framegrabber*. A framegrabber is a computer card that converts the output of the camera to digital format and places the image in computer memory so that it may be processed by the machine vision software.

The *software* will typically take several steps to process an image. Often the image is first manipulated to reduce noise or to convert many shades of gray to a simple combination of black and white. Following the initial simplification, the software will count, measure, and/or identify objects in the image. As a final step, the software passes or fails the part according to programmed criteria. If a part fails, the software may signal a mechanical device to reject the part; alternately, the system may stop the production line and warn a human worker to fix the problem that caused the failure.

Though most machine vision systems rely on black-and-white cameras, the use of color cameras is becoming more common. It is also increasingly common for machine vision systems to include digital camera equipment for direct connection rather than a camera and separate framegrabber.

Advantages of Machine Vision

A machine vision system that has been carefully engineered to meet a well-defined set of requirements will perform well and be very cost effective. Specific advantages of such a system include the following:

1. *Precision*: The well-designed vision system is capable of measuring dimensions to one part in a thousand or better. Because these measurements do not require contact, there is no wear or danger to delicate components.

2. *Consistency*: Because vision systems are not prone to the fatigue suffered by human operators, operational variability is eliminated. Furthermore, multiple systems can be configured to produce identical results.

3. *Cost effectiveness*: With the price of computer processing dropping rapidly, machine vision systems are becoming increasingly cost effective. A $10,000 vision system can easily replace three human inspectors who each earn $20,000 per year or more. Furthermore, the operating and maintenance costs of vision systems are low.

4. *Flexibility*: Vision systems can make a wide variety of measurements. When applications change, the software can be easily modified or upgraded to accommodate new requirements.

Applications of Machine Vision

The applications of MV include:

- Large-scale industrial manufacture
- Short-run unique object manufacture
- Safety systems in industrial environments
- Inspection of pre-manufactured objects (e.g., quality control, failure investigation)
- Visual stock control and management systems (counting, barcode reading, store interfaces for digital systems)
- Control of automated guided vehicles (AGVs)
- Automated monitoring of sites for security and safety
- Monitoring of agricultural production
- Quality control and refinement of food products
- Retail automation
- Consumer equipment control
- Medical imaging processes (e.g., interventional radiology)
- Medical remote examination and procedures

ROBOTICS AND MACHINE VISION

The machine vision system provides robots with eyes. The intelligent sensing system can be used to detect locations, identify items, and even measure the dimensions of processable targets. The need for standard robot programming in robot welding applications is reduced and accurate 3D CAD information is no longer required. Machine vision enables robots to navigate orchards and battlefields or to follow precisely the contours of a fighter aircraft. Image analysis and pattern recognition algorithms locate tumors in the body, identify customers at bank automatic teller machines (ATMs), and detect hidden cracks in machined parts.

Here is an example of machine vision used together with robotics to perform a task better than a human can.

"A manufacturer of anti-lock brake sensors experienced recall problems due to the high failure rate of components. The main cause of the failures was the wires to the sensors, which were damaged. The cause of the damage was the fatigue of the worker, who tucked the wires around the body of the sensor. A machine vision system was installed and connected to a robot. It was able to perform the task using a three-dimensional machine vision system, which controlled the robot. The failure rate has been reduced to zero and the system works three shifts a day without breaks. The solution has eliminated a tedious and repetitive task, which often leads to an unpleasant working environment."

Reasons for Increased Adoption of Vision-guided Robotics

- Production demands for higher speed and higher flexibility.
- Increased use of vision guided robots has become a standard part of production equipment and are designed into the application from the beginning.
- The need for more flexibility.
- The need for higher automation to save labor costs.
- The need for higher accuracy in production.
- The increasing acceptance of the technology.
- Decreasing investment for simple applications.
- More user friendly MMI.

Applications of Vision-guided Robots

Industries of all types are embracing vision-guided robots because of the benefits the technologies provide: lower costs, flexibility, reliability, and safety, just to name a few. Every industry with high production costs or critical production steps are potentially interested in vision-guided robotics. Some of the industries using vision-guided robots are:

- Automotive, automotive suppliers: part handling, assembly, gauging, robot-guided inspection, and logistics
- Electronics: pick & place
- Packaging: pick & place
- Machining: loading & unloading
- Press shops: loading & unloading
- Aerospace: rivet-robot
- Ceramics: part handling
- Consumer goods
- Packaging of consumer goods
- Warehouse logistics and object handling

Machine Vision Technologies

- Image processing
- Scene analysis
- Pattern recognition
- Model-based vision
- Sensor fusion
- Motion analysis
- Infrared image analysis
- Dimensional inspection
- Industrial inspection and process control

ROBOTIC ACCIDENTS

Robot accidents occur during programming, program touchup, maintenance, repair, testing, setup, or adjustment rather than under normal operating conditions. The operator, programmer, or corrective maintenance worker may temporarily be within the robot's working envelope where unintended operations can result in injury. For example:

1. A robot's arm functioned erratically during a programming sequence and struck the operator.

2. A materials-handling-robot operator entered a robot's work envelope during operations and was pinned between the back end of the robot and a safety pole.

3. A fellow employee accidentally tripped the power switch while a maintenance worker was servicing an assembly robot. The robot's arm struck the maintenance worker's hand.

Causes of Robot Accidents

Unsafe Acts

- Placing oneself in hazardous positions while programming or performing maintenance within the robot's work envelope.

- Inadvertently entering the envelope because of unfamiliarity with the safeguards in place or not knowing if they are activated.

- Making errors in programming, interfacing peripheral equipment, and connecting input/ output sensors.

Unsafe Conditions

- Mechanical failure.

- Safeguards deactivated.

- Hazards from pneumatic, hydraulic, or electrical power that can result from malfunction of control or transmission elements of the robot's power system such as control valves, voltage variations, or voltage transients disrupting the electrical signals to the control and/or power supply lines.

- Electrical shock and release of stored energy from accumulating devices that can result in injury to personnel.

Prevention of Accidents

System components must be designed, installed, and secured so that the hazards associated with stored energy are minimized. Adequate room must be provided for a robot's movement as well as for workers. There must be a means for controlling the release of stored energy in all the robotic systems and for shutting off power from outside the restricted envelope. A detailed risk assessment should be performed to ensure the safety of workers who operate, service, and maintain the robotics system.

ROBOTICS AND SAFETY

Safety is everyone's responsibility. The present-day industrial robot is a dumb, senseless idiot. Therefore, safety in robotics must be managed by humans. In matters of robot safety, the safety of humans should come first, then the safety of the robot, and finally the safety of other related equipment. To prevent hazards to the system, the robot systems operator should ensure that all appropriate safeguards are established for all robot operations. The most dangerous situation in which a human must work with a robot is when repairing it. The next most dangerous situation in which a human must work with a robot is during the robot's training or programming. The least dangerous situation in which humans must work with a robot is during the robot's normal operation. Both workers and visitors need to be protected from robots. Robots can injure people in many ways either through bodily impact or by pinning the human against some structure. All workers should be educated about the safety issues involved in working with robots or other equipment. When arc-welding robots are used, shields or curtains should be placed around the welding area to protect passerby from the bright light of the arc. Whenever repair personnel works with a live robot, they should know the location of nearest emergency stop button.

Safeguarding Devices

Personnel should be safeguarded from hazards associated with the restricted envelope (space) through the use of one or more safeguarding devices:

- Mechanical limiting devices;
- Non mechanical limiting devices;
- Presence-sensing safeguarding devices;
- Fixed barriers (which prevent contact with moving parts); and
- Interlocked barrier guards.

ROBOTS MAINTENANCE

The user of a robot or robot system should establish a regular and periodic inspection and maintenance program to ensure safe equipment operations. This program should include, but not be limited to, the recommendations of the robot manufacturer and the manufacturers of other associated robot system equipment such as conveyer mechanisms, part feeders, tooling, gages, and sensors. These recommended maintenance programs are essential for minimizing the hazards that can result from component malfunction, breakage, and unpredicted movements or actions of the robot or other system equipment. To ensure that robots are safely and adequately maintained, it is recommended that periodic maintenance be conducted and documented, including the names of the personnel who perform maintenance and the names of the independent verifiers.

ROBOTS INSTALLATION

A robot or robot system should be installed by the users in accordance with the manufacturer's recommendations and in conformance to acceptable industry standards. Temporary safeguarding devices and practices should be used to minimize the hazards associated with the installation of new equipment. The facilities, peripheral equipment, and operating conditions that should be considered are:

- Installation specifications;
- Physical facilities;
- Electrical facilities;
- Action of peripheral equipment integrated with the robot;
- Identification requirements;
- Control and emergency stop requirements; and
- Special robot operating procedures or conditions.

EXERCISES

1. How are robots classified based on geometry?
2. Identify various types of robot control.
3. Distinguish between point-to-point and continuous path motion of a robot.
4. What is a work envelope of a robot?
5. Define accuracy.
6. Define precision.
7. What is the importance of the work envelope of a robot?
8. Classify robots based on their geometry.

9. Distinguish between accuracy and repeatability of a robot.

10. Classify robots on the basis of path movement.

11. By means of a sketch, explain the construction of a polar type robot.

12. List the parameters used for selection of a robot for a particular application.

13. State the major subsystems of a robot and their functions. Show these subsystems on a sketch.

14. Describe briefly the various types of motion controls possible in robots.

15. Define machine vision. What are the components of machine vision?

16. Name two methods by which path is controlled by a robot controller.

17. Explain the anatomy of a robot manipulator in detail?

18. Distinguish between accuracy and repeatability.

19. What do you understand by degree of freedom of robot?

20. Define speed of response.

21. What are the characteristics of an industrial robot?

22. Classify robots based on their geometry. Explain the application pertaining to each class.

23. With the help of sketches, discuss the common robot configurations.

24. Briefly explain the various components of an industrial robot. State the main function of each of the components.

25. Define a robot.

26. List the advantages and disadvantages of robots.

27. Explain the following terms with reference to a robot: repeatability and accuracy.

28. Explain briefly non-servo controlled and servo controlled robots.

29. List the various types of joints used in robots.

30. Sketch the revolving joint and show the relative joint motions.

31. What are the laws of robotics?

32. What are the types of wrist motions?

33. What is a robotic joint? Distinguish between prismatic and revolute joints.

34. What are the various technologies of machine vision?

35. How is robotic vision sensed? What are the component systems used in most common vision based applications?

36. What are the causes of robotic accidents? How can accidents be prevented?

37. Draw a sketch showing all the subsystems of an articulated type of robot.

38. Discuss the anatomy of a robot and explain the important parts of a robot with a sketch.

39. Explain the following terms:
    - Accuracy
    - Degree of freedom
    - Repeatability
    - Speed

40. Sketch an articulated arm type robot and label its parts.

41. Differentiate the four common types of robot configurations with the help of sketches.

CHAPTER 14

ROBOTIC SENSORS

INTRODUCTION

Sensors serve as a robot's sight, hearing, touch, taste, and smell. Without sensors, a robotic device would not be able to discern anything about current surroundings. A signal is returned from the sensor to the robot CPU and is applied to the current situation or saved for later analysis. Without sensors, a robot is just a machine. Sensors provide feedback to the control systems and give the robots more flexibility. As robots are used in hazardous areas and industrial processes, they are fitted with sensors to gather information about the nearby environment and objects around, to come out with precise results of the work. Robots need sensors to deduce what is happening in their world and to be able to react to changing situations. Sensors are used to gain information about the world, which is useful for guiding a robot to achieve a particular goal. The control of a manipulator or industrial robot is based on the correct interpretation of sensory information. This information can be obtained from a robot either internally (for example, joint positions and motor torque) or externally using a wide range of sensors.

For a robotic device to be useful, it needs to have some information of what is currently happening in the environment around it. Whether this information is the temperature of the air, the distance from an object, or how much pressure a robotic arm is applying, the sensor is the piece of equipment providing this information. Depending on the job requirements of the robot, sensors are needed to provide extremely accurate information. As with most things, higher performance comes at a cost. A robot built by a hobbyist may employ sensors costing less than a dollar, while a robot built for industrial use may utilize sensors costing hundreds of dollars.

TYPES OF SENSORS IN ROBOTS

The basic function of a sensor is to measure some feature of the world, such as light, sound, or pressure and convert that measurement into an electrical signal, usually a voltage or current. Typical sensors respond to stimuli by changing their resistance (photocells), changing their current flow (phototransistors), or changing their voltage output (the sharp IR sensor). The electrical output of a given sensor can easily be converted into other electrical representations. The control of the manipulator or industrial robot is based on correct interpretation of sensory information. Sensing includes activities such as seeing, hearing, touching, smelling, and measuring distance. A wide variety of devices are used by robots to obtain information. They include not only transducers for physical quantities such as microphones for sounds, but also data processing input devices such as keyboard for textual information and specialized sensors. Sensors in robots can be classified as:

- Exteroceptors or External Sensors (for the measurement of robot's environmental parameters).

- Proprioceptors or Internal Sensors (for the measurement of robots internal parameters).

EXTEROCEPTORS OR EXTERNAL SENSORS

Exteroceptors are sensors that measure the positional or force-type interaction of the robot with its environment. These sensors are added to robots to perceive the world in which they operate and interact with the environment outside the robot. External sensors can be categorized as:

- Contact sensors
- Non contact sensors

Contact Sensors

Contact sensing is one of the most basic requirements for any manipulator interacting physically with the environment in a non-structured manner. It is the ability of the robot to determine the shape, size, weight, or even the surface texture of an object by touching it. The contact sensors are situated in the design of the gripper of the robot to provide the robot with information about the forces in the wrist or the joint of the robot and also in the objects that are to be handled. It is very important for robots to determine this information especially in fettling and grinding operations and in assembly operations. The functions of contact sensors in controlling manipulation may be classified into the following basic material handling and assembly operations:

- *Searching*—detecting a part by sensitive touch sensors on the hand exterior without moving the part.

- *Recognition*—determining the identity, position, and orientation of a part, again without moving it, by sensitive touch sensors with high spatial resolution.

- *Grasping*—acquiring the part by deformable, roundish fingers, with sensors mounted on their surfaces.

- *Moving*—placing, joining, or inserting a part with the aid of force sensors.

   Most important types of robotic sensors of contact type are:

- Tactile sensors
- Force sensors

Non-contact Sensors

These sensors are used to give the robot information about the process or the environment without the use of physical contact. Non-contact sensors include:

- Pneumatic sensors, which detect part presence by air, flow disturbance.

- Ultrasonic sensors that analyze sound waves reflected from a part.

- Proximity sensors that register the approach, arrival, or removal of parts.

- Optical sensors utilizing interrupted light beams across the path of an incoming part.

- Machine vision systems that use visual sensors, usually video cameras, to provide data that allows the robot to make intelligent decisions regarding parts.

TACTILE SENSORS

For a robot to accomplish light, delicate tasks, the end effector must possess human hand like qualities. It must have a sense of touch. Touch is of particular importance for close-up assembly work and for providing feedback necessary to grip delicate objects firmly, without causing damage. Tactile sensors are used to identify and to control directly the interaction between the robot end effector and the environment. The interaction with the object may occur by direct contact. These sensors either detect when the hand touches something, or they measure some combination of force and torque components that the hand is exerting on an object. These sensors sense the external condition of an object, the roughness or smoothness of its surface, its slipperiness, its elasticity, etc. This sort of information is necessary in order to position the robot manipulators on an object and to compute the force and pressure that must be applied, say to grasp and lift the object. Tactile sensors find their use in various locations of a robot system including joints between the robot links, wrist between arm and end effector, the finger points and the fingertip. Tactile sensors are important for assembly and for quality. Sensors useful in these areas include pressure sensors used for part orientation and gripping force control, piezoelectric sensors for continuous pressure, etc. Tactile sensors can be classified into:

- Touch sensors
- Stress sensors

Touch Sensors

Touch sensors produce a binary output signal, depending upon whether or not they are in contact with something. Touch sensors may be mounted on the outer and inner surfaces of each finger. The outer sensors may be used to search for an object and possibly determine its identity, position, and orientation. The inner mounted sensors may be used to obtain information about an object before it is acquired and about grasping forces and workpiece slippage during acquisition. The simplest kind of touch sensors require no specific sensor device at all if the objects they are going to touch are electrically conductive. Microswitchs are the most commonly used and least expensive kind of touch sensor which either turns on or off as contact is made. A microswitch is shown in Figure 14.1.

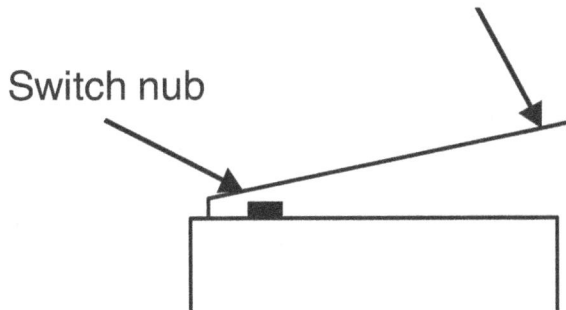

Switch nub

FIGURE 14.1 A Typical Microswitch.

Often, the switch is simply mounted on a robot so that when the robot runs into something, the switch is pressed, and the microprocessor can detect that the robot has made contact with some object and take appropriate action. The bumper skirt on the robot shown in Figure 14.2 is an example of a touch sensor. When the robot runs into a wall, the bumper skirt hits a micro switch which lets the robot controller know that the robot is up against a wall. Other types of touch sensors are used internally to let the robot know when an arm is extended too far and it should be retracted.

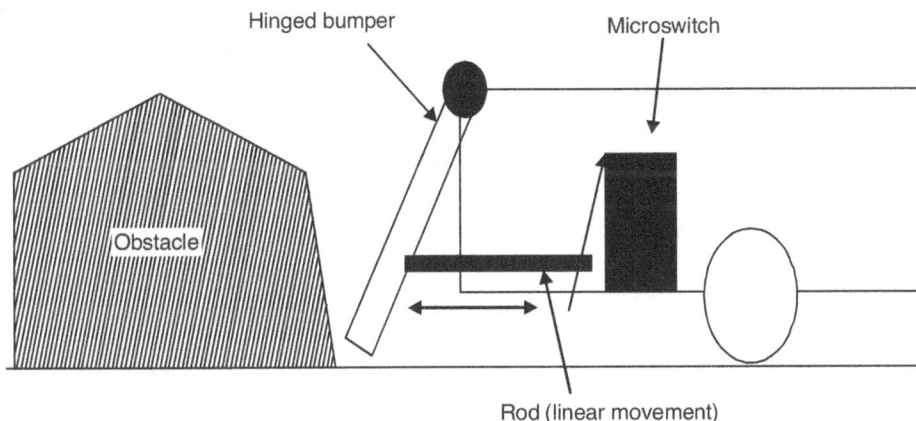

Hinged bumper   Microswitch

Obstacle

Rod (linear movement)

FIGURE 14.2 Robotic Platform Employing Bumper Coupled to Touch Sensor.

Touch sensors can also serve as *limit switches* to determine when some movable part of the robot has reached the desired position. For example, if a motor drives a robot arm, perhaps using a gear rack, touch switches could detect when the arm reached the limit of travel on the rack in each direction.

Force/Torque Sensors

Force/torque sensors measure the amount of force and torque exerted by the mechanical hand. These sensors are normally used to measure the robotic system's forces as it performs various operations. Forces, which the robot uses in manipulation and assembly, are usually of great importance. The major parts of these devices are transducers, which measure force/torque. The interaction forces and torques, which appear during mechanical assembly operations at the robot hand level, can be measured by sensors mounted on the joints or on the manipulator wrist. Individual stress sensors usually respond only to force in one direction on them. However, the combination of two or more can report forces as well as torques in two or three directions. Strain gauges are used to make force sensors, torque sensors, and sensors that can measure both kinds of stress simultaneously.

The simplest type of tactile sensor is a gripper that is equipped with an array of miniature micro switches. This type can only determine the presence or absence of an object at a particular point or array of points. A more advanced type of tactile sensor uses arrays of pressure-sensitive piezoelectric material. This material conducts an electrical current when stressed. The more pressure applied to the material, the more electrical current is produced. This aspect allows the sensor to perceive changes in force and pressure. The tactile sensor can also be used in force feedback applications. This is essential where the gripper is handling delicate, fragile objects so that the robot will not apply too much force and crush the object it is holding. For tactile sensing to be truly useful in adaptive assembly tasks, the sensor must be capable of sensing physical quantities like the human hand. These qualities include pressure, direction of forces, temperature, vibration, and texture.

PROXIMITY SENSORS (POSITION SENSORS)

A proximity sensor is a device that senses and indicates the presence or absence of an object without requiring physical contact. This helps the workstation controller to safely move the end effectors quickly towards that object even if its position is not precisely known. The signal from the proximity detector would give the workstation controller the warning it would need in order to slow down and avoid a collision. Most proximity sensors indicate only the presence or absence of an object within their sensing region but some can give some information about the distance between the object and the sensor as well. Proximity sensors are classified according to their operating principle: inductive, Hall effect, capacitive, ultrasonic, and optical. These noncontact sensors have widespread use, such as for high speed counting, protection of workers, indication of motion, sensing presence of ferrous materials, level control, and noncontact limit switches. Some of the common proximity sensors are explained below:

(a) *Optical proximity sensors* operate on either visible or invisible light. These sensors measure the amount of light reflected from an object.

(b) A *photoelectric proximity sensor* uses a light-beam generator, a photodetector, a special amplifier, and a microprocessor. The light beam reflects from an object and is picked up by the photodetector. The light beam is modulated at a specific frequency, and the detector has a frequency sensitive amplifier that responds only to light modulated at that frequency. This prevents false imaging that might otherwise be caused by lamps or sunlight. If the robot is approaching a light-reflecting object, its microprocessor senses that the reflected beam is getting stronger. The robot can then steer clear of the object.

(c) An *acoustic proximity sensor* works on the same principle as sonar. An oscillator generates a pulsed signal, having a frequency somewhat above the range of human hearing. This signal is fed to a transducer that emits ultrasound pulses at various frequencies in a coded sequence. These pulses reflect from nearby objects and are returned to another transducer, which converts the ultrasound back into high-frequency pulses. The return pulses are amplified and sent to the robot controller. The delay between the transmitted and received pulses is timed, and this will give an indication of the distance to the obstruction.

(d) A *capacitive proximity sensor* uses a radio-frequency oscillator, a frequency detector, and a metal plate connected into the oscillator circuit. The oscillator is designed so that a change in the capacitance of the plate, with respect to the environment, causes the frequency to change. This change is sensed by the frequency detector, which sends a signal to the apparatus that controls the robot. In this way, a robot can avoid bumping into things. Objects that conduct electricity to some extent, such as house wiring, animals, cars, or refrigerators, are sensed more easily by capacitive transducers than are things that do not conduct, like wood-frame beds and dry masonry walls.

(e) *Magnetic field sensors* are excellent proximity sensors. Inductive sensors are based on the change of inductance due to the presence of metallic objects. Hall effect sensors are based on the relation, which exists between the voltage in a semiconductor material and the magnetic field across that material. Inductive and Hall effect sensors detect only the proximity of ferromagnetic objects.

RANGE SENSORS

A range sensor is a device that can provide precise measurement of the distance from the sensor to an object. Range sensors are useful for locating objects within the workstation area and for controlling a manipulator. Range sensing is accomplished by means of television cameras or sonar transmitters and receivers. They are used for robot navigation, obstacle avoidance, or to recover the third dimension for monocular vision. The main goal of any ranging system is to repeatedly obtain accurate range information of surroundings. Ranging systems are usually used in automated guided vehicles or where the robot has a large workspace. Range-imaging sensors have been applied primarily to object recognition. However, they are also very suitable for other tasks, such as finding a factory floor or a road, detecting obstacles and pits, and inspecting the completeness of subassemblies.

Range sensors are based on one of these two principles: *time-of-flight* and *triangulation*. *Time-of-flight sensors* estimate the range by measuring the time elapsed between the transmission and return of a pulse. Laser range finders and sonar are the best-known sensors of this type. *Triangulation sensors* measure range by detecting a given point on the object surface from two different points of view at a known distance from each other. Knowing this distance and the two view angles from the respective points to the aimed surface point, a simple geometrical operation yields the range. The range sensors are classified into two categories:

■ *Passive devices*, such as stereoscopic vision systems and,

■ *Active devices*, such as ultrasonic ranging systems.

(a) *Stereoscopic vision system*: the main problem with range sensing occurs when the transmitter does not sense all objects in the workspace. The use of additional transmitters helps to reduce, but not eliminate the problem. In the case of a stereoscopic vision system, a single projector with multiple cameras is used to view the target area from different angles. The system senses more data than a single-camera system, but requires more hardware, software, and computing time to work.

(b) *Ultrasonic ranging systems*: Ultrasonic ranging systems like the one used on the automatic-focusing polaroid camera have been widely used to give environmental awareness to a mobile robot. An ultrasonic sensor determines the range by measuring the elapsed time between the transmission of certain frequencies and their detected echoes. Different discrete frequencies are used because surface characteristics could cancel a single waveform, preventing detection.

MACHINE VISION SENSORS

Robot vision is a complex sensing process. It involves extracting, characterizing, and interpreting information from images in order to identify or describe objects in the environment. The repeatability and accuracy of a vision system, the ability to produce approximately the same results when given approximately the same inputs, are its greatest virtues. It can perform simple tasks such as monitoring and inspection faster and more reliably. A vision sensor (camera) converts the visual information to electrical signals, which are then sampled and quantized by a special computer interface electronics yielding a digital image. Solid-state CCD image sensors have many advantages over conventional tube-type sensors as: small size, lightweight, more robust, better electrical parameters, which recommend them for robotic applications. Virtually all-existent vision sensors are designed for television, which is not necessarily best, suited for robotic applications. Because of the reduced resolution, and robot hand obstructing the field of view, the common wisdom approach of placing camera above the working area is of questionable value for many robotic applications.

Mounting the vision sensor in the robot hand may be a better solution, which eliminates these problems. Illumination is a very important component of the image acquisition. Controlled illumination offers expedient solutions to many robotic vision problems. Vision sensors are used extensively in inspection handling and acquisition of parts, say from an assembly belt. In addition, vision is used in geometric analysis of shape to determine the points of a part that have to be grasped by the robot.

A general vision system consists of a light source, an image sensor, an image digitizer, a system control computer, and some form of output.

The image sensor of a machine vision system is defined as an electro optical device that converts an optical image to a video signal. The image sensor is usually a vacuum tube TV camera or a solid state-sensing device. *Tube-type cameras*: Vidicons are the most common tube-type cameras. The image is focused on a photosensitive surface where a corresponding electrical signal is produced by an electron beam scanning the photosensitive surface. The electron beam passes easily through the photo sensor at a highly conductive point caused by very intense light. Fewer electrons pass through the photo sensor where lower light levels have made it less conductive. Scanning the electron beam carefully across the entire surface produces electrical information about the entire image.

VELOCITY SENSORS

They are used to estimate the speed with which a manipulator is moved. The velocity is an important part of the dynamic performance of the manipulator. The DC tachometer is one of the most commonly used devices for feedback of velocity information. The tachometer, which is essentially a DC generator, provides an output voltage proportional to the angular velocity of the armature. This information is fed back to the controls for proper regulation of the motion.

PROPRIOCEPTORS OR INTERNAL SENSORS

From a mechanical point of view, a robot appears as an articulated structure consisting of a series of links interconnected by joints. Each joint is driven by an actuator, which can change the relative position of the two links connected by that joint. Proprioceptors are sensors measuring both kinematic and dynamic parameters of the robot. Based on these measurements, the control system activates the actuators to exert torques so that the articulated mechanical structure performs the desired motion. The most common joint (rotary) position transducers are: potentiometers, synchros and resolvers, encoders, RVDT (rotary variable differential transformer), etc.

Encoders are digital position transducers, which are the most convenient for computer interfacing.

Incremental encoders are relative-position transducers, which generate a number of pulses proportional with the traveled rotation angle. They are less expensive and offer a higher resolution than the absolute encoders. As a disadvantage, incremental encoders have to be initialized by moving them in a reference ("zero") position when power is restored after an outage.

Absolute shaft encoders are attractive for joint control applications because their position is recovered immediately and they do not accumulate errors as incremental encoders may do. Absolute encoders have a distinct n-bit code (natural binary, Gray, BCD) marked on each quantization interval of a rotating scale. The absolute position is recovered by reading the specific code written on the quantization interval that currently faces the encoder reference marker.

Joint position sensors are usually mounted on the motor shaft. When mounted directly on the joint, position sensors allow feedback to the controller with the joint backlash and drive train compliance parameters. Angular velocity is measured (when not calculated by differentiating joint positions) by tachometer transducers. A tachometer generates a DC voltage proportional to the shaft rotational speed. Digital tachometers using magnetic pickup sensors are replacing traditional, DC motor-like tachometers, which are too bulky for robotic applications.

Strain gages mounted on the manipulator's links are sometimes used to estimate the flexibility of the robot's mechanical structure. Strain gages mounted on specially profiled (square, cruciform beam or radial beam) shafts are also used to measure the joint shaft torques.

ROBOT WITH SENSORS

Because sensors are any devices that provide input of data to the robot controller a wide verity of sensors exist. Figure 14.3 shows the some of the basic types of sensors used in robot. These are:

- *Light sensors*, which measure light intensity.
- *Heat sensors*, which measure temperature.
- *Touch sensors*, which tell the robot when it bumps into something.
- *Ultra sonic rangers*, which tell the robot how far away objects are.
- *Gyroscopes*, which tell the robot which direction is up.

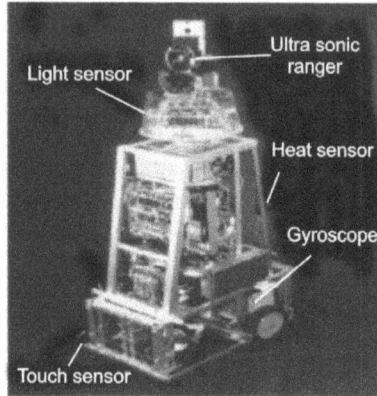

FIGURE 14.3 Sensors in Robots.

Light Sensors

Light sensors are used to detect the presence and intensity of light. These can be used to make a light-seeking robot and are often used to simulate insect intelligence in robots.

Heat Sensors

Heat sensors help robots determine if they are in danger of overheating. These sensors are often used internally to make sure that the robot's electronics do not breakdown.

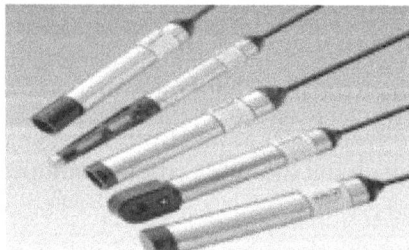

Ultra Sonic Sensors

Ultra sonic sensors are used to determine how far a robot is away from object. They are often used by robots that need to navigate complicated terrain and cannot risk bumping into anything.

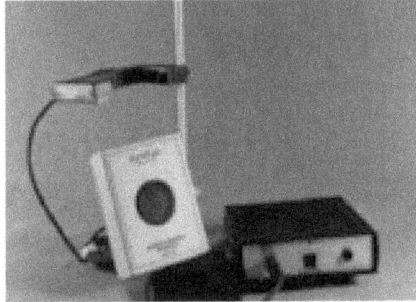

Gyroscopes

Gyroscopes are used in robots that need to maintain balance or are not inherently stable. Gyroscopes are often coupled with powerful robot controllers that have the processing power necessary to calculate thousands of physical simulations per second.

EXERCISES

1. What is a sensor?

2. Identify various types of sensors used in robots.

3. What is a proximity sensor? Name three techniques for designing proximity sensors.

4. Differentiate between contact and non-contact sensors in robots with the help of examples. Describe the construction and working principle of one contact and one non-contact type of sensor.

5. Differentiate between a proximity and a range sensor.

6. What are touch sensors? Describe its different types along with their advantages and disadvantages.

7. What are machine vision sensors?

8. What are the classifications of proximity sensors?

9. What do you understand by tactile sensors?

10. Define range sensors and its types.

11. Explain the principle of working of the following sensors used in robots:

- Range sensor
- Proximity sensor
- Tactile sensor
- Magnetic sensor

CHAPTER 15

ROBOT END EFFECTORS

INTRODUCTION

The function of a robot is to interact with the surroundings. The robot does this by manipulating objects and tools to fulfill a given task. The robot end effector becomes a bridge between the computer controlled arm and the world around it. Earlier industrial robots were used as stand alone machines for painting, spot welding, or pick and place work in which parts were moved from one location to another without much attention paid to how the parts were picked up and put down. Nowadays, robots are put to work in more challenging applications. The tasks the robot perform may involve assembling parts or fitting them into clamps and fixtures. These tasks place greater demands on the accuracy of the arm and the end effector. The actions of the robot and the gripper determine whether the assembly will go smoothly or the parts will get damaged in the process.

END EFFECTOR

One of the most important areas in the design of robot systems is the design of end effectors. Most of the problems that occur in production are caused by badly designed tooling and not by faults in the robots.

In robotics, an *end effector* is a device or tool connected to the end of a robot arm. Basically, it is a tool to grip, hold, and transport objects and position them in a desired location. *End effectors* are the devices through which a robot interacts with the world around it, grasping and manipulating parts, inspecting surfaces, and working on them. As such, end effectors are among the most important elements of a robotic application and an integral component of the overall tooling, fixturing, and sensing strategy. An end effector attached to the robotic arm is shown in Figure 15.1.

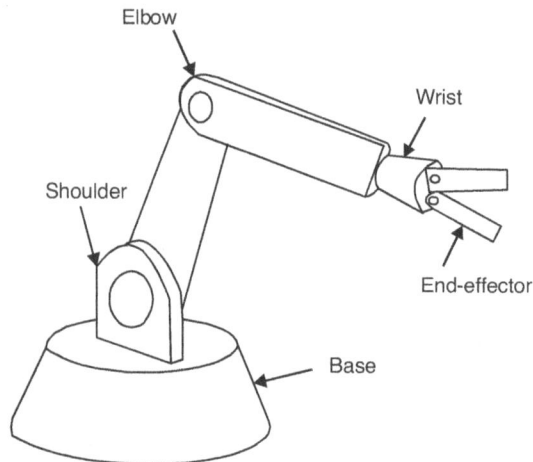

FIGURE 15.1

An end effector is often different from a human hand. It could be a tool such as a gripper, a vacuum pump, tweezers, scalpel, blowtorch, etc. or just about anything that helps to do the job. A robotic end-effector is any object attached to the robot flange (wrist) that serves a function. This would include robotic grippers, robotic tool changers, robotic collision sensors, robotic rotary joint, robotic press tooling, compliance device, robotic paint gun, robotic deburring tool, robotic arc welding gun, etc. Robot end effectors are also known as robotic peripherals, robotic accessories, robot tools or robotic tools, *end of arm tooling (EOA)*, or end-of-arm devices. Some robots can change end-effectors, and be reprogrammed for a different set of tasks. If the robot has more than one arm, there can be more than one end-effector on the same robot, each suited for a specific task.

End-of-arm tooling is defined as the subsystem of an industrial robot system that links the mechanical option of the robot (manipulator) to the art being handled or worked on. An industrial robot is essentially a mechanical arm with a flat tool mounting plate at its end that can be moved to any spatial point within its reach. End-of-arm tooling in the form of specialized devices to pick up parts or hold tools to work on parts is physically attached to the robot's tool mounting plate to link the robot to the workpiece. A robot can become a production machine only if an end effector has been attached to its mechanical arm by means of the tool mounting plate as shown in Figure 15.2. Tool mounting plate is for the interface between end effector and the controller.

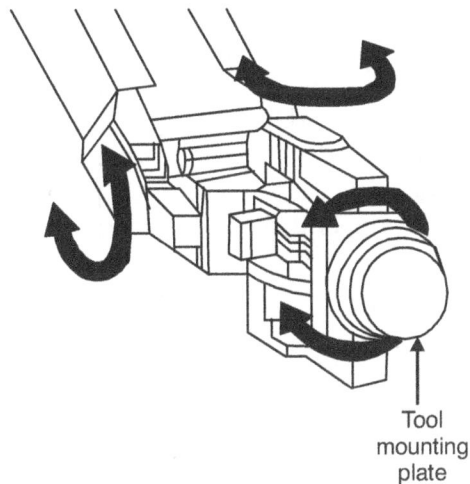

Tool
mounting
plate

FIGURE 15.2

The structure of an end effector, and the nature of the programming and hardware that drives it, depends on the intended task. If a robot is designed to set a table and serve a meal, then robotic hands, more commonly called grippers, are the most functional end effectors. The same or similar gripper might be used, with greater force, as pliers or a wrench for tightening nuts or crimping wire.

CLASSIFICATION OF END EFFECTORS

An end effector covers a wide range of different tools and devices that can be classified into two main categories:

- Grippers
- Tools for process applications

GRIPPERS

Grippers are end efectors that actually grip a part for transfer operations in the work envelope of the robot. A *gripper* is a component of the robot used to manipulate an object loose from the robot itself. This can be a component it needs to pick up, or the waste it is programmed to find and dispose of. Grippers include mechanical hands and also anything like hooks, magnets, and suction devices, which can be used for holding or gripping. Grippers take advantage of point-to-point control (exact path that the robot takes between what it is picking up and where it is placing it). Grippers should be designed so that it requires the minimum amount of maneuvering in order to grip the workpiece. Grippers are used in applications for material handling, machine loading, and assembling. In many grippers, the mechanism is activated through a pneumatic piston that moves the gripper's fingers. A robot gripper is shown in Figure 15.3.

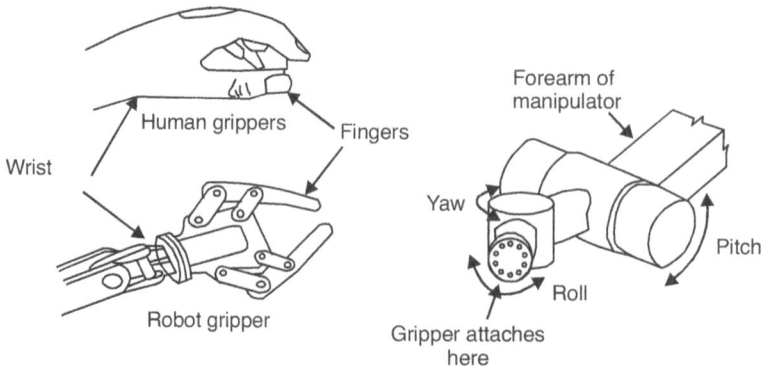

FIGURE 15.3 Robot Gripper.

There are four main categories, which makes use of a gripper:

1. *No gripping*: In this situation, the workpiece is held in a jig (a specially designed purpose built holder) and the robot performs an activity on it. Jobs that use no gripping can include spot welding, flame cutting, and drilling.

2. *Coarse gripping*: In this case, the robot holds the workpiece but the gripping does not have to be precise. Jobs that use coarse gripping include handling and dipping castings, unloading furnaces, stacking boxes, or sacks.

3. *Precise gripping*: A robot holds the workpiece, which requires accurate positioning, for example, unloading and loading machine tools.

4. *Assembly*: The robot is required to assemble parts, which require accurate positioning, and some form of sensory feedback to enable the robot to monitor and correct its movements.

SELECTION OF GRIPPER

The selection of gripper is usually made by examining the geometry of the part, its orientation, the space available, and the manufacturing treatment to be performed. External gripping is the most widely used type, where the closing force is utilized to clench the part. An internal grip makes way for unobstructed access to the outside surface of the part, which is necessary for polishing/buffing, grinding, or painting applications. The opening force of the gripper is used to hold the part. There are numerous types of grippers both in style and power source. Determining which is the best type to use is an important issue that robotics users must face. Selection of the gripper is based upon a number of factors that may need consideration:

- Source of power
- Gripping force
- Gripping style
- Weight
- Environmental capabilities
- Sensor capabilities
- Number of jaws
- Other factors

Source of Power

According to source of power, grippers are classified as pneumatic, hydraulic, and electrical.

(a) *Pneumatic grippers* have no motors or gears. So it is simple to translate the power of a piston/cylinder system into a gripping force. Most manufacturing facilities already have compressed air, so little effort is required to bring it to a gripper in a cost efficient manner. Pneumatics put out a high amount of gripping force in a small and light package. This is critical when limited space is a factor.

There are two main types of pneumatic grippers: *angular* and *parallel*. The main difference between them is how their jaws move. The jaws of an angular gripper swing open and closed on pivots. The jaws of a parallel gripper slide open and closed in tracks. Either style can have two or three jaws. Jaw movement is usually synchronized, but some models can be modified so the jaws can move independently. Parallel grippers can grab a part from the outside or the inside. The jaws can be actuated directly by the piston or indirectly through a cam or wedge mechanism. The former have longer stroke lengths, but the latter produce more gripping force.

Angular grippers cost less than parallel grippers, which require more precise machining and more expensive bearings for the jaws. They can only grasp parts from the outside. Compared with parallel grippers, angular grippers can't handle a wide range of part sizes, and they don't resist side loads as well. On the plus side, the jaws on angular grippers can swing out of the way of incoming parts. This can save time when picking parts off a conveyor, because the actuator does not have to move as much.

(b) *Hydraulic grippers* are a cost effective alternative to pneumatic grippers. The primary advantage of a hydraulic actuator is its gripping power. Hydraulic power is used when there is a need for extra gripping force. Because hydraulic fluid is not a compressible medium as is air, there is more gripping power available. Hydraulic grippers have disadvantages also. Hydraulic grippers are more costly and they are generally less accurate than pneumatic or electric grippers. Hydraulic grippers have the disadvantage of being not suited for clean room applications. Pneumatic grippers are common because they are clean, especially when compared to hydraulics.

There might be more strength in a hydraulic gripper, but they are messy and much more costly to maintain.

(c) Electricity is the other major option when looking for a source of power for robotic grippers. *Electric grippers* are a fairly new technology. They have not been cost effective until recently. Electric grippers are suitable for applications that require high speeds and for those requiring light or moderate grip forces. Electric grippers are cleaner than either pneumatic or hydraulic grippers. Electric grippers do not put out any dirt or particulate, so they are good for clean room operations. The major advantage of electric powered grippers is control. Pneumatic grippers are not very controllable. Typically the valve, which controls the gripper, is fully opened or closed. With electric, it is relatively easy to add a stepper motor to control the gripper as to how far it opens and closes. This adds flexibility and the capability of varying grip force with ease. The disadvantage of electric is the grippers tend to be a bit larger because you have to work the motor in, and they tend to have less force than pneumatic.

Gripping Force

When considering the gripping force required, a number of factors must be considered. Not only must the mass of the object to be gripped be taken into account but also the accelerations imposed on it by the robot must be taken into account. The coefficient of friction between the gripper and the object may also be an important factor. This can often be increased by using one of the special rubber based materials that have been developed. The use of these materials can create maintenance issues, however, as they have a finite life. One other way of reducing the gripping force required is to use fingers designed for the form of the component. This reduces the flexibility of the gripper but dramatically increases the weight carrying capacity.

Gripping Style

The configurations of grippers include *parallel and angular styles*, and those with two or three jaws. Parallel jaws move in a motion parallel in relation to the gripper body, while angular jaws open and close around a central pivot point, moving in an arcing motion. Determining which gripper to use is dependent on the specific needs of an application.

An *angular gripper* is used when one needs to get the tooling out of the way. Angular grippers are usually the most economical. However, because of the sweeping action of an angular gripper, it is not always practical for some applications, and it is more difficult to design.

A *parallel gripper* is used for pulling a part down inside a machine because the fingers fit into small areas better. Space constraints might lead to the use of parallel over angular. Two jaw parallel grippers are the most widely used. They are the most popular type because two jaw parallel grippers are easier to design and program as there is only axis of motion.

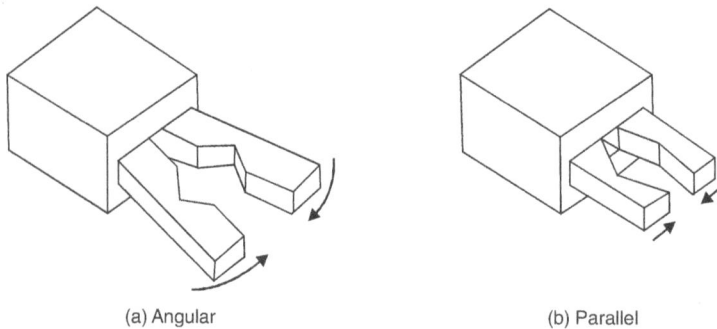

(a) Angular   (b) Parallel

FIGURE 15.4 Angular and Parallel Grippers.

Angular and parallel grippers are shown in Figure 15.4.

A gripper can also be *external or internal*. The external gripper is used to grasp the exterior surface of objects with closed fingers, whereas the internal gripper grips the internal surface of objects with open fingers. External and internal grippers are shown in Figure 15.5.

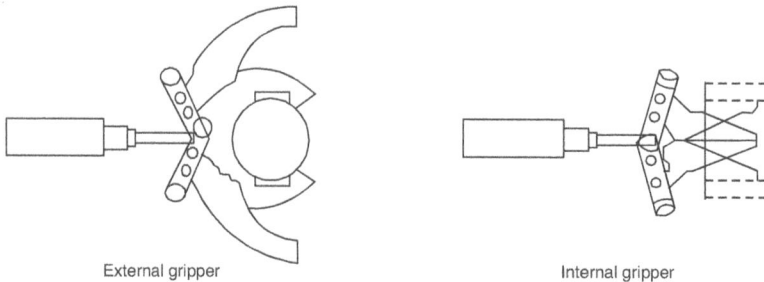

External gripper   Internal gripper

FIGURE 15.5 External and Internal Grippers.

Weight

Industrial robots have fixed lifting capabilities and the combined weight of the gripper and gripped component may be important. Even when this weight is within the capability of the robot, it may cause an unacceptable increase in the cycle time of the operation. The distance between the robot flange and the center of mass may also be important and this should be kept to a minimum.

Environmental Capabilities

End effectors are often required to work in hostile environments. High temperatures, dust, or the presence of chemicals will require special materials or designs to be used.

Sensor Capabilities

For certain applications, some degree of sensory feedback from the gripper is necessary. This may be measurement of insertion or gripping forces or may simply be a proximity sensor to say if anything is between the jaws of the gripper. Some standard grippers are provided with feedback to show the separation of the jaws but most grippers have no feedback.

Number of Jaws (Two or Three Jaw Gripper)

In two jaw units, which come in both parallel and angular styles, the grasper affords two mounting spots for the fingers to hold the part. The jaws move in a simultaneous motion, opening and closing toward the central axis of the gripper body. Three jaws furnish more contact with the part to be clasped and are more accurate in centering than are two jaws.

Others

Other factors to be considered include the speed of the gripper jaws and the range of sizes of the components they can grip. The amount of maintenance required is also important though most modern mechanisms require little or no maintenance. For some situations, the behavior of the gripper on power failure may be critical. Some but not all use either springs to apply the gripping force or non-return valves to ensure that pressure is maintained.

GRIPPING MECHANISMS

A gripper is specifically end of arm tooling that uses a mechanical mechanism and actuator to grasp a part with gripping surfaces. Fingers are designed to:

- Physically mate with the part for a good grip.
- Apply enough force to the part to prevent slipping.

Typical mechanisms

- Linkage actuation
- Gear and rack
- Cam
- Screw
- Rope and pulley
- Miscellaneous, e.g., bladder, diaphragm

TOOLS

Tools are devices, which robots use to perform operations on an object, for example, drills, paint sprays, grinders, welding torches, and any other tool, which get a specific job done. Tools take advantage of continuous path control (the path the end effector takes needs to careful, steady and continuously controlled at every moment). Take the case of a spray gun as a tool. If it moves too quickly, the paint will be too thin. On the other hand, if it moves too slowly, the paint will be too thick or in blobs. Any tool required can be fitted to the end of the robotic arm and can be programmed to select and change tools without human intervention.

TYPES OF TOOLS

At times, a robot is required to manipulate a tool to perform an operation on a work part. In such applications, the end-effector is used as a gripper that can grasp and handle a variety of tools and the robot has multitool handling function. However, in most robot applications in which only one tool is to be manipulated, the tool is directly mounted on the wrist. Here the tool itself acts as the end effector. In such applications, the end effector is a tool itself. Some of the tools used are:

- Spot-welding tools
- Arc-welding tools
- Spray-painting nozzles
- Rotating spindles for drilling
- Rotating spindles for grinding
- Deburring tools
- Pneumatic tools

CHARACTERISTICS OF END-OF-ARM TOOLING

End-of-arm tooling in a robot work cell should have the following characteristics:

- The tooling must be capable of gripping, lifting, and releasing the part or family of parts required by the manufacturing process.
- The tooling must sense the presence of a part in the gripper, using sensors located either on the tooling or at a fixed position in the work cell.
- Tooling weight must be kept to a minimum because it is added to part weight to determine maximum payload.
- Containment of the part in the gripper must be ensured under conditions of maximum acceleration at the tool plate and loss of gripper power.
- The simplest gripper that meets the first four criteria should be the one implemented.

ELEMENTS OF END-OF-ARM TOOLING

End-of-arm tooling is commonly made up of four distinct elements, which provide for:

1. Attachment of the hand or tool to the robot tool mounting plate.
2. Power for actuation of tooling motions.
3. Mechanical linkages.
4. Sensors integrated into the tooling.

Mounting Plate

The means of attaching the end-of-arm tooling to an industrial robot is provided by a tool mounting located at the end of the last axis of motion on the robot. This tool mounting plate contains either threaded or clearance holes arranged in a pattern for attaching tooling. For a fixed mounting of a gripper or tool, an adapter plate with a hole pattern matching the robot tool mounting plate can be provided. The remainder of the adapter plate provides a mounting surface for the gripper or tool at the proper distance and orientation from the robot tool mounting plate. If the task of the robot requires it to automatically interchange hands or tools, a coupling device can be provided. An adapter plate is attached to each of the grippers or tools to be used, with a common lock-in position for the pickup by the coupling device. The coupling device may also contain the power source for the grippers or tools and automatically connects the power when it picks up the tooling.

Power

Power for actuation of tooling motions can be pneumatic, hydraulic, or electrical, or the tooling may not require power, as in the case of hooks or scoops. Generally, pneumatic power is used where possible because of its ease of installation and maintenance, low cost, and light weight. Higher-pressure hydraulic power is used where greater forces are required in the tooling motions. However, contamination of parts due to leakage of hydraulic fluid often restricts its application as a power source for tooling. Although it is quieter, electrical power is used less frequently for tooling power, especially in part-handling applications, because of its lower applied force. Several light payload assembly robots utilize electrical tooling power because of its control capability. In matching a robot to end-of-arm tooling, consideration should be given to the power source provided with the robot. Some robots have one for tooling power, especially in part-handling robots, and it is an easy task to tap into this actuation for tooling functions.

Mechanics

Tooling for robots may be designed with a direct coupling between the actuator and workpiece. Take the case of an air cylinder that moves a drill through a workpiece, or use indirect couplings to gain mechanical advantage, as in the case of a pivot-type gripping device. A gripper may also have provisions for mounting interchangeable fingers to conform to various part configurations. In turn, fingers attached to grippers may have provisions for interchangeable to conform to various part configurations.

Sensors

Sensors are incorporated in tooling to detect various conditions. For safety considerations, sensors are normally designed into tooling to detect workpiece or tool retention by the robot during operation. Sensors are also built into tooling to monitor the condition of the workpiece or tool during an operation, as in the case of a torque sensor mounted on a drill to detect when a drill has broken. Sensors are also used in tooling to verify that a process is completed satisfactorily as wire-feed detectors in arc welding torches and flow meters in dispensing heads. More recently, robots specially designed for assembly tasks contain force sensors (strain gauges) and dimension sensors in the end-of-arm tooling. Tools are the devices that actually perform the task, *e.g.*, gluing guns, drills, welding torches, arc welding guns, spray guns, automatic screw drivers.

TYPES OF GRIPPERS

Robotic end effectors include everything from simple two-fingered grippers and vacuum attachments to elaborate multifingered hands. The grippers are classified as:

1. Passive grippers
2. Active grippers

Passive Grippers

Most end effectors in use today are passive, *i.e.*, they emulate the grasps that people use for holding a heavy object or tool, without manipulating it in the fingers. However, a passive end effector may be equipped with sensors, and the information from these sensors may be used in controlling the robot arm. Passive grippers can hold parts, but cannot manipulate them or actively control the grasp force. Passive grippers include:

1. *Nonprehensile end effectors*

These end effectors are called nonprehensile because they neither enclose parts nor apply grasp forces across them. These include vacuum, electromagnetic, and Bernoulli-effect end effectors.

2. *Wrap end effectors*

These grippers hold a part in the same way that a person might hold a heavy hammer. In such applications, humans use wrap grasps in which the fingers envelop a part, and maintain a nearly uniform pressure so that friction is used to maximum advantage. These include blades and linkages.

3. *Pinch end effectors*

The next type of the end effector includes common two-fingered grippers. These grippers employ a strong "pinch" force between two fingers, in the same way that a person might grasp a key when opening a lock. Most such grippers are sold without fingertips because they are the most product-specific part of the design. The fingertips are designed to match the size of components, the shape of components (*e.g.*, flat or V-grooved for cylindrical parts), and the material (*e.g.*, rubber or plastic to avoid damaging fragile objects). Pinch end effectors include 2 or 3 fingers, parallel or angular motion, fingertip styles like flat, V, etc. Because two-fingered end effectors typically use a single air cylinder or motor that operates both fingers in unison, they will tend to center parts that they grasp.

Active Grippers

These include active servo grippers and *dexterous* robot hands found in research laboratories and teleoperated applications. Here the distinctions depend largely on the number of fingers and the number of joints or degrees of freedom per finger. Servo-controlled end effectors provide advantages for fine-motion tasks. In comparison to a robot arm, the fingertips are small and light, which means that they can move quickly and precisely. The total range of motion is also small, which permits fine resolution position and velocity measurements. When equipped with force sensors such as strain gauges, the fingers can provide force sensing and control, typically with better accuracy than can be obtained with robot wrist or joint mounted sensors. A servo gripper can also be programmed either to control the position of an unconstrained part or to accommodate to the position of a constrained part. The sensors of a servo-controlled end effector also provide useful information for robot programming. For example, position sensors can be used to measure the width of a grasped component, thereby providing a check that the correct component has been grasped. Similarly, force sensors are useful for weighing grasped objects and monitoring task related forces. For most industrial applications, the robot's hand or end effectors must be designed to perform a specific job. Therefore, when a robot changes tasks, it needs a new gripper. Here are a few of the many available types of grippers.

Some of the commonly used grippers are:

- Finger grippers
- Mechanical grippers where friction or the physical configuration of the gripper retains the object
- Suction or vacuum cups used for flat objects
- Magnetized gripper devices used for ferrous objects

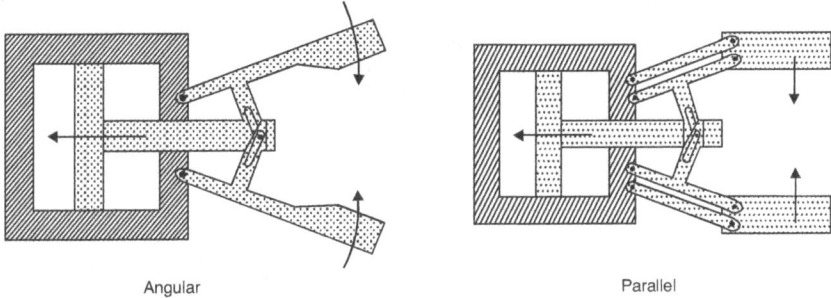

Angular   Parallel

FIGURE 15.6

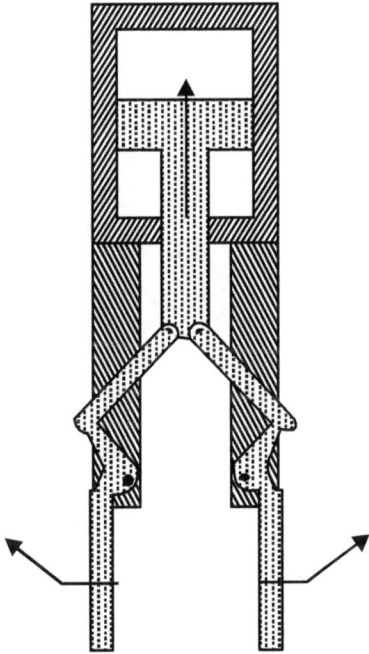

FIGURE 15.7

FINGER GRIPPERS

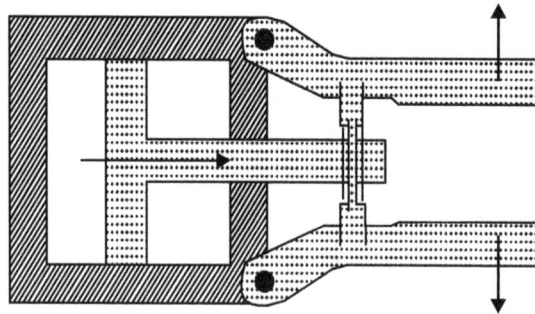

FIGURE 15.8

The most commonly used grippers are finger grippers. These will generally have two opposing fingers or three fingers like a lathe chuck. The

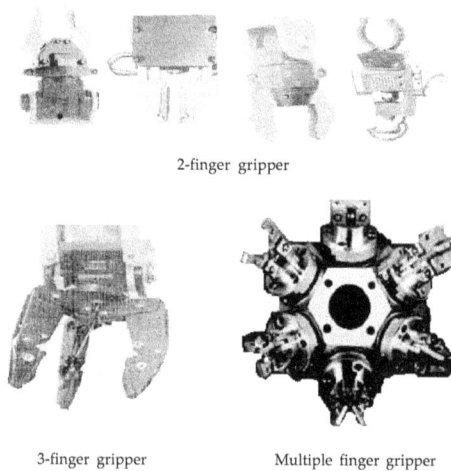

2-finger gripper

3-finger gripper   Multiple finger gripper

FIGURE 15.9 Types of Grippers.

fingers are driven together such that once gripped, any part is centered in the gripper. This gives some flexibility to the location of components at the pick-up point. Two finger grippers can be further split into parallel motion or angular motion fingers as shown in Figure 15.6.

Two fingered pneumatic actuated: As the cylinder is actuated, it translates to the fingers opening or closing. The extra links help increase holding force. (Refer to Figure 15.7.)

Two-finger internal gripper: As the cylinder is actuated, the fingers move outward. (Refer to Figure 15.8.)

Other types of grippers with two, three, and multiple fingers are shown in Figure 15.9.

MECHANICAL GRIPPERS

One can think of a mechanical gripper as a robot hand. The mechanical grippers use mechanical fingers actuated by a mechanism to grasp an object. A basic robot hand will have only two or three fingers. A mechanical hand that wraps around an object will rely on friction in order to secure the object it is holding. Friction between the gripper and the object will depend on two things, first is the type of surface whether it be metal on metal, rubber on metal, smooth surfaces, or rough surfaces and the second is the force which is pressing the surfaces together.

There are different types of mechanism: pneumatic, electrical, mechanical, or hydraulic. Some kinds of fingers can be detachable. In most cases, two fingers are enough to grasp the target in good condition. The main problem often consists of countering the gravity problem when the gripper has to grasp something. The solution is given by applying an opposite force that one needs to lift. Then one can consider the physical constriction and the friction methods. The first one is accomplished by designing the contacting surfaces of the fingers to be in the approximate shape of the part geometry. In the second method, the fingers applied a strong enough friction force to retain the part against gravity. This method is usually cheaper because of the design of the fingers.

Mechanical grippers are often fitted with some type of pad usually made from polyurethane as this provides greater friction. Pads are less likely to damage the workpiece. Pads are also used so to have a better grip, as the polyurethane will make contact with all parts of the surface when the gripper is closed. Mechanical grippers can be designed and made for specific purposes and adjusted according to the size of the object. They can also have dual grippers. Robots benefit from having dual grippers as they can increase productivity, be used with machines that have two work stations where one robot can load two parts in a single operation, and where the cycle time of the robot is too slow to keep up with the production of the other. Some mechanical grip devices contain three moving fingers that simultaneously close to grasp a part or tool. These hands are particularly useful in handling cylindrical-shaped parts, as the three-point contact centers around parts of varying diameters on the centerline of the hand.

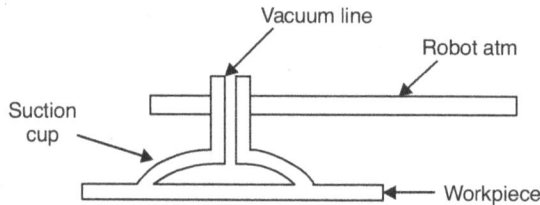

FIGURE 15.10 Vacuum Grippert.

VACUUM/SUCTION GRIPPERS

Suction grippers are of two types:

1. *Devices operated by a vacuum*: The vacuum may be provided by a vacuum pump or by compressed air.
2. *Devices with a flexible suction cup*: This cup presses on the workpiece. Compressed air is blown into the suction cup to release the workpiece.

The advantage of the suction cup is that if there is a power failure it will still work, as the workpiece will not fall down. The disadvantage of the suction cup is that they only work on clean, smooth surfaces. The suction cups are made of rubber-like materials (silicone, neoprene). Depending on the design and the material, suction cups are suitable for workpiece temperatures upto 200°C. The number of grippers (cups) determines the size and weight of object to be grasped. Figure 15.10 shows a vacuum gripper.

Advantages

▪ Requires only one surface of a part to grasp.
▪ A uniform pressure can be distributed over some area, instead of concentrated at a point.
▪ The gripper is lightweight.
▪ Many different materials can be used.
▪ No danger of crushing fragile objects.

Disadvantages

▪ The maximum force is limited by the size of suction cups.

▪ Positioning may be somewhat inaccurate.

▪ Time may be needed for the vacuum in the cup to build up.

▪ WThe robot system must include a form of a pump for air and the level of noise can cause annoyance in some circumstances.

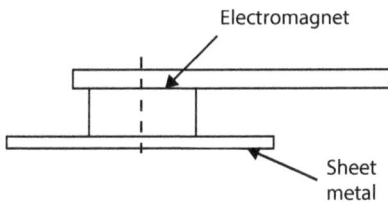

FIGURE 15.11 Magnetic Gripper.

MAGNETIC GRIPPERS

Magnetic grippers are considered when the part to be handled is of ferrous content. The principle is the same as of vacuum grippers except that magnetic grippers replace the cups. This system has the advantage that it can quickly handle metal parts with holes. For maximum effect, the magnet needs to have complete contact with the surface of the metal to be gripped. Any air gaps will reduce the strength of the magnetic force; therefore flat sheets of metal are best suited to magnetic grippers. If the magnet is strong enough, a magnetic gripper can pick up an irregular shaped object. In some cases, the shape of the magnet matches the shape of the object. A magnetic gripper is shown in Figure 15.11.

A disadvantage of using magnetic grippers is the temperature. Permanent magnets tend to become demagnetised when heated and so there is the danger that prolonged contact with a hot workpiece will weaken them to the point where they can no longer be used. The effect of heat will depend on the time the magnet spends in contact with the hot part. Most magnetic materials are relatively unaffected by temperatures up to around 100 degrees. A unique feature that sets the magnetic gripper apart from other grippers is that it leaves little to no residual magnetic charge when turned off, therefore releasing the part instantly. The magnetic gripper is able to release its load quickly, and despite its compact size, can sustain enormous holding power.

Electromagnets can be used instead and are operated by a DC electric current. If an electromagnet is used, a power failure will cause the part to be dropped immediately, which may produce an unsafe condition. Although electromagnets provide easier control and faster pickup and release of parts, permanent magnets can be used in hazardous environments requiring explosion-proof electrical equipment and lose nearly all of their magnetism when the power is turned off. Permanent magnets are also used in situations where there is an explosive atmosphere and sparks from electrical equipment would cause a hazard.

Advantages

- Variation in part size can be tolerated. The grippers do not have to be designed for one particular work part.
- Ability to handle metal part with holes.
- Pickup times are very fast.
- Requires only one surface for gripping.

Disadvantages

- Residual magnetism that remains in the workpiece.
- Possible side slippage.
- More than one sheet will be lifted by the magnet from a stack.

EXERCISES

1. Explain the working principle of an end effector with the help of a sketch and give its important applications.

2. How are the robot end effectors classified?

3. List the different end effectors used on robots.

4. What are the various types of grippers used in robots?

5. Explain the construction of different end effectors for different types of applications.

6. Identify the different types of end effectors used in robots and their applications.

7. Define end effectors.

8. Discuss the working of any three types of end effectors.

9. What are the characteristics of end-of-arm tooling?

10. Briefly explain the elements of end-of-arm tooling.

11. Discuss the various factors to be considered in selecting a gripper.

12. Show the following grippers with the help of a diagram:
    (i)   Two finger gripper
    (ii) Internal and external gripper
    (iii) Angular and parallel gripper

13. Distinguish between magnetic and vacuum grippers.

14. List the advantages and disadvantages of vacuum grippers.

15. What are the advantages and disadvantages of magnetic grippers?

16. Explain the construction of different end effectors for different types of applications. Describe a simple servo control system for a robot.

17. Explain the following with a sketch:
    - Magnetic grippers
    - Vacuum grippers
    - Mechanical grippers

CHAPTER 16

ROBOT PROGRAMMING

INTRODUCTION

Robots are becoming more powerful, with more sensors, more intelligence, and cheaper components. As a result robots are moving out of controlled industrial environments into uncontrolled service environments such as homes, hospitals, and workplaces where they perform tasks ranging from delivery services to entertainment. It is this increase in the exposure of robots to unskilled people that requires robots to become easier to program and manage. The flexibility of a robot system comes from its ability to be programmed. How the robot is programmed is a main concern of all robot users. A good mechanical arm can be underutilized if it is too difficult to program. Earlier robot programming was easy because it only required guiding the robot through the sequence of desired movements. To execute complete tasks of the type found in assembly, robot-programming languages had to be introduced. Although the introduction of robot programming languages has represented an important breakthrough in industrial robotics, currently available languages are not easy to use. The programmer must define all the movements very precisely. Programming the robot is not the same as programming the computer.

ROBOT PROGRAMMING

A *robot program* can be defined as a path in space to be followed by the manipulator, combined with peripheral actions that support the work cycle. A robot can perform complex tasks under the control of stored programs, which can be modified at will. The process of robot programming includes teaching it the task to be performed, storing the program, executing the program, and debugging it. Robotic programming is similar to real-time programming in that the programs must be interrupt driven and take account of limited resources. Robot programs can be simple or complex. Simple programs will have a set of pre-programmed responses to expected events. For example, if the robot is hit, it may move in a particular direction. Complex robot programs may include a way to learn from past events and actions and predict what will happen. For example, a robot programmed to move left when it reaches a barrier will always move in this direction. However, a robot might be programmed to remember which direction has the fewest barriers and will move in that direction. This element of learning is more likely found in robot programs than real-time programs. Examples of the peripheral actions include opening and closing the gripper, performing logical decision making, and communicating with other pieces of equipment in the robot cell.

ROBOT PROGRAMMING TECHNIQUES

A number of different techniques are used to program robots. The principal task of robot programming is to control the motions and actions of the manipulator. A robot is programmed by entering the programming commands into its controller memory. The methods of entering the commands are:

1. Online programming
2. Lead-through programming

3. Walk-through programming
4. Offline programming
5. Task programming

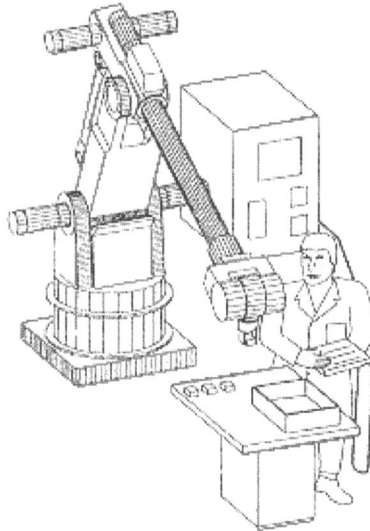

FIGURE 16.1 Online Programming.

ONLINE PROGRAMMING

Online programming takes place at the site of production itself and involves the workcell. Online programming systems (Refer to Figure 16.1) uses a *teach pendant* to direct the robot's movement which allows trained personnel to physically lead the robot through the desired sequence of events by activating the appropriate pendant button or switch. Position data and functional information are "taught" to the robot, and a new program is written. Taught data is stored in the pendant's memory then transferred to the robot's controller. The teach pendant can be the sole source by which a program is established, or it may be used in conjunction with an additional programming console and/or the robot's controller. When using this technique of teaching or programming, the person performing the teach function can be within the robot's working envelope, with operational safeguarding devices deactivated or inoperative. There are a number of teach pendant types available, depending on the type of application for which they will be used. If the goal is simply to monitor and control a robotics unit, then a simple control box style is suitable. If additional capabilities, such as on the site programming are required, more sophisticated boxes should be used.

Online programming is a convenient and easy method of programming when tasks are simple and revisions or adjustments can be made on the spot. However, the production line must be stopped during the programming and there are safety issues to consider, as the programmer must work within the robot's work envelope.

Advantages

- Easily accessible.
- The robot is programmed in concordance with the actual position of equipment and pieces.

Disadvantages

- Unavailability of robot for production during programming phase.
- Slow movement of the robot while programming.
- Time consuming process.
- Program logic and calculations are hard to program.
- Suspension of production while programming.
- Cost equivalent to production value.
- Inefficient in flexible manufacturing system.

The Teach Pendant (Manual Control Pendant)

The teach pendant has the following primary functions:

- Serve as the primary point of control for initiating and monitoring operations.
- Guide the robot or motion device, while teaching locations.
- Support application programs.

The teach pendant as shown in Figure 16.2 is used with a robot or motion device primarily to teach robot locations for use in application programs. The teach pendant is also used with custom applications that employ "teach routines," that pause execution at specified points and allow an operator to teach or re-teach the robot locations used by the program. There are two styles of teach pendants: *the programmer's pendant*, which is designed for use while an application is being written and debugged, and *the operator's pendant*, which is designed for use during normal system operation.

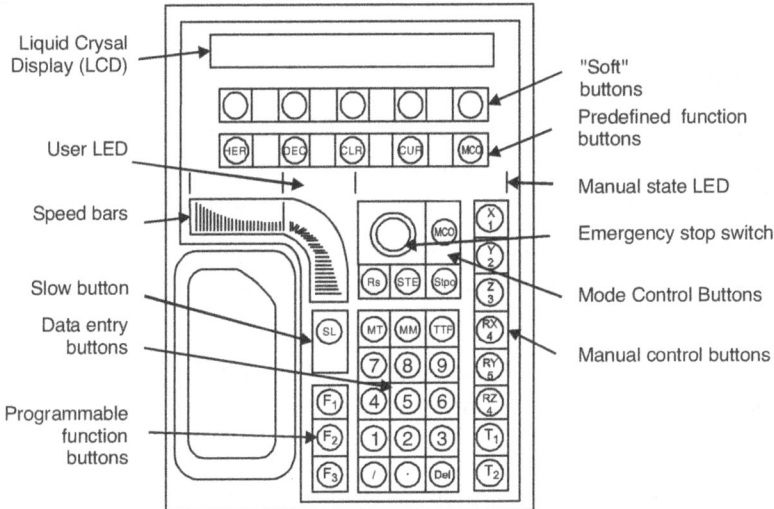

FIGURE 16.2 Teach Pendant.

The operator's pendant has a palm-activated switch, which is connected to the remote emergency stop circuitry of the controller. Whenever this switch is released, arm power is removed from the motion device. To operate the teach pendant, the left hand is put through the opening on the left-hand side of the pendant and the left thumb is used to operate the pendant speed bars. The right hand is used for all the other function buttons.

The major areas of the teach pendant are:

1. Liquid crystal display (LCD)

2. *Data entry buttons*: The data entry buttons are used to input data, normally in response to prompts that appear on the pendant display. The data entry buttons include YES/NO, DEL, the numeric buttons, the decimal point, and the REC/DONE button, which behaves like the Return or Enter key on a normal keyboard. In many cases, application programs have users press the REC/DONE button to signal that they have completed a task.

3. *Emergency stop switch*: The emergency stop switch on the teach pendant immediately halts program execution and turns off arm power.

4. *User LED*: The pendant is in background mode when the user LED is not lit and none of the predefined functions are being used. The user LED is lit whenever an application program is making use of the teach pendant.

5. *Mode control buttons*: The mode control buttons change the state being used to move the robot, switch control between the teach pendant and the application programs, and enable arm power when necessary.

6. *Manual control buttons*: When the teach pendant is in manual mode; these buttons select which robot joint will move, or the coordinate axis along which the robot will move.

7. *Manual state LEDs*: The manual state LEDs indicates the type of manual motion that has been selected.

8. *Speed bars*: The speed bars are used to control the robot's speed and direction. Pressing the speed bar near the outer ends will move the robot faster, while pressing the speed bar near the center will move the robot slower.

9. *Slow button*: The slow button selects between the two different speed ranges of the speed bars.

10. *Predefined function buttons*: The predefined function buttons have specific, system-wide functions assigned to them, like display of coordinates, clear error, etc.

11. *Programmable function buttons*: The programmable function buttons are used in custom application programs, and their functions will vary depending upon the program being run.

12. *Soft buttons*: The "soft" buttons have different functions depending on the application program being run, or the selection made from the predefined function buttons.

LEAD-THROUGH PROGRAMMING

Lead-through programming requires the operator to move the robot arm through the desired motion path during a teach procedure, thereby entering the program into the controller memory for subsequent playback. There are two methods of performing the lead-through teach procedure:

- Powered lead-through
- Manual lead-through

The difference between the two is the manner in which the manipulator is moved through the motion cycle.

Powered lead-through is commonly used as the programming method for playback robots with point-to-point control. It involves the use of a teach pendant (hand-held control box) which has toggle switches or contact buttons for controlling the movement of the manipulator joints. Using the toggle switches or buttons, the programmer drives the robot arm to the desired positions, in sequence, and records the positions into memory. During subsequent playback, the robot moves through the sequence of positions under its own power.

Manual lead-through is convenient for programming playback robots with continuous path control in which the continuous path is an irregular motion pattern such as in spray painting. This programming method requires the operator to physically grasp the end-of-arm or tools attached to the arm and manually move through the motion sequence, recording the path into memory. Because the robot arm itself may have significant mass and would therefore be difficult to move, a special programming device often replaces the actual robot for the teach procedure. The programming device has a similar joint configuration to the robot, and it is equipped with a trigger handle (or other control switch), which is activated when the operator wishes to record motions into memory. The motions are recorded as a series of closely spaced points. During playback, the path is recreated by controlling the actual robot arm through the same sequence of points.

Powered lead-through is the most common programming method in industry at this time.

Advantages

- Easy to program: shop personnel can readily learn it and does not require deeper programming experience.

Disadvantages

- Interruption in production.
- Teach pendant have limitations in the amount of decision making logic that can be incorporated in the program.
- No interface to other computer subsystems in the factory.

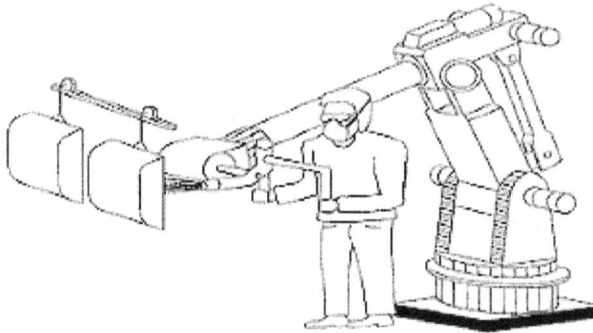

FIGURE 16.3 Walk-through Programming Method.

WALK-THROUGH PROGRAMMING OR TEACHING

In walk-through programming, the teacher physically moves ("walks") the robot through the desired positions within the robot's working envelope (Refer to Figure 16.3). During this time, the robot's controller may scan and store coordinate values on a fixed-time interval basis. These values and other functional information are replayed in the automatic mode. This may be at a different speed than that used in the walk-through.

This type of walk-through programming uses triggers on manual handles that move the robot. When the trigger is depressed the controller remembers the position. The controller would then generate the movement between these points when the program is played. The walk-through methods of programming require the teacher to be within the robot's working envelope with the robot's controller energized at least in the position sensors. This may also require that safeguarding devices be deactivated. A person doing the teaching has physical contact with the robot arm and actually gains control and walks the robot's arm through the desired positions within the working envelope. With the walk-through method of programming, the person doing the teaching is in a potentially hazardous position because the operational safeguarding devices are deactivated or inoperative. The walk-through method is appropriate for spray painting and welding robots.

OFFLINE PROGRAMMING

Offline programming is accomplished on computers located away from the robot station (Refer to Figure 16.4). Using simulation software, data

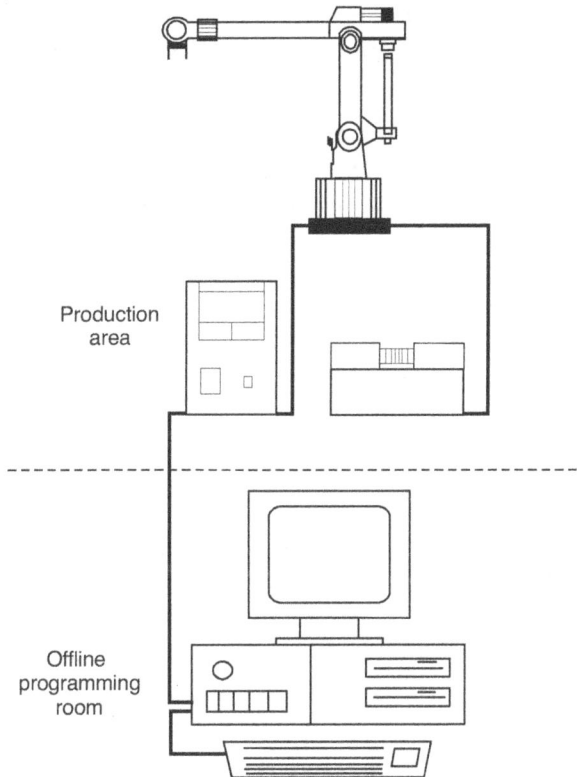

FIGURE 16.4. Offline Programming.

is generated and then sent to the robot's controller where it is translated into instructions. Additionally, the software contains modeling data, which assists selection of the best robot configuration for a particular application. The advantage of this programming method is that programming can be done while the robot is still in production on the preceding job, thus production time of the robot is not lost to delays in teaching the robot a new task. This ensures higher utilization of the robot.

Advantages

- Effective programming of program logics and calculations with state of the art debugging facilities.
- Locations are built according to models and this can mean that programmers will have to fine tune programs online or utilize sensors.
- Effective programming of locations.

- Verification of program through simulation and visualization.
- Well documented through simulation model with appropriate programs.
- Reuse of existing CAD data.
- Cost independent of production. Production can continue while programming.
- Process support tools, for instance, selection of welding parameters.

  Disadvantages

- Demands investment in an offline programming system.
- Needs extensive training.

TASK LEVEL PROGRAMMING

Task level programming requires specifying goals for the position of objects, rather than the motions of a robot needed to achieve those goals. A task level specification is meant to be totally robot independent, no positions or paths that depend on robot geometry are specified by the user. Task level programming systems requires complete geometric models of the environment and of the robot as input.

In the task level language, robot actions are specified by their effects on objects. For example, a user would specify that a job should be placed in a pallet rather than specifying the sequence of manipulator motions needed to perform the insertion. A task planner would transfer the task level specifications into robot level specifications.

MOTION PROGRAMMING

Motion programming with today's robot languages requires a combination of textual statements and lead-through techniques. Accordingly, this method of programming is sometimes referred to as *online/offline programming*. The textual statements are used to describe the motion, and the lead-through methods are used to define the position and orientation of the robot during and/or at the end of the motion.

The lead-through methods provide a very natural way of programming motion commands into the robot controller. In manual lead-through, the operator simply moves the arm through the required path to create the program. In powered lead-through, the operator uses a teach pendant to drive the manipulator. The teach pendant is equipped with a toggle switch or a pair of contact buttons for each joint. By activating these switches or buttons in a coordinated fashion for the various joints, the programmer moves the manipulator to the required positions in the workspace.

Coordinating the individual joints with the teach pendant is sometimes an awkward way to enter motion commands to the robot. For example, it is difficult to co-ordinate the individual joints of a jointed-arm robot to drive the end-of-arm in a straight-line motion. Therefore, many of the robots using powered lead-through provide two alternative methods for controlling movement of the manipulator during programming, in addition to individual joint controls. With these methods, the programmer can control the robot's wrist end to move in straight-line paths. The names given to these alternatives are (1) world co-ordinate system, and (2) tool co-ordinate system. Both systems make use of a Cartesian co-ordinate system. In the *world co-ordinate system*, the origin and frame of reference are defined with respect to some fixed position and alignment relative to the robot base. In the *tool co-ordinate system*, the alignment of the axis system is defined relative to the orientation of the wrist faceplate (to which the end effector is attached). In this way, the programmer can orient the tool in a desired way and then control the robot to make linear moves in directions parallel or perpendicular to the tool.

The speed of the robot is controlled by means of a dial or other input device located on the teach pendant and/or the main control panel. Certain portions of the program should be performed at high speeds (*e.g.*, moving parts over substantial distances in the workcell), while other parts of the program require low-speed operation (*e.g.*, moves that require high precision in placing the work part). Speed control also permits a given program to be tried out at a safe slow speed, and then for a higher speed to be used during production.

OVERVIEW OF ROBOT PROGRAMMING LANGUAGES

Virtually all robots are programmed with some kind of robot programming language. These programming languages are used to command the robot to move to certain locations, signal outputs, and read inputs. The programming language is what gives a robot its flexibility and allows a robot to react to its environment (through the use of sensors), and to make decisions. Robot programming languages fall into three basic categories:

1. Specialized robot languages

These languages have been developed specifically for robots. The commands found in these languages are mostly motion commands with minimal logic statements available. Most of the early robot languages were of this type, although many still exist today. VAL is an example of this.

2. Robot library for a new general-purpose language

These languages were created by first creating a new general-purpose programming language and then adding robot-specific commands to it. These languages are generally more capable than a specialized language, because they tend to have better logic testing capabilities. KAREL language is an example of this type.

3. Robot library for an existing computer language

These languages are developed by creating extensions to popular computer programming languages. Consequently, these languages closely resemble traditional computer programming languages, and benefit from the power of existing, widely used languages. Robot script is an example of this type of language.

Generally, all the robot programs have three basic modes of operation. They are:

- *Monitor mode*: In this mode, the user defines the various positions to be used in the program. These positions will be taught using the teach pendant and stored into the memory to be used in the program.

- *Editor mode*: This mode is used by the user to write new programs and edit the existing ones. In this mode, the syntax checking will be done.

- *Run mode*: This is the mode in which the robot is actually executing the program. The robot is actually performing the sequence of motions in the run mode.

REQUIREMENTS FOR A STANDARD ROBOT LANGUAGE

Because of the different methods of developing robot languages, many different types of languages are available. Currently, there are no standards for robot languages, and each robot manufacturer has developed their own, each with their own syntax and data structures. Some factories have robots from multiple robot manufacturers, thus, multiple languages are running on these control systems. This requires robot programmers to be proficient in many languages, or for the robot programmers to specialize in certain languages. The result of this variety is a demand for a common language that can be used on any type of robot. To develop a new robot programming language, the deficiencies of the existing languages, as well as the requirements of a new language, need to be identified.

As robot workcells are becoming more complex, the robot has to perform more complicated moves and interact with more sensors and other peripheral devices. Robots often have to decide which part is coming down the line, and determine the proper program to run. Existing robot-programming languages, especially the specialized languages, often have limited ability to use subroutines and do logic testing. Some of the early generation controllers just don't have the memory capacity to hold such large programs.

Most robot languages require the user to compile a program before running it on the robot controller. Compiling a program converts the human-readable source code into a machine-readable form. This is usually done to increase the speed of the program execution, because the program is already in the easiest form for the computer to read. Interpreters, on the other hand, convert the human-readable code to computer-readable code on the fly, as the program executes. Usually, this results in a significant amount of computer processing time being used to do the conversion. In the early days of robotics, compiling was required because the processors were not fast enough to interpret a program. Many robot controllers still require the programs to be compiled before running, even though there is now plenty of computer power to interpret programs.

ROBOT LANGUAGES

A language is a system of communication, which usually is connected to human spoken language and which is based on an arbitrary system of symbols. The most important feature of a language is its ability to produce messages. In a computer, the executable control program is formed of a sequence of machine-language commands. A machine-language command consists of a numerical code, which contains the type of the command and the source and destination addresses of the information. To make programming easier, several high-level programming languages have been developed. Instead of numbers and addresses, the developer can now use words and names. Before the use of such a high-level control program, it must be compiled to machine-language code. This is done by compilers, which have been developed for each language.

Different languages have different aims and are suitable for different purposes. For example, MATLAB is a mathematical language, which has been developed to solve mathematical problems. It has built-in functions for powerful mathematical analysis, but it is not suitable for real-time control of a mobile robot. HTML is a markup language to describe how information appears in web browsers, but it is not suitable to solve mathematical problems. Some of the basic types of commands in programming languages are:

1. Motion and sensing functions (*e.g.*, MOVE, MONITOR)
2. Computation functions (*e.g.*, ADD, SORT)
3. Program flow control functions (*e.g.*, RETURN, BRANCH)

TYPES OF ROBOT LANGUAGES

The earliest methods for training a robot like mechanical setup, point-to-point path recording, and task lead through did not use word-based languages. Some of the high-level computer languages now used to program robots are: Wave, AL, ACL, AML, APT, ARCL, ZDRL, HELP, Karel, CAP 1, MML, RIPL, MCL, RAIL, RPL, ARMBASIC, Androtext, VAL, IBL, and Ladder Logic.

Wave

Wave was the first high-level language created for programming a robot. Standford Artificial Laboratory developed it in 1973.

AL

The AL (Arm Language) high-level programming language was developed at the robotics research center of Stanford University.

ACL

The Advanced Command Language (ACL) is a robot language that employs a user-friendly conversational command environment. Yaskaua robots use it.

AML

AML is the programming language used for the control of robots produced by IBM. AML is intended to provide a complete interpreted computer language along with all of the programming support typically associated with high-level programming languages.

APT

The Automatically Programmed Tools (APT) language is a computer language dealing with motion. Electronic Systems Laboratory of MIT developed it in 1956.

ARCL

ARCL (A Robot Control Language) was based on Pascal-like syntax. It was a compiled language and the developed cross-compiler required three passes before the executable code was ready to be downloaded and executed in a robot. This language has a Pascal-like syntax with sensory-control and motion control commands. An example of an ARCL-language command is MOVA (GRIP, HI, CONT, MED), which opens the gripper on the robot. ARCL emphasized sensory-based programming rather than planned trajectory motion and was designed for educational robot.

ZDRL

ZDRL (Zhe Da Robot Language) was a motion-oriented robot language. It was an interpretative system and the language was composed of system commands and program instructions. System commands were used to prepare the system for execution of user-written programs. ZDRL included 32 system commands and 37 program instructions, and it contained capabilities for program editing, file management, location-data teaching, program executing, and program debugging.

HELP

HELP is a high-level programming language developed for use with General Electric's Allegro assembly robot.

Karel

Karel, Karel 2, and Karel 3 are robot control languages used by some FANUC robot controllers.

CAP 1

The Conversational Auto Programming 1 (CAP 1) robot language is used by the FANUC 32-18-T Robot Controller.

MML

MML was a model-based mobile robot language that was developed at the University of California. It is a high-level offline programming language, which contains functions for high-level sensor functions, geometric model description and path planning, and others. This language contains an important concept of slow and fast functions, which architecture is essential for real-time control of robots. A slow function is executed sequentially, while a fast function is executed immediately. The second important concept is the separation of the reference and current posture, which makes precise and smooth motion control and dynamic posture correction possible.

RIPL

RIPL (Robot Independent Programming Language) is based on an object-oriented Robot Independent Programming Environment [RIPE]. The RIPE computing architecture consists of a hierarchical multiprocessor approach, which employs distributed general and special-purpose processors. This architecture enables the control of diverse complex subsystems in real time while co-coordinating reliable communications between them.

MCL

MCL is short for Manufacturing Control Language and was developed by McDonnell Douglas for the U.S. Air Force's ICAM project.

RAIL

RAIL is a high-level programming language developed by Automatix for use with robots and vision systems.

RPL

RPL is a high-level programming language developed by SRI and is used to configure automated manufacturing systems.

ARMBASIC

ARMBASIC is an extension of the hobbyist computer language BASIC. It was used with the Microbot Mini-Mover 5 educational robot.

Androtext

Androtext is a high-level computer language developed by Robotronic Corporation to make commanding a personal robot easier.

VAL

VAL stands for Victor's Assembly Language. VAL is a high-level programming language developed for PUMA lines of robots. The programming language is similar to BASIC. It has a complete set of vocabulary words for writing and editing robot programs.

IBL

IBL (Instruction Based Learning) is a method to train robots using natural language instructions. IBL uses unconstrained language with a learning robot system. A robot is equipped with a set of primitive sensory-motor procedures such as turn left or follow the road that can be regarded as an execution-level command language. The user's verbal instructions are converted into a new procedure and that procedure becomes a part of the knowledge that the robot can use to learn increasingly complex procedures. With this procedure, the robot should be capable of executing increasingly complex tasks. Because errors are always possible in human-machine communication, IBL verifies whether the learned subtask is executable. If it is not, then the user is asked for more information.

Ladder Logic

Ladder logic is a programming language designed to be used by electricians. It closely resembles the relay logic that appears on the inside lids or doors of dishwashers and washing machines. The only robots that use ladder logic programming are those that come without a controller or with a programmable logic controller.

EXAMPLE OF A ROBOT PROGRAM USING VAL

The VAL language is the most advanced commercial language designed for use with Unimation, Inc. industrial robots. VAL stands for Victor's Assembly Language. It is basically an offline language in which the program defining the motion sequence can be developed off line, but the various point locations used in the work cycle are most conveniently defined by leadthrough. To demonstrate the VAL language, let us assume that the robot must pick up objects from a chute and place them in successive boxes. One possible sequence of robot activity is as follows:

1. Move to a location above the part in the chute.
2. Move to the part.

3. Close the gripper jaws.
4. Remove the part from the chute.
5. Carry the part to a location above the box.
6. Put the part into the box.
7. Open the gripper jaws.
8. Withdraw from the box.

The corresponding VAL program is as follows:

EDIT DEMO. 1

- PROGRAM DEMO. 1

  1. ? APPRO PART, 50 @
  2. ? MOVES PART @
  3. ? CLOSE I @
  4. ? DEPARTS 150@
  5. ? APPROS BOX, 200@
  6. ? MOVE BOX@
  7. ? OPENI @
  8. ?DEPART@

The exact meaning of each line is:

1. Move to a location 50 mm above the part in the chute.
2. Move along a straight line to the part.
3. Close the gripper jaws.
4. Withdraw the part 150 mm from the chute along a straight-line path.
5. Move along a straight line to a location 200 mm above the box.
6. Put the part into the box.
7. Open the gripper jaws.
8. Withdraw 75 mm from the box.

When the program is executed, it causes the robot to perform the steps which describe the task.

Most articulated robots perform by storing a series of positions in memory, and moving to them at various times in their programming. For example, a robot that is moving items from one place to another might have a simple program like this:

Define points P1–P5:

1. Safely above workpiece
2. 20 cm above bin A
3. At position to take part from bin A
4. 20 cm above bin B
5. At position to take part from bin B

Define program:

- Move to P1
- Move to P2
- Move to P3
- Close gripper
- Move to P4
- Move to P5
- Open gripper
- Move to P1 and finish

For a given robot, the only parameters necessary to locate the end effector (gripper, welding torch, etc.) of the robot completely are the angles of each of the joints. However, there are many different ways to define the points for the robot to move to, some of which can be much more efficient, depending on the task to be accomplished.

EXERCISES

1. Name some programming languages for robot programming.
2. Explain the use of teach pendant for robot programming.
3. Identify any four robot programming languages. Name the important requirement of programming languages.
4. What is meant by teach pendant programming of robots?
5. What is a teach pendant?
6. Give three methods for programming a robot.
7. Give an example of a robot program.

8. List the names of three robot-programming languages.

9. What do you mean by task level programming?

10. What are the different classes of robot languages? Discuss.

11. Compare online and offline programming of robot.

12. Discuss the lead through programming technique used in robot programming.

13. List at least four languages used for programming of robots.

14. Write notes on languages available, features required of programming language, and classification of commands in respect of robot programming languages.

15. Explain the following
    - Manual programming
    - Lead through programming

CHAPTER 17

APPLICATIONS OF ROBOTS

INTRODUCTION

The present-day applications for robots are much broader than most people realize. While the emphasis in robot development is on industrial robots, factories are not the only place where robots are used. Small, medium, and large companies in just about every industry are taking a fresh look at robots to see how this powerful technology can help them solve manufacturing challenges. In business offices and elsewhere, robots serve as mail delivery carts, promotional or show robots, laboratory assistants, hospital orderlies, and window washers. In general, industrial robots are best used for jobs that are dirty, dull, dangerous, or difficult, the types of jobs that humans do most poorly.

Robots are used in a wide range of industrial applications. The first commercial application of an industrial robot took place in 1961, when a robot was installed to load and unload a die-casting machine. This was an unpleasant task for human operators. Many robot applications took place in areas where a high degree of hazard or discomfort to humans existed, such as in welding, painting, and foundry operations. The earliest applications were in materials handling, spot welding, and spray painting. Robots were initially applied to jobs that were hot, heavy, and hazardous such as die-casting, forging, and spot welding. The reasons for using robots are:

1. *Reduced costs*—robots can perform tasks more economically than humans.

2. *Improved productivity*—robots are not only less expensive than human labor, but also have higher rates of output. This increase in productivity is due to robot's slightly faster work pace but much is the result of the robot's ability to work almost continually, without lunch breaks and rest periods.

3. *Better quality*—robots have a distinct advantage of being able to perform repetitive tasks with a higher degree of consistency, which in turn leads to improved product quality.

4. Elimination of hazardous tasks.

ROBOT CAPABILITIES

The three important capabilities of robots that make them useful for applications are:

- Transport
- Manipulation
- Sensing

Transport

Material handling is one of the basic operations, which is performed on an object as it passes through the manufacturing process. The object is moved from one location to another to be stored, machined, assembled, or packaged. The robot's capability to acquire an object, move it through space, and release it makes it ideal for transport operations. Simple tasks such as part transfer from one conveyor to another may only require one- or two-dimensional movements, which are often performed by non-servo robots.

Other parts handling operations such as machine loading and unloading and packaging may be more complicated and require varying degrees of manipulative capability. Servo-controlled robots perform these operations.

Manipulation

Another basic operation performed on an object as it is transformed from raw material to a finished product is processing, which generally requires some types of manipulation, *i.e.*, workpieces are inserted, oriented, or twisted to be in proper position for machining, assembly, or some other operation. A robot's capability to manipulate both parts and tooling make it very suitable for processing applications. Examples in this regard include spot and arc welding, and spray painting.

Sensing

A robot's ability to react to its environment by means of sensory feedback is also important, particularly in applications like assembly and inspection. These sensory inputs may come from a variety of sensor types, including proximity switches, force sensors, and machine vision systems.

In each application, one or more of the robot's capabilities of transport, manipulation, or sensing is employed. These applications make a robot ideal for many applications now performed manually.

APPLICATIONS OF ROBOTS

Manufacturing Applications

- Arc and spot-welding
- Spray painting
- Machine loading and unloading
- Machining
- Die casting
- Forging
- Investment casting
- Parts transferring
- Plastics molding
- Finishing
- Assembly

- Inspection

Materials Handling

- Transport goods
- Pick and place
- Palletizing

Space Industry

- Robot arms used as manipulator to handle bulky telescopes
- Mounted on space shuttle or repair craft

Military

- Remote bomb detonation
- Smart bombs

Medical Applications

- Intelligent wheelchairs
- Robot arms used to manipulate and handle patients

MANUFACTURING APPLICATIONS

Welding

Perhaps the most popular applications of robots are in industrial welding. The repeatability, uniformity quality, and speed of robotic welding are unmatched. A robot performing a welding operation is shown in Figure 17.1. The two basic types of welding are spot welding and arc welding, although laser welding is done. Some environmental requirements should be considered for a successful operation. The automotive industry is a major user of robotic spot welders. The other major welding task performed by robots is arc or seam welding. In this application, two adjacent parts are joined together by fusing them, thereby creating a seam.

Why should robots be used for welding?

A welding process that contains repetitive tasks on similar pieces might be suitable for automation. The number of items of any type to be welded determines how difficult automating a process will be or not. If parts normally need adjustment to fit together correctly, or if joints to be welded are too wide or in different positions from piece to piece, automating the procedure will be difficult or impossible. Robots work well for repetitive tasks or similar pieces that involve welds in more than one axis. The most

FIGURE 17.1 Robot Performing Welding Operation.

prominent advantages of automated welding are precision and productivity. Robot welding improves weld repeatability. Once programmed correctly, robots will give precisely the same welds every time on workpieces of the same dimensions and specifications.

Arc Welding

Arc or fusion welding is considered a major growth area for the application of robotics (Refer to Figure 17.2). The process is very hostile to the operator, generating noise, fumes, and intense light. Automation produces high quality welds with greater consistency and at a faster rate. In general, equipment for automatic arc welding is designed differently from that used for manual arc welding. Automatic arc welding normally involves high-duty

FIGURE 17.2 Robots for Arc Welding.

cycles, and the welding equipment must be able to operate under those conditions. In addition, the equipment components must have the necessary features and controls to interface with the main control system.

A special kind of electrical power is required to make an arc weld. A welding machine, also known as a power source, provides the special power. All arc-welding processes use an *arc welding gun* or *torch* to transmit the welding current from a welding cable to the electrode. They also provide for shielding the weld area from the atmosphere. The nozzle of the torch is close to the arc and will gradually pick up spatter. A *torch cleaner* (normally automatic) is often used in robot arc welding systems to remove the spatter. All of the continuous electrode wire arc processes require an *electrode feeder* to feed the consumable electrode wire into the arc. The process is applied to automobile subassemblies mainly for reasons of strength, low distortion, high-speed applications where one-sided access is required, and sealing.

Spot Welding

Automatic welding imposes specific demands on resistance welding equipment. Often, equipment must be specially designed and welding procedures developed to meet robot-welding requirements. The spot welding robot (Refer to Figure 17.3) is the most important component of a robotized spot welding installation. Welding robots are available in various sizes, rated by payload capacity and reach. Robots are also classified by the number of axes. Spot welding involves applying a welding tool to some object, such as a car body, at specified discrete locations. A spot welding gun applies appropriate pressure and current to the sheets to be welded. This requires the robot to move its hand (end effector) to a sequence of positions with sufficient accuracy to perform the task properly. It is desirable to move at

FIGURE 17.3 Spot-welding Gun.

a high speed to reduce cycle time, while avoiding collisions and excessive wear or damage to the robot.

There are different types of welding guns, used for different applications, available. An automatic weld-timer initiates and times the duration of current. A robot can repeatedly move the welding gun to each weld location and position it perpendicular to the weld seam. It can also replay programmed welding schedules. A manual-welding operator is less likely to perform as well because of the weight of the gun and monotony of the task. Spot welding robots should have six or more axes of motion and be capable of approaching points in the work envelope from any angle. This permits the robot to be flexible in positioning a welding gun to weld an assembly. A robot easily performs some movements that are awkward for an operator, such as positioning the welding gun upside down.

Typical components of an integrated robotic spot welding cell are:

- Spot welding robot
- Spot welding gun
- Weld timer
- Electrode tip dresser
- Spot welding swivel

Electron Beam Welding

Electron beam welding (EBW) is a fusion joining process that produces a weld by impinging a beam of high-energy electrons to heat the weld joint. Electrons are elementary atomic particles characterized by a negative charge and an extremely small mass. An electron beam-welding gun uses a high intensity electron beam to target a weld joint. The weld joint converts the electron beam to the heat input required to make a fusion weld. The electron beam is always generated in a high vacuum. The use of specially designed orifices separated a series of chambers at various levels of vacuum permits welding in medium and non-vacuum conditions. Although, high vacuum welding will provide maximum purity and high depth to width ratio welds. An electron beam robot welding system benefits the customer with a low contamination vacuum, narrow weld zone, uses low filler metal, and has low distortion.

MIG Welding

Gas metal arc welding (GMAW) is frequently referred to as MIG welding. MIG welding is a commonly used high deposition rate welding process. Wire is continuously fed from a spool. MIG welding is therefore referred to as a semiautomatic welding process. Robotic systems are integrated towards MIG welding applications on a consistent basis. With advances in technology, and the benefits of a GMAW robotic cell, factories and job shops large and small are investing in a robot. Return on investment of a robotic system is possible after just a few years. M I G welding robots, or GMAW cells have many benefits for customers. Robots are all position capable, have higher deposit rates than SMAW, need less operator skill, can perform longer welds without stopping, and have minimal post weld cleaning.

TIG Welding

Gas tungsten arc welding (GTAW) is frequently referred to as TIG welding. TIG welding is a commonly used high-quality welding process. TIG welding has become a popular choice of welding processes when high quality, precision welding is required. In TIG welding, an arc is formed between a non-consumable tungsten electrode and the metal being welded. Gas is fed through the torch to shield the electrode and molten weld pool. If filler wire is used, it is added to the weld pool separately. A TIG welding robot system has many benefits to customers. A robot produces high-quality welds, welds can be made with or without filler metal, variable precise controls, low distortion, and free of spatter.

Spray Painting

Another popular and efficient use for robots is in the field of spray painting. The consistency and repeatability of a robot's motion have enabled near perfect quality while at the same time wasting no paint. The spray painting applications seems to epitomize the proper applications of robotics, relieving the human operator from a hazardous, albeit skillful job, while at the same time increasing work quality, uniformity, and cutting costs. In spraying applications, the robot manipulates a spray gun, which is used to apply some material such as paint, stain, or plastic powder, to either a stationary or moving part. These coatings are applied to a wide variety of parts, including automotive body panels, appliances, and furniture. The spray-painting environment constitutes a fire hazard and is also dangerous to human worker's respiration. An early paint-spraying machine was built by Pollard in the 1930s. Today, this machine would be called an industrial robot. Robots perform painting, coating, and dispensing jobs in many industries today. Companies making products such as motorcycles, bicycles, boats, jet skis, and cars are using painting automation to their advantage. Painting robots are generally equipped with six axes, three for the base motions and three for applicator orientation. Some units incorporate machine vision for guidance or to check application quality. Typically, these painting robots are electrically driven, rather than hydraulically or pneumatically powered. Advantages of using spray-painting robots (Figure 17.4) include:

FIGURE 17.4 Robots for Spray Painting.

1. Humans are removed from a hostile environment.
2. Less energy is needed for fresh air requirements and the need for protective clothing is reduced.
3. The quality of the painting is improved, reducing works and warranty costs.
4. Less paint and other materials are used.
5. Direct labor costs are reduced.

Machine Tending

Machine loading and unloading is a more complex application than basic material handling; for this application, the robot provides both manipulative and transport capabilities (Refer to Figure 17.5). Robots can be used to grasp a workpiece from a supply point (*e.g.*, a conveyor belt), transport it to a machine, orient it, and then insert it into the machine work holder. This may require that the robot signal the machine tool when the workpiece is in the correct position, so that tvhe part can be secured in the work holder. The robot then releases the part and withdraws the

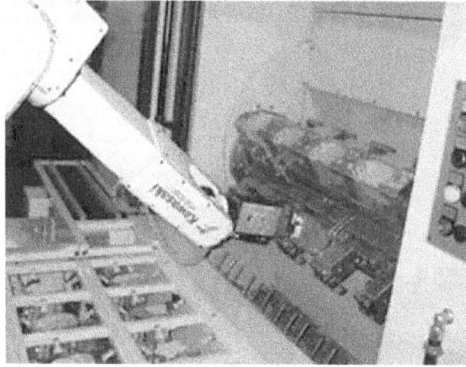

FIGURE 17.5 Machine Tending Operation.

arm so that machining can begin. Upon completion of the machining, the robot unloads the workpiece and transfers it to another machine or conveyor. In a robotic cell, a single robot can service several machines. The single robot may be used to perform other operations while the machines are performing their primary functions. This may require that the robot be able to exchange end-effectors. Examples of machine tending functions include the following:

- Exchanging machine tools, such as lathe and machining centers.
- Stamping press loading and unloading.
- Tending plastic injection molding machines.
- Holding a part for a spot welding operation.
- Loading hot billets into forging presses.
- Loading auto parts for grinding.
- Loading gears onto CNC milling machines.

Forging

Robots have been applied in many different types of forging applications such as hammer forge operations, upsetter operations, roll forges, hot forming presses, and draw bench applications. In some cases, a robot acts as a forging machine operator or as a role of forge helper.

Forge hammers: Forging hammers are either hydraulic, steam hydraulic, or air driven. One-half of the forming die is on the anvil, and the other half of the die is on a ram that moves up and down, either under force by air, steam, or gravity. Under the control of an operator, the hammer is allowed to strike the part that is lying between the two dies a certain number of times. Depending on the observation of the forging operator, the operator determines when to take a part out of one die and moves it into the next. The function of the robot in this application can be to act as a forge helper. When working heavier parts, the robot can be used to load and unload furnaces and process the billets to the forging bed, where the operator can take over and process the parts through the various forming cycles. The robot then can maneuver the finished product to a trim operation.

Die Casting

Die casting involves forcing nonferrous metals into dies under high pressure to form parts of a desired shape. A typical die-casting task involves unloading a part from the die-casting machine, quenching the part, and then disposing of the part on a conveyor belt or into a bin. A possible robot task cycles in die-casting could include any of the following:

1. Alternately loading two or more die casting machines.

2. Unloading, quenching, trimming, and disposing of the part.

3. Unloading the die casting machine and preparing the dye for the next casting cycle. (This would require another gripper or attachment to spray the die.)

4. Loading an insert into the die casting machine and unloading the finished part.

Plastic Molding

The plastic molding process is typically used for thermoplastic materials. The material to be molded is supplied in a granule form and moved from a hopper to a cylinder, from which a plunger forces the granules through a heat chamber into the mold. Then, the mold half opens, and the products are withdrawn. Many automotive parts are injection molded today, as well as many parts utilized in household appliances. The robot is typically employed to remove the part from the mold either by grasping a sprue and runner assembly. A robot is typically used at an injection molding machine workcell where parts must not be dropped because of fragility, or where runs are so short that it is not economic to build a totally automatic mold to drop the part through the bottom of the machine.

Assembly

This is the process of robot manipulation of components, resulting in a finished assembled product. Assembly (Refer to Figure 17.6) is the process

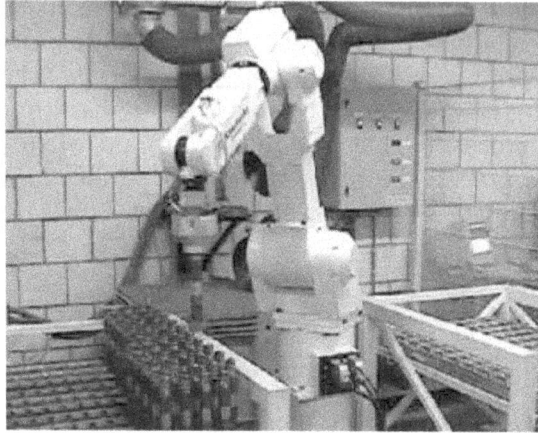

FIGURE 17.6 Robot Doing Assembly Operation.

of fitting and holding together parts and assemblies; generally performed by means of nuts, bolts, screws, fasteners, or snap-fit joints. Examples of assembly operations include:

- Assembly of computer hard drives.
- Insertion of lamps into instrument panels.
- Insertion and placement of components onto printed circuit boards.
- Automated assembly of small electric motors.
- Furniture assembly.

Sealing/Dispensing

For dispensing applications, the robot manipulates a dispenser or gun to apply a material such as paint, adhesive, sealant, or washing solution to a stationary or moving part. Additional equipment used to complete a dispensing system may include material containers, pumps, and regulators. It is very important that, for operations that involve the application of material that produces flammable or explosive fumes, the robot be sealed and have a system that purges the robot's internal cavities. If the robot is not sealed and does not have a purging system, the possibility of the ignition of flammable or explosive fumes by the arcing of the robot's internal electrical components (*i.e.*, motors, electronic components, and electrical connections) exists.

In applications where the part is on a conveyor line, the robot's motion is coordinated with the motion of the conveyor. The manipulative capability of the robot is the primary function that makes a robot especially well suited for dispensing applications. The major benefit of using a robot (Figure 17.7) for dispensing applications is that a robot provides uniform application of material (repeatability). Other benefits include a reduction of labor costs, coating material waste, and exposure of workers to hazardous materials.

FIGURE 17.7 Robot Performing Sealing Operation.

Examples of dispensing applications include:

- Painting parts on an automated line.
- Application of adhesive and sealant to car bodies.
- Application of thermal material to rockets.
- Washing.

Inspection and Measurement

With a growing interest in product quality, the focus has been on zero defects. However, the human inspection system has somehow failed to achieve its objectives. Robot applications of vision systems have provided services in part location, completeness and correctness of assembly products, and collision detection during navigation. Machine vision applications require the ability to control both position and appearance in order to become a productive component of an automated system.

Material Removal

Robotic material removal is a new application that has many uses in industrial automation. The robot can grind, roll, and file metal parts to precision. The robotic system can remove material with quality every single time it is being used. The robot can work long hours and produce more throughputs. Robotics has become more affordable and many factories are looking to buy a robotic deburring cell for their operation. With expert engineering, any material removal application can bring your company a good return on investment (ROI). A material removal robot can be very beneficial to an operation because of the robot's precision and quality. With a material removal robot system technology can be as precise as removing spots on jewellery. Material removal can be described as a deburring process, and can be integrated to the most precise applications. Companies can automate a material removal process and gain many benefits.

Deburring

Robotic deburring is the use of a robot to remove burrs, sharp edges, or fins from metal parts. The robot can grind, roll, and file metal parts to precision (Refer to Figure 17.8). The robotic system can produce a deburring

FIGURE 17.8 Robot Performing Deburring Operation.

application, *i.e.*, quality every single time it is being used. The robot can work long hours and produce more throughputs. Robotics has become more affordable and many factories are looking to buy a robotic deburring cell for their operation. A robotic deburring cell can work long hours without fatigue removing rough edges. The quality of the process, as well as the long hours of a robotic cell is unmatched.

Grinding

Manual grinding is tough, dirty, and noisy work. The metal dust produced by grinding is harmful to employee's eyes and lungs. Grinding robots remove excess material from the surface of machined parts/products quickly and efficiently. A robot can perform work with more consistency and higher quality results.

Drilling

With the automation of a drilling robot system, companies can improve accuracy and repeatability. Drilling robots can work 24 hours a day without worries or fatigue thus enhancing operational output. (Refer to Figure 17.9.)

FIGURE 17.9 Robot Performing Drilling Operation.

MATERIAL HANDLING APPLICATIONS

Using robots for handling materials are an essential component of today's automated manufacturing systems. Material handling and logistics is the movement, protection, storage, and control of materials and products throughout the process of their manufacture and distribution, consumption, and disposal. It is a repetitive operation; carried out often under unpleasant and hard working conditions, which requires little skill. This application makes use of the robot's capability to transport objects. Figure 17.10 shows a robot doing material handling operation.

FIGURE 17.10 Material Handling Robot.

The term "material handling" covers a lot of ground in the world of robotics: tiny work pieces that people can't handle very well, if at all; large, heavy parts like engine blocks and wheels; bulky items like bags and boxes; delicate and expensive electronic components; medical equipment. The list is extensive. Robotic material-handling applications range from tending injection-molding machines and machine tools, to reorienting parts between processes, to packaging and palletizing.

By fitting the robot with an appropriate end-effector (*e.g.*, gripper), the robot can grasp the object that needs to be moved. The robot may be mounted either stationary on the floor or on a traversing unit, enabling it to move from one workstation to another. The robot can also be ceiling mounted. The primary benefits of using robots for material handling are reduction of direct labor costs and removal of humans from tasks that may be hazardous, tedious, or fatiguing. Also, the use of robots for moving fragile objects results in less damage to parts during handling. Robots that are used for material handling can interface with other material handling equipment such as containers, conveyors, guided vehicles, monorails, and automated storage/retrieval systems. A handling process consists of eight sequences:

(1) Transfer of the robot arm up the workpiece

(2) Fine motion approaching the workpiece

(3) Grasping

(4) Fine motion uprising the workpiece

(5) Transfer the workpiece to the desired position

(6) Fine motion down to the destination position

(7) Release the workpiece

(8) Fine motion upward

The gripping sequence is the more delicate part of handling material. The gripper's and the robot's positions have to be checked for collision with other objects in the environment. The limitation of the robot's workspace can also be a problem for all other planning sequences of the handling.

The following are examples of material handling applications:

- Transferring parts from one conveyor to another.
- Transferring parts from a processing line to a conveyor.
- Loading bins and fixtures for subsequent processing.
- Moving parts from a warehouse to a machine.
- Transporting explosive devices.
- Transfer of parts from a machine to an overhead conveyor.

Parts Transferring

Parts transferring refer to removing parts from pallets and placing them in bins or on conveyor belts, or removing parts from bins and conveyor belts and placing them on pallets. Part transfer applications are generally referred to as material handling robot applications. With the advancements in end-of-arm tooling, and technology companies are taking a closer look into part transfer robotics. As robots become more affordable, and competition becomes fierce, many companies will look into a part transfer robot to automate their press operation. An integrated part transfer robot (Refer to Figure 17.11) system is easy to install and brings many benefits to our customers. A robot moves a part in and out of a press, with ease and no fatigue. The robot can transfer a part 24 hours a day, giving more flexibility to the company.

Palletizing

Palletizing is the act of loading or unloading material onto pallets. A robotic palletizing system (Refer to Figure 17.12) allows more flexibility to run more products for longer periods of time. With the advancements in end-of-arm tooling, robot-palletizing workcells have been integrated in many factories. The use of robots in palletizing is a popular material-handling

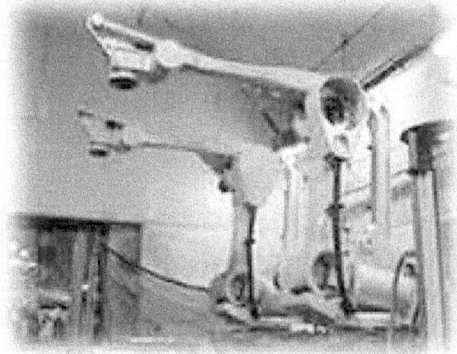

FIGURE 17.11 Robot used for Parts Transferring.

FIGURE 17.12 Robots Performing Palletizing Operation.

operation, particularly where more than one type of packaging is being handled. The newspaper industry has been particularly hard hit by increased labor costs. Part of the solution to this problem was to use robots like Cincinnati Milacron Robot being used to palletize advertising inserts for a newspaper. Robotic palletizing technology can help with productivity and profitability. The robotic workcells can be integrated towards any project. The savings of using a robot for longer hours without fatigue is always a consideration for plant managers. Robots have become more affordable in recent years and can be paid off in just a few years worth of work. Many factories, food processing plants, and palletizing plants have automated their application with a palletizing robot.

Pick and Place Operations

Industrial robots also perform what are referred to as pick and place operations. Pick and place is the name commonly given to the operation of picking up a part and placing it appropriately for subsequent operations. Pick-and-place operations have some requirements. The part must not be dropped. It must be held securely enough to prevent it from slipping in the gripper but gently enough to avoid damage. In addition, care must be taken to avoid disturbing the part during approach and departure. A pick and place robot is a material handling robot that can work 24 hours a day without worries or fatigue. The consistent output of a robotic system along with quality and repeatability are unmatched. Among the most common of these operations is loading and unloading pallets, used across a broad range of industries. This requires relatively complex programming, as the robot must sense how full a pallet is and adjust its placements or removals accordingly. Robots have been vital in pick and place operations in the casting of metals and plastics. In the die casting of metals, for instance, productivity using the same die-casting machinery has increased up to three times, the result of robot's greater speed, strength, and ability to withstand heat in parts removal operations.

Pick and place robotic cells are being integrated into factory floors all over the world. They are a more favorable solution to production lines because they move at faster speeds and have more flexibility than ever before. Managers are realizing the long-term savings with a pick and place robotic workcell rather than the operation they are currently doing. An increase in output with a material handling robotic system has saved factories money. With the advancements in technology, and robotics becoming more affordable pick and place robotic cells are being installed for many different pick and place automation applications.

Machine Tool Loading and Unloading

Machine loading and unloading applications are generally referred to as material handling. With the advancements in end-of-arm tooling, and technology, companies are taking a closer look into the benefits of machine loading robotics. As robots become more affordable, and competition becomes fierce, many companies will look into machine loading robotics to automate their operation. A robot can be given a longer reach than a human worker and might be able to load and unload more then one machine. An integrated machine-loading robot is easy to install and brings many benefits to our customers. A robot loads a part into the press, with ease and no fatigue. The robot can load parts 24 hours a day, giving more flexibility to the company.

Order Picking

A robot order picking system can be programmed to do multiple tasks. With the advancements in end-of-arm tooling, order-picking application can be used in most companies. A robotic system benefits a firm with flexibility, repetitive quality and no fatigue. A robot order picking process would most likely be associated with material handling.

CLEANROOM ROBOTS

Cleanroom robots are specifically sealed and isolated from dust or any kind of air particles to perform tasks in an isolated atmosphere. Mostly used in medical or lab application to handle parts, tend machinery, and dispense drugs. A robot is a valuable tool in a clean room setting because it can provide the same quality results without error for a long period of time. A robotic system allows laboratories and other clean room robot applications much flexibility. (Refer to Figure 17.13.)

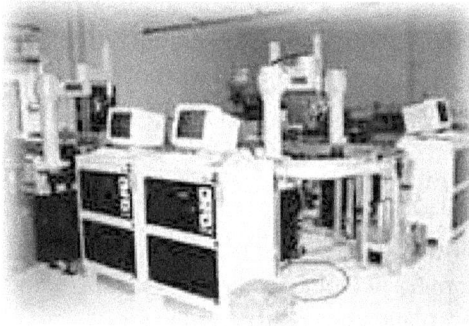

FIGURE 17.13 Clean Room Robot Cell.

EXERCISES

1. Name some industrial applications of robots.

2. What are the reasons for successful application of robots in manufacturing industries?

3. Identify some typical applications of robots in the industry.

4. Write an explanatory note on the industrial applications of robots.

5. Explain the use of a robot for industrial material handling with the help of some examples.

6. Explain the use of robot for machine loading/unloading.

7. Prepare a list of any four industrial operations that can be performed by robot.

8. List various types of assembly tasks, which can be performed by robots.

9. What are the reasons for using robots in industries?

10. List the capabilities of a robot.

11. Discuss the material handling application of robots.

12. What is machine tending in industrial robots?

13. Write a short note on sealing/dispensing application of robots.

14. Write an essay on the robots used in spray painting applications.

15. Write a short note on the following:
    - Material handling.
    - Welding operation.

CHAPTER 18

ROBOTS USING REAL-TIME EMBEDDED SYSTEMS

(The material in this chapter appears and was adapted from *Real-Time Embedded Components and Systems with Linux and RTOS* by S.Siewert and J. Pratt. Mercury Learning and Information, 2016. ISBN: 978-1-942270-04-1.)

INTRODUCTION TO ROBOTS USING REAL-TIME EMBEDDED SYSTEMS

Modern robotic applications can be operated as real-time embedded systems. Real-time systems are computer systems that can monitor, respond to, or control their environment making decisions without active human control. A robot as part of a real-time embedded system can use sensors, actuators and other inputs and outputs to affect objects in the real world and must do so within the real-time physical constraints of environments that humans often operate in as well. This chapter reviews basic concepts important to the design and implementation of basic real-time robotic systems.

Real-time applications also might have deadlines that are beyond human ability. A real-time system must simply operate within an environment to monitor and/or control a physical process at a rate required by the physics of the process. In the case of robotics, this is a distinct advantage that robotics has over human labor, the ability to keep up with a process that requires faster response and more accuracy than is humanly possible. Furthermore, robots can perform repetitive tasks for long hours without tiring. Figure 18.1 shows an industrial robotic assembly line.

FIGURE 18.1 Industrial Robots on an Assembly Line.

It is important to note that robotics is often deployed in controlled environments such as assembly lines rather than in uncontrolled environments where humans often operate better, at least presently.

ROBOTIC ARM

The robotic arm approximates the dexterity of the human arm with a minimum of five degrees of rotational freedom including: base rotation, shoulder, elbow, wrist, and a gripper. The gripper can be a simple claw or have the dexterity of a human hand. In general, the gripper is often called an end effector to describe the broad range of devices that might be used to manipulate objects or tools. These basic arms are available as low cost hobby kits and can be fit with custom controllers and sensors for fairly advanced robotics projects. The main limitation of low-cost hobby arms is that they are unable to grip and move any significant mass, offer less accurate and repeatable positioning, and are less dexterous than industrial or research robotic arms. Most industrial or research robotic arms have six or more degrees of freedom (additional wrist motion and complex end effectors) and can manipulate masses from one to hundreds of kilograms. Robotic arms are often combined with machine vision, with cameras either fixed in the arm or with views of the arm from fixed locations. A basic five degree of freedom arm with end effector vision can be used to implement interesting tasks including search, target recognition, grappling, and target relocation. Figure 18.2 shows the OWI-7 robotic trainer arm with a reference coordinate system.

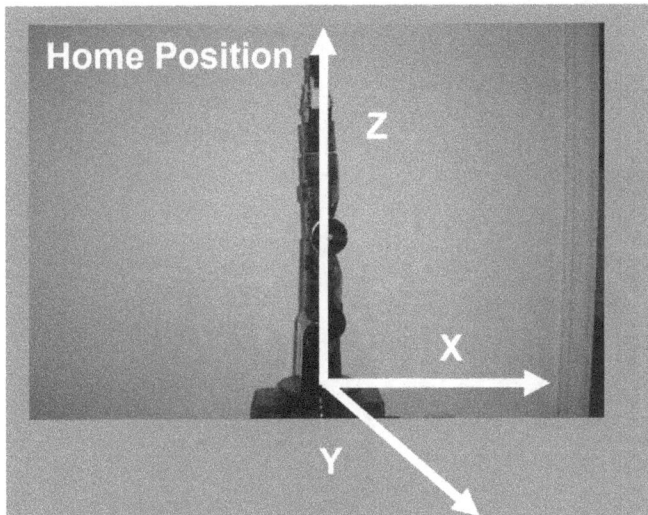

FIGURE 18.2 Robotic Arm Coordinates and Home Position.

With a reference coordinate system with an origin at the fixed base for the arm, the reach-ability of an arm can be defined based upon the arm mechanical design and kinematics. Figure 18.2 shows the OWI arm with elbow rotation such that the forearm is held parallel to the base surface. In this position, the base can be rotated to move the end effector over a circular trace on the surface.

The five degree of freedom arm is capable of tracing out reachable circles around its base. Figure 18.4 shows the innermost ring of reach-ability for the OWI arm.

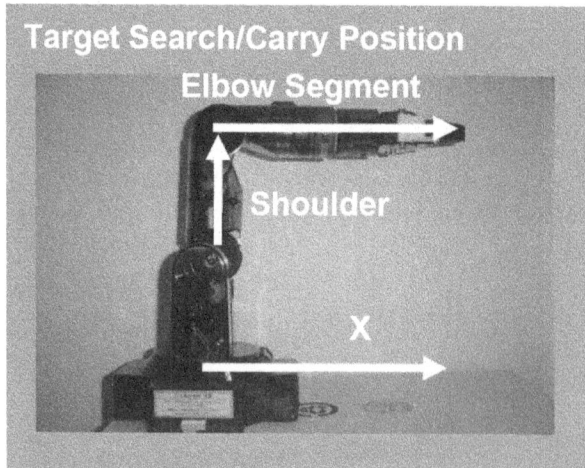

FIGURE 18.3 Robotic Elbow Rotation Only.

FIGURE 18.4 Innermost Surface Ring Reach-ability.

FIGURE 18.5 Intermediate Surface Ring Reach-ability.

Combined rotation of the shoulder and elbow allows the OWI arm end effector to reach locations on circular arcs around the base at various radii from the innermost ring. Figure 18.5 shows an intermediate ring of reach-ability.

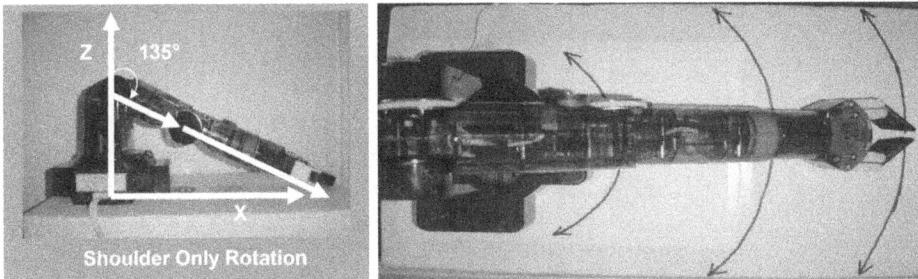

FIGURE 18.6 Outermost Surface Ring Reach-ability.

Finally, the outermost ring of reach-ability for the OWI arm is defined by arm length with no elbow rotation and shoulder rotation such that the end effector reaches the surface. Figure 18.6 shows the limit of outermost reach-ability.

This basic analysis only considers the surface reach-ability of the OWI arm on its X, Y base plane. More sophisticated tasks might require three dimensional reach-ability analysis. Once the *kinematics* (the mathematical representation of the mechanics of objects in motion) and reach-ability analysis has been completed so that the joint rotations are known for moving the end effector to and from desired target locations, now an actuation and control interface must be designed.

ACTUATION

Actuation and end effector control is greatly simplified when the target object masses that the end effector must work with are negligible. Significant target mass requires more complex active joint motor torque control. Moving significant mass requires motor controllers with DAC output (Digital to Analog) and stepper motors with active feedback control channels for each degree of freedom. For the OWI arm and negligible payload mass, the actuation can be designed using relays or simple H-bridge types of motor controllers. The motors must be reversible. The simplest circuit for reversing a motor can be implemented with switches to change the polarity across the motor leads as shown in Figure 18.7. With changing polarity, the motor is controlled in forward and reverse directions.

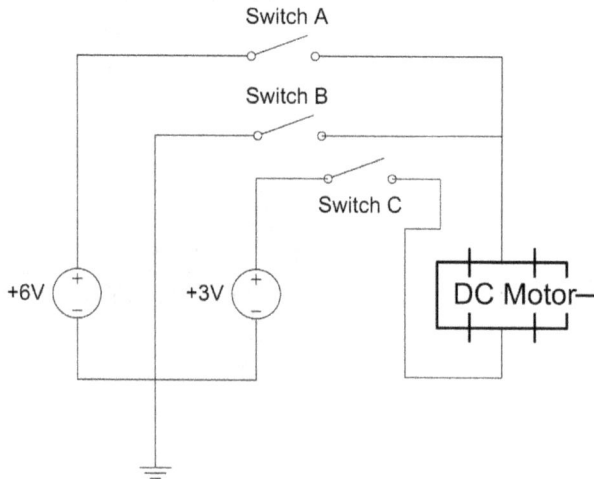

FIGURE 18.7 Three Switch Reversible Motor.

Each switch state of "on" or "off" combine to result in one of three types of operation: 1) no motion or "off"; 2) forward; and 3) reverse. The possible switch states and the operation that results, known as the actuation, are enumerated in Table 18.1.

Setting three switches is not very practical since this would require three relays and therefore fifteen total for a five degree of freedom arm. Figure 18.8 shows how two relays can be used to implement the three switch reversible motor circuit by making use of relays that include normally open and normally closed poles.

TABLE 18.1 3 Switch Reversible Motor Control.

SW-A	SW-B	SW-C	MOTOR
Off	X	Off	Off
Off	On	On	Forward
On	Off	On	Reverse

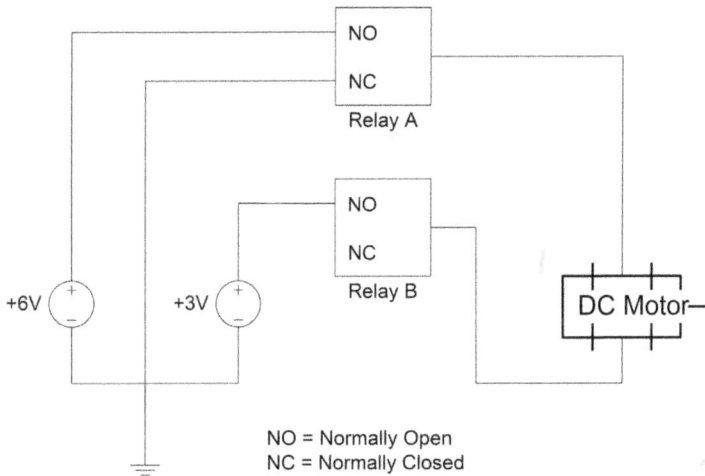

FIGURE 18.8 Two Relay Reversible Motor.

This simplifies the relay reversible motor actuation to ten relays required for a five degree of freedom arm. Table 18.2 summarizes the motor actuation as a function of the relay setting for this design.

TABLE 18.2 2 Relay Reversible Motor Control.

RLY-A	RLY-B	MOTOR
Off	Off	Off
Off	On	Forward
On	Off	Off
On	On	Reverse

This is scaled to actuate a five degree of freedom arm as shown in Figure 18.9.

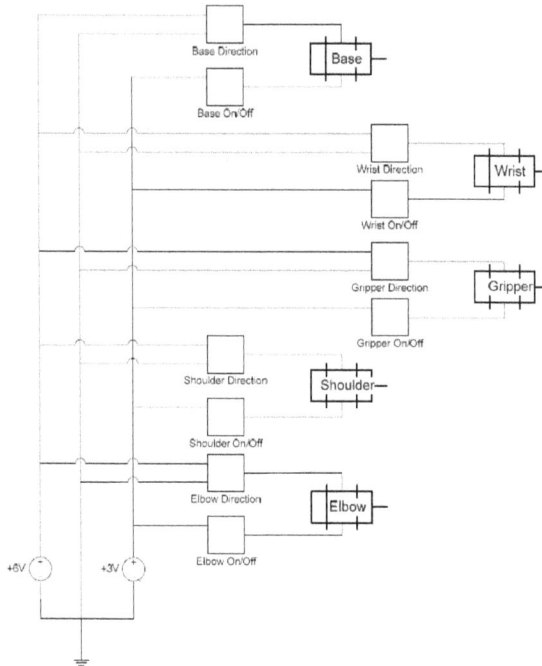

FIGURE 18.9 Five Degree of Freedom Robotic Arm Relay Circuit.

Actuation of a robotic arm can lead to mechanical arm failure if the motors are allowed to over drive the joint rotations based upon the arm mechanical limits of rotation for each joint. To avoid gear damage a mechanical clutch or slip system can be employed, a feature of the OWI arm, however reliance upon a clutch or slip system is still not ideal. Joints designed with mechanical slip or clutches can slip under the weight of the arm and cause positioning errors if too loose, and if too tight, over driving a joint will still cause gear damage. A better approach is to integrate hard and soft limit switches so that electrical and software protection mechanisms prevent over rotation of joints. Figure 18.10 shows a circuit design for hard limit switches.

The limit switches shown in Figure 18.10 must be mounted on the arm such that the joints cause the switch to be activated at each limit of mechanical motion. The downside to this circuit is that the arm joint which hits a limit remains inoperable until it is manually reset.

A better approach is to use soft limits monitoring with a software service that periodically or on an interrupt basis samples the output of a switched circuit through an A/D (Analog to Digital) converter so that software can disable motors that hit a limit. This design is shown in Figure 18.11.

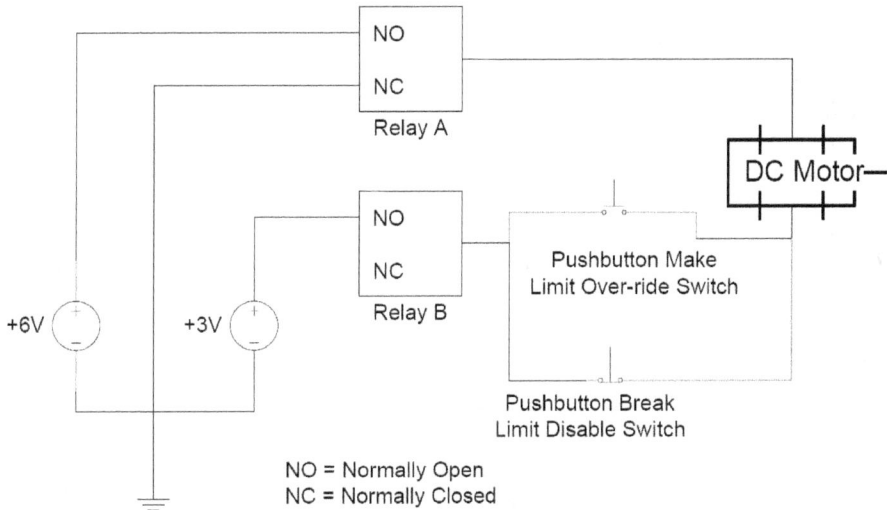

FIGURE 18.10 Use of Hard Limit Switches for Arm Joint Motor Control.

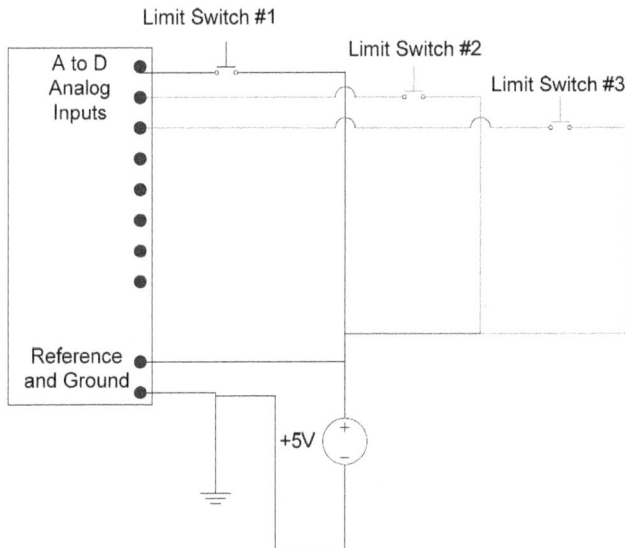

FIGURE 18.11 Use of Soft Limit Switches for Motor Control Safing.

Each of the soft limit switches can be mechanically integrated so they will trigger before the hard limit switches, allowing software to safe (disable) a potentially over rotated joint, decide whether a limit over-ride for recovery is feasible, and then recover by commanding rotation back to the operable range. If software limits monitoring fails or the software controller fails, the hardware limits will continue to protect the arm from damage. The relay actuation design with hard and soft limit switches provides basic arm actuation, but only with binary on/off motor control.

The concept of reversible motor poles can be generalized using relays in an H-bridge, providing more motor control states than the two relay design. The H-bridge relay circuit is shown in Figure 18.12.

Inspection of the relay H-bridge states shows that the H-bridge also provides additional control features.

The braking features of an H-bridge can provide the basis for torque and over-shoot control so that the motor controller can ramp up torque and ramp it down while positioning. The ramp-up can be provided by a DAC (Digital to Analog Converter) and the ramp-down braking can be provided by the H-bridge braking states. The short-circuit states of the H-bridge, "fuse tests," must specifically be avoided by H-bridge controller logic.

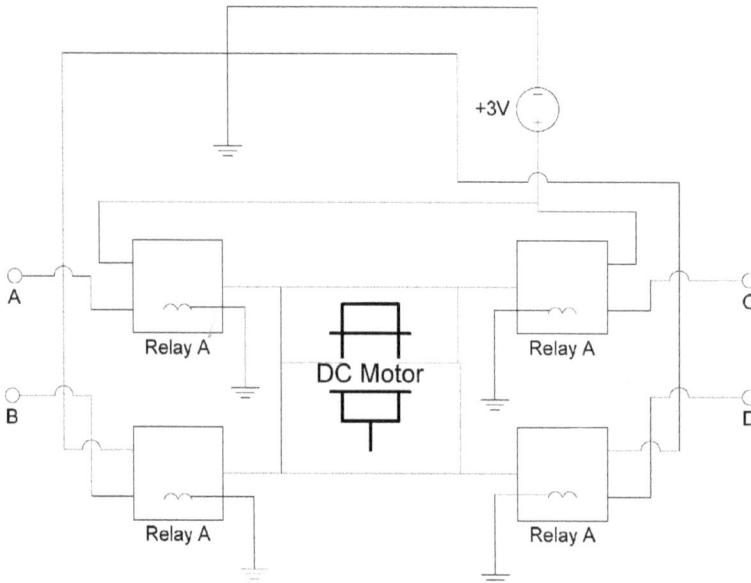

FIGURE 18.12 Relay H-Bridge Motor Control.

Relay actuation provides only on and off motor control and requires the use of electro-mechanical relay coils which are create noise, dissipate significant power, and take up significant space even for compact reed relays. Figure 18.13 shows the same H-bridge controller design as Figure 18.12, but using solid state MOSFETs (Metal Oxide Substrate Field Effect Transistors).

TABLE 18.3 Relay H-Bridge Motor Control States.

A	B	C	D	MOTOR
0	0	0	0	Off
0	0	1	1	Brake
0	1	0	1	Fuse Test
0	1	1	0	Reverse
1	0	0	1	Forward
1	0	1	0	Fuse Test
1	1	0	0	Brake

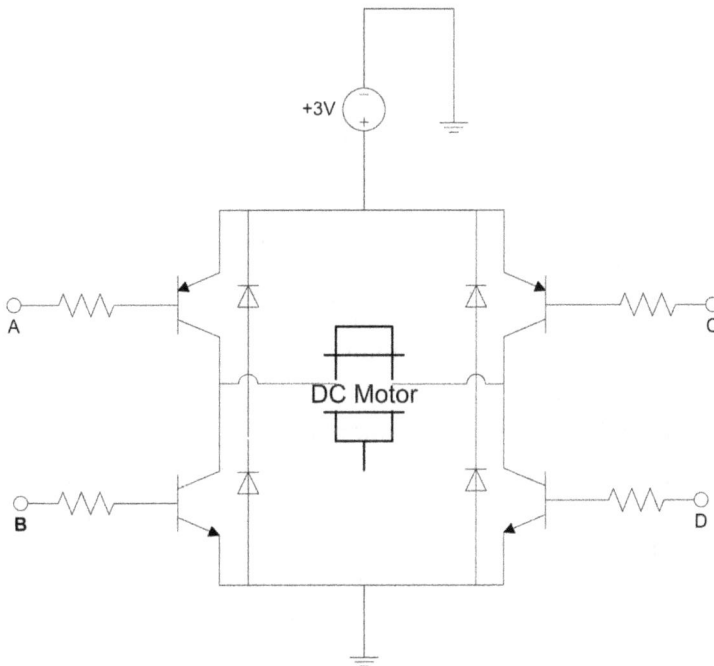

FIGURE 18.13 MOSFET H-Bridge Motor Control.

The MOSFET design provides the same states and control as the four relay H-bridge, but much more efficiently with less power required compared to electromagnetic coil actuation.

END EFFECTOR PATH

The ability to actuate an arm still does not provide the ability to navigate the arm's end effector to and from specific reachable locations. This must be done by path planning software and by end effector guidance. The simplest path planning and end effector guidance function implements arm motion dead-reckoning and single joint rotation sequences. Dead-reckoning simply turns on a joint motor for a period of time based upon a rotation rate that is assumed constant, perhaps calibrated during an arm initialization sequence between joint limits. Using a dead-reckoning estimation for joint rotation will lead to significant positioning error, but can work for positioning tasks where significant error is tolerable. Moving one joint at a time is also tolerable for paths which do not need to optimize the time, energy, or distance for the motion between two targets. Much more optimal paths between two targets can be implemented using position feedback and multiple concurrent joint rotation. Concurrent joint rotation is simple to do with a relay or H-bridge controller, although the kinematics describing the path taken are more complex. Motion feedback requires active sensing during joint rotation.

SENSING

Joint rotation sensing can be provided by joint position encoders and/or machine vision feedback. Position encoders include:

1. Electrical (multi-turn potentiometer)
2. Optical (LED and photodiode with light-path occlusion and counting)
3. Mechanical switch (with a momentary switch counter)

Position encoders provide direct feedback during arm positioning. This feedback can be used to drive the feedback in a control loop when the arm is moved to a desired target position. This assumes the desired target is known, either pre-programmed or known through additional sensing such as machine vision. Figure 18.14 shows the basic feedback control design for a position encoded controlled process to move an arm from one target position to another.

Figure 18.14 can be further refined to specifically show an actuation with feed-back design using relays and a potentiometer position encoder feedback channel. The main disturbance to constant rotation will come from stick/slip friction in the joint rotation and motor ramp up and down characteristics in the motor/arm plant. Figure 18.15 shows this specific feed-back control design.

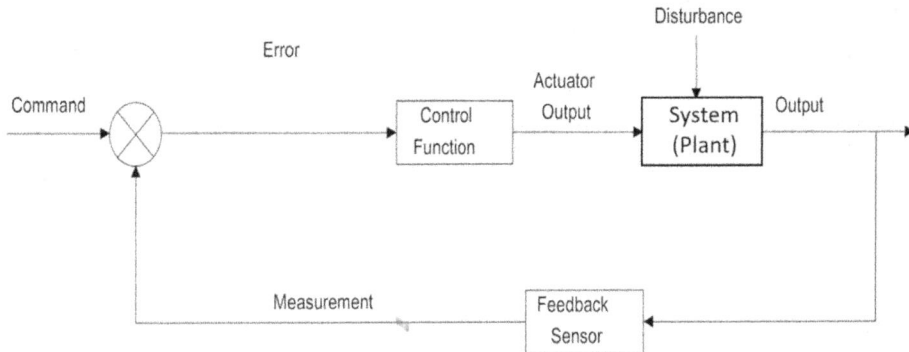

FIGURE 18.14 Basic Feedback Control Arm Positioning.

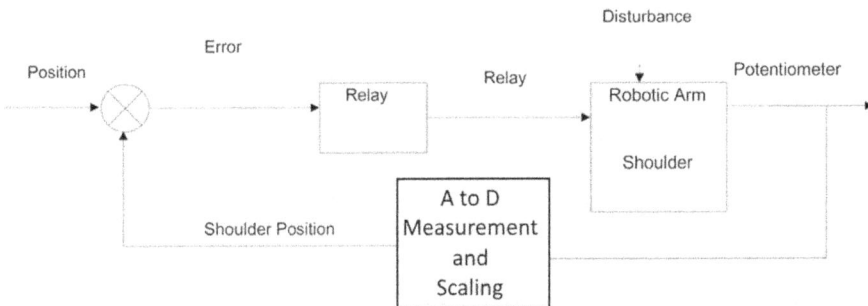

FIGURE 18.15 Relay Control with Position Encoder Feedback Through and A/D Converter.

Closer inspection of the design in Figure 18.15 reveals that the control loop has an analog and a digital domain as shown in Figure 18.16.

This mixed signal control loop design requires sampling of the feedback sensors and a digital control law. The control law can be implemented for each joint on an individual basis as a basic PID (Proportional, Integral, Differential) process control problem. For the PID approach a proportional gain, an integral gain, and a differential gain are used in the control transfer function which is:

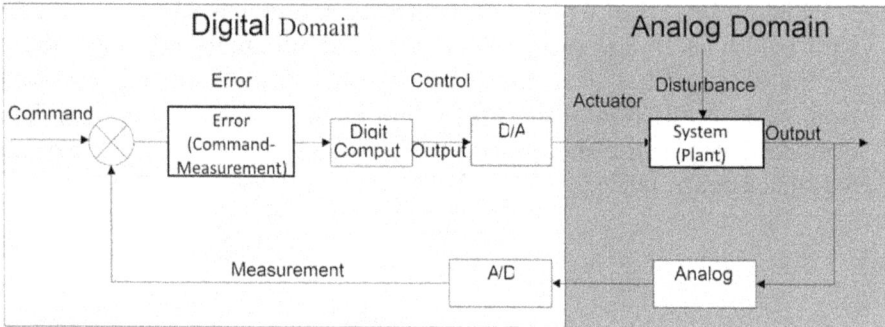

FIGURE 18.16 Arm Positioning with Feedback Digital and Analog Domains.

$$G_c(s) \quad K_p \quad \frac{K_i}{s} \quad K_d s \quad \frac{K_p s \quad K_i \quad K_d s^2}{s}$$

The transfer function defined by a Laplace transform is depicted as a control loop block diagram by Figure 18.17.

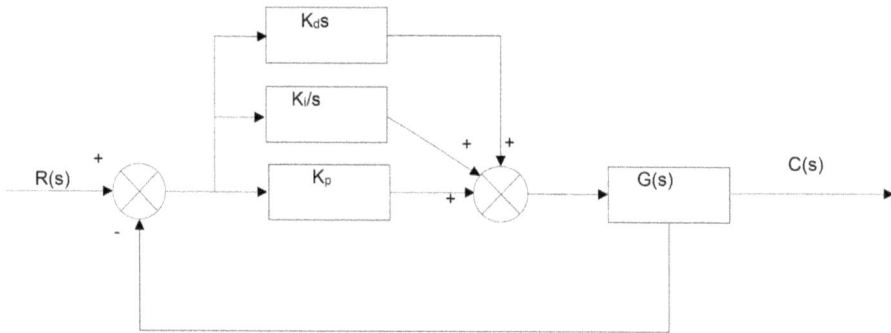

FIGURE 18.17 PID Control Loop.

The Laplace transform for the PID control law makes traditional stability analysis simple, but to implement a PID control law on a digital computer, a state space or time domain formulation for the PID control law must be understood. Furthermore, a relationship between the measured error and the control function output must be known in terms of discrete samples. This time domain relationship is:

$$u(t) \quad K_p y(t) \quad K_i \quad y(t) \quad K_d \frac{y}{t}$$

This equation can then be used to design the control loop as shown in Figure 18.8.

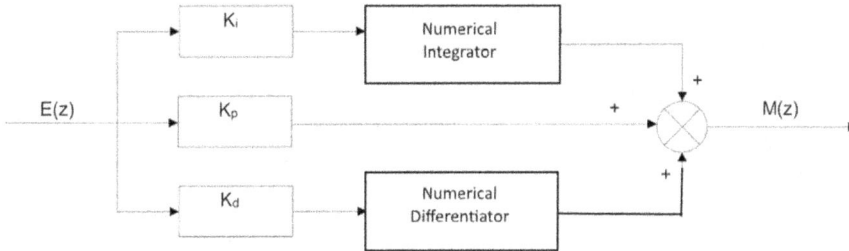

FIGURE 18.18 PID Digital Control Loop.

The numerical integration can be done using an algorithm such as forward integration, trapezoidal, or Runge-Kutta over series of time samples. Likewise, differentiation over time samples can be approximated as a simple difference. The proportional, integral, and differential gains must then be applied to the integrated and differentiated functions and summed with the proportional for the next control output. Applying these three components of the digital control law with appropriately tuned gains leads to quick rise time, minimum overshoot, and quick settling time as shown in Figure 18.19.

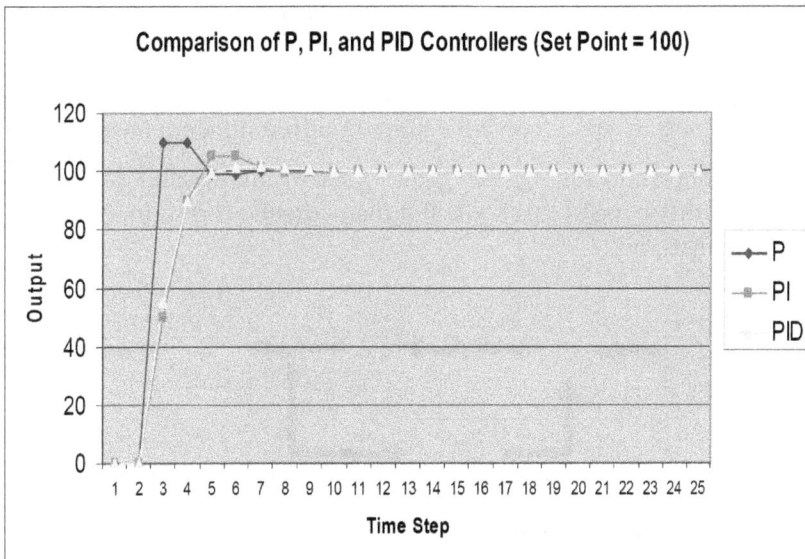

FIGURE 18.19 PID Time Response.

Figure 18.19 shows proportional control alone, proportional with integral, and finally the full PID. Tuning the gains for a PID control law can be accomplished as summarized by Table 14.4.

TABLE 18.4 PID Gain Tuning Rules.

Parameter	Rise Time	Overshoot	Settling Time
K_p gain increase	Decreases	Increases	Small Change
K_i gain increase	Decreases	Increases	Increases
K_d gain increase	Small Change	Decreases	Decreases

The PID controller provides a framework basic single-input, single output control law development. More advanced control can be designed using the modern control state space methods for multiple-inputs and outputs. State space control provides a generalized method to analyze and design a control function for a set of differential equations based upon the kinematics and mechanics of a robotic system.

A *system matrix,* **B** *input matrix*

y Cx Du

y *output vector,* **u** *input vector*

C *ouput matrix,* **D** *feed forward matrix*

Figure 18.20 shows the feed-back control block diagram for the generalized set of state space control system of differential equations.

A detailed coverage of state space control analysis and design methods is beyond the scope of this book, but many excellent resources are available for further study.

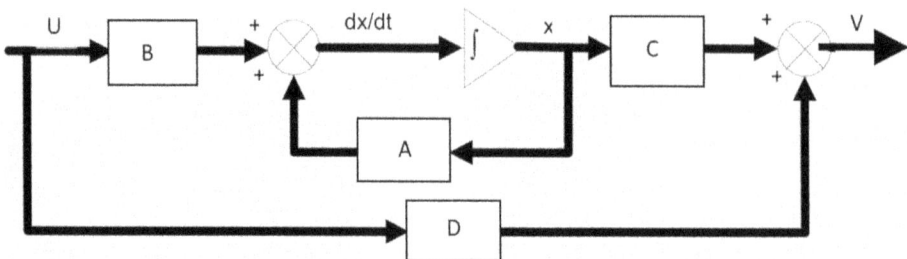

FIGURE 18.20 State Space Feed-back Control.

TASKING

The robotic arm must be commanded and controlled at a higher level than actuation and feed-back control in order to provide the basic capabilities of:

1. Searching, identifying and acquiring a visual target for pick up
2. Path planning and execution to pick up a target
3. Object grappling and grapple feedback
4. Carry path planning and execution to relocate the object to a new target location

These four tasks compose the larger task of pick and place. The OWI arm command and control software and hardware actuation and feed-back control system can be designed so that a single high level command can initiate pick and place tasks within the arm reach-ability space. The target search, identification, and acquisition task is a fundamental machine vision task. It requires an overhead, side, or front/back fixed camera system or a simpler embedded camera in the end effector. With an embedded camera, the arm can start a search sequence to sweep the camera field of view over the concentric reach-ability rings in Figures 18.4, 18.5, and 18.6. While the camera is being swept over the rings, frame acquisition of NTSC frames at 30 frames/sec or less can format the digital camera data into a digital frame stream. The digital frame stream images can be processed so that each frame is enhanced to better define edges and to segment objects for comparison matching with a target object description. The target object description can include target geometry and color (invariants) as well as target size (variant). Once the target is seen (matched with a segmented object in the field of view) then the arm can start to track and close in on the target.

Closing in on a visual target in the reach-ability space of the OWI requires constant video visual object centering based upon computation of the object's centroid in the field of view. The arm can use the centroid visual feedback to rotate the base to control XY plane errors in order to keep the object centered as the arm is lowered toward the XY plane. The kinematics require that the arm shoulder and elbow be lowered to approach a target on the XY plane and to control the X translation of the end effector and embedded camera. Simultaneously the base rotation can be controlled to coordinate the target Y translation to keep the target centered as the arm is lowered. This basic task requires actuation and control of three degrees of freedom at a minimum. The OWI wrist only has a single degree of freedom which rotates about the forearm. More sophisticated robotic arms also include rotation about the other two axes of the wrist joint. So, a fixed camera embedded in the OWI arm may need independent tilt/pan control so that the camera angle can be maintained perpendicular to the XY plane as the arm is lowered. More sophisticated wrist degrees of freedom would also provide this fine camera pointing.

Target pick up requires the arm to use position and limits feed-back so that the arm knows when it has intersected the XY plane and acquired the XY plane located target object. Furthermore, the grappler should be in the fully open position at this time. Once this ready to grapple position has been achieved with feedback machine vision and positioning, the grappler can be closed around the target object. Positive indication of successful target grappling can be provided by sensing switches built into the end effector fingers. Most often brass contacts separated by a semi-conducting foam (IC packing foam) or micro-switches with interface plates on each finger provide good feedback for grappling.

Once the target has been successfully grappled, the arm can now switch to a carry path planning task to guide it to the drop off target location. This drop off path planning might once again involve search or a pre-determined target position at a relative offset from the target acquisition location. Either way this is essentially identical to the acquire path planning and execution except that the arm is raised to a carry height at a desired reach-ability ring, then carried with base translation. The raise and carry sequence can be concurrent for a more optimal path in terms of time and distance traveled. The arm is again lowered to the drop-off target and the target object is released using the grappler feedback to ensure that the fingers no longer sense a grip on the object.

This overall sequence is a high level automation using tasking. It is still commanded by the operator, but the robotic arm performs the sub-tasks composing the overall task autonomously. Two major architectural concepts in the field of robotics have emerged which provide a framework for robot tasking and planning interfaced to lower level controls and with or without human interaction. These are *shared control* and the *degree of autonomy* that a robotic system has along with the subsumption architecture. In the next section, these architectural concepts are briefly introduced.

AUTOMATION AND AUTONOMY

Figure 18.21 shows command and control loops for a robotic system that range from fully autonomous to telerobotic.

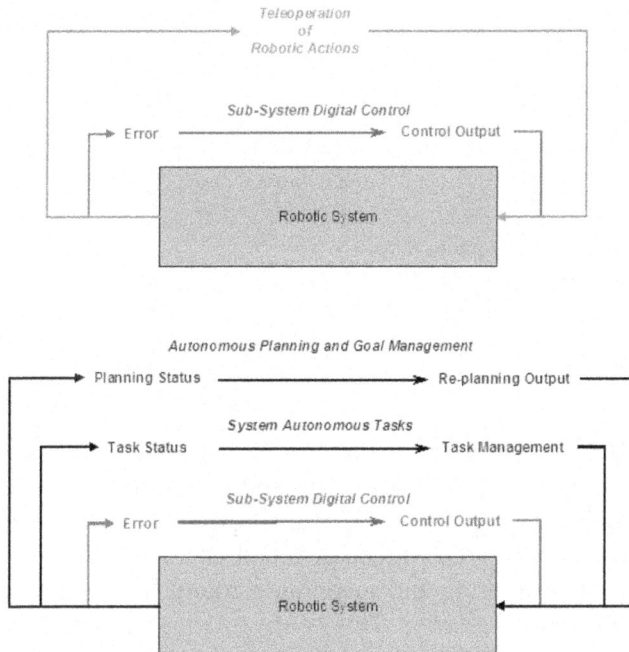

FIGURE 18.21 Teleoperated and Fully Autonomous Robotic Task Control Loops.

An intermediate level of control between fully autonomous and telerobotic is called shared control. In shared control some aspects of robotic tasking are autonomous, some are telerobotic, and others are automated, but require operator concurrence to approve action or must be initiated by the operator. The concept of shared control is shown in Figure 18.22.

Telerobotic systems may still have closed-loop digital control, but all tasking and decision making comes from the operator, and in the extreme case, literally every movement of the robot is an extension of the user's movements and no actuation is initiated autonomously. Once the robotic action is commanded by the user controlled motion may be maintained by closed-loop digital control. This is similar to concepts in virtual reality where a user input is directly replicated in the virtual model and the concept of

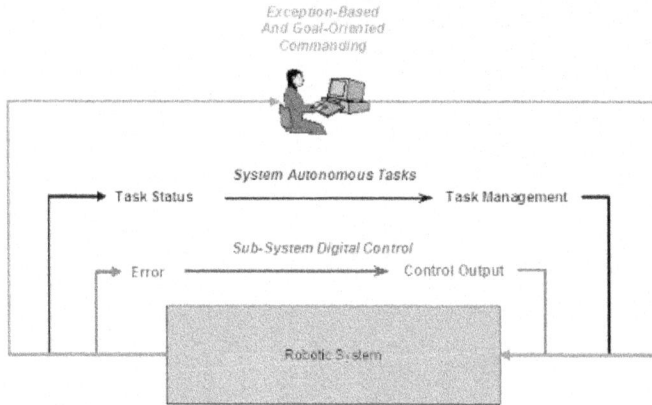

FIGURE 18.22 Shared Control of a Robotic System.

remotely piloting aircraft. For example, it would be possible to set up an OWI interface so that the OWI arm attempts to match the movement and position of the human arm based upon arm flexure measurements.

Robotics requires a hierarchy of automation, the *subsumption architecture*, where low level component actuation and sensing is interfaced to higher level sub-system control. Sub-systems in turn are tasked with goals and configured with behaviors. For example, a rolling robot may be tasked with exploring and mapping a room, but configured to have a collision avoidance behavior. This control, behavior, and tasking must be coordinated by an intelligent human operator or by artificially intelligent goal-based planning. This chapter provides an introduction to robotic system architecture and design from a bottom-up viewpoint providing practical examples for how to control and task a five degree of freedom robotic arm. As the reader can see, building upon these concepts will allow a robotics system that can operate in real-time through the use of feedback from sensing and instructions programmed into the robot's Artificial Intelligence.

EXERCISES

1. What is a Real-Time Embedded System?

2. With a reference coordinate system with an origin at the fixed base for the arm, the _____ of an arm can be defined based upon the arm mechanical design and kinematics.

3. Why are soft and hard limit switches used in robotic embedded systems?

4. Dead-reckoning control of a joint motor requires calibration of the _____,

5. Motion feedback requires active sensing during joint rotation: give an example of one method of providing this feedback?

6. Provide the four tasks necessary for a robot to perform in pick and place operation.

7. An intermediate level of control between fully autonomous and telerobotic is called _____ control.

8. Define subsumption architecture.

CHAPTER 19

THE USE OF UNMANNED AERIAL VEHICLES (UAVS) IN INDUSTRIAL AUTOMATION

INTRODUCTION

The drone is more formally known as an Unmanned Aerial Vehicle (UAV) or an Unmanned Aircraft System (UAS) or a Remotely Piloted Aircraft (RPA) that can fly autonomously or guided remotely without a human in control (Glover, 2014). A drone is like a male honey bee. Several names are available in the literature, like UAV, Unmanned Air Vehicle, Unmanned Aircraft Vehicle, Unmanned Aerospace Vehicle, Uninhabited Aircraft Vehicle, Unmanned Airborne Vehicle, Unmanned Autonomous Vehicle, Upper Atmosphere Vehicle, Aerial Robots, etc. In this book, both terms, drone and UAV have been used interchangeably.

The UAV is defined as an aircraft that operates without an onboard pilot. Essentially, a UAV is a flying robot that can be remotely controlled or fly autonomously through software-controlled flight plans in their embedded systems (Austin, 2010). It can work in conjunction with several onboard sensors, including Inertia Measurement Unit (IMU) and Global Positioning System (GPS). The UAV is a lightweight flying device similar to an aircraft; however, different from an aircraft since it is not operated by a pilot onboard (Gupta *et al.*, 2018). Based on how its movement is controlled, it can fly with a human remotely controlling the UAV, or in the most advanced cases, fly itself without any human intervention.

The terms UAV and UAS are closely related; however, the distinction should be made. The UAS is a much broader term describing the entire system that enables UAV operations. The UAS terminology is frequently used by the Federal Aviation Administration (FAA) of the United States, the European Aviation Safety Agency, and the UAV Systems Association (Dawson, 2018). A UAS consists of a UAV platform, in addition to the ground control station (GCS), communication, data link, piloting, sensing, flight planning, and other critical components needed for the operation of a UAV. A simple generic UAS system is shown in Figure 19.1.

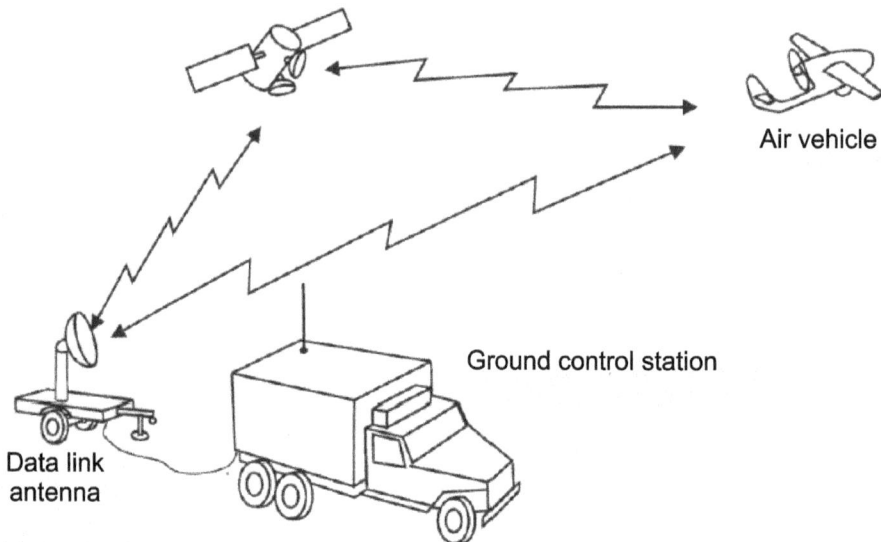

FIGURE 19.1 A typical UAS system.

The UAV is a component of the UAS since it refers to only the vehicle/aircraft itself. The UAV-related topics cannot be covered completely without reference to the relevant UAS components. A UAS unit may consist of more than one UAV; for example, a US Air-force Predator unit contains four UAVs. Most UASs have been developed from the military world, but in the last decades, UASs have penetrated a variety of civilian applications (Boucher, 2015). This is mainly due to the lowering of price and the development of sensors, imaging technologies, and UAV components. The UAS is an emerging sector of the aerospace industry with great opportunity and market demands ($\approx$70% of global growth and market share is in the US) that can be leveraged to high profitability in the near future (PwC, 2018).

The International Civil Aviation Organization (ICAO) employs the RPA system as these systems, based on cutting-edge developments in aerospace technologies, offer advancements for civil-commercial applications and improvements to the safety & efficiency of the entire civil aviation (ICAO, 2015). Many pilots prefer RPA because flying certain types of UAVs requires more skill. The RPA is essentially interchangeable with UAV. Taking control of an RPA requires more than simple handheld controls (Dawson, 2018).

The UAVs carry cameras, sensors, communications equipment, or other payloads. Remotely operated UAVs are actively controlled by a pilot on the ground, while autonomous UAVs perform functions without active human involvement. Varying degrees of automation are available in most modern UAVs (Singhal *et al.*, 2018). For example, a UAV may be remotely operated by a pilot but may use automation to maintain its position without commands from the pilot. The relative autonomy of a specific platform will also depend on the software used. The UAVs are opening up many new opportunities, from pilots for UAVs to electronics, sensors, and cameras. UAVs' increase in awareness and mission capabilities are driving innovations and new applications worldwide (White Paper, 2018).

HISTORICAL DEVELOPMENTS

Hot air balloons and explosives were first used for military reconnaissance in the French Revolutionary Wars to attack the enemies. The history of drones traces back to 1849 in Italy when Venice was fighting for its independence from Austria. Austrian soldiers attacked Venice with hot air, hydrogen- or helium-filled balloons equipped with bombs (Newcome, 2004). The kite was also used to collect information. In 1883, Douglas Archibald, an Englishman, attached an anemometer to a kite line and measured wind velocity at altitudes up to 1,200 ft. Mr. Archibald attached cameras to kites in 1887, providing one of the world's first reconnaissance UAVs. William Eddy took hundreds of photographs from kites during the Spanish-American War, which may have been one of the first uses of UAVs in combat (Dronethusiast, 2020).

During World War I, interest in UAVs was created in 1916 when Hewitt-Sperry Automatic Airplane was developed, and the Dayton-Wright Airplane Company invented an unmanned aerial torpedo in 1917 that carried explosives to a target at a preset time (Keane and Carr, 2013). This led to full-sized airplanes with primitive controls capable of stabilizing and navigating without a pilot onboard. The conversion of manned airplanes to target drones continued during the 1920s and 1930s (NIAS, 2018). During World War II, a large number of radio-controlled, radar-controlled, and television-controlled glide bombs were invented and used. During World War II, the first large-scale production of drones was made by Reginald Denny Industries (IWM, 2018). They manufactured nearly 15,000 drones for the US Army to be used for training anti-aircraft gunners. The Cold War also contributed to the evolution of drones. The most famous was the US Navy TDR-1, the US Army AZON, and the German Fritz X.

In 1917, Charles Kettering (of General Motors) developed a biplane UAV for the Army Signal Corps. It took him about three years to develop, which was called the Kettering Aerial Torpedo, but is better known as the "Kettering Bug" or just the "Bug." The Bug could fly nearly 40 miles at 55 mph and carry 180 lb of high explosives (Blom, 2011). The air vehicle was guided to the target by preset controls and had detachable wings released when it was over the target, allowing the fuselage to plunge to the ground like a bomb (Keane and Carr, 2013). In 1917, Lawrence Sperry developed a UAV, similar to Kettering, for the Navy, called the Sperry-Curtis Aerial Torpedo. It made several successful flights out of Sperry's Long Island airfield but was not used in the war.

The first used drone appeared in 1935 as a full-size retooling of the de Havilland DH82B "Queen Bee" biplane, which was fitted with a radio and servo-operated controls in the back sea (NIAS, 2018). The term *drone* dates to this initial use, a plane on the "Queen Bee" nomenclature. The plane, launched in 1941 (shown in Figure 19.2), could be conventionally piloted from the front seat, but generally, it flew unmanned, and was shot at by artillery gunners in training. The first target drone converted for a battlefield unmanned aerial photo reconnaissance mission was a version of the MQM-57 in 1955 (Dronethusiast, 2020).

FIGURE 19.2 Launch of a DH.82 Queen Bee (mother of drones) target drone (1941) (Dronethusiast, 2020).

Several pioneers developed important parts of the UAV system. For example, Archibald Montgomery Low (Professor Low of England), known as the "Father of Radio Guidance Systems," developed the first data link and solved interference problems caused by the UAV engine. On 3 September, 1924, he made the world's first successful radio-controlled flight. In 1933, the British flew three refurbished Fairey Queen biplanes by remote control from a ship (IWM, 2018). Two of them crashed, but the third one flew successfully, making Great Britain the first country to fully appreciate the value of UAVs, especially after they decided to use one as a target but could not shoot it down. In 1937, another Englishman, Reginald Leigh Denny, and two Americans, Walter Righter and Kenneth Case, developed a series of UAVs called RP-1, RP-2, RP-3, and RP-4. They formed a company in 1939 called the Radioplane Company, which later became part of the Northrop-Ventura Division. Radioplane Company built thousands of target drones during World War II. (One of their early assemblers was Norma Jean Daugherty, later known as Marilyn Monroe). Of course, the Germans used lethal UAVs (V-1s and V-223s) during the later years of the war, but it was not until the Vietnam War era that UAVs were successfully used for reconnaissance (Sholes, 2007).

The development of UAVs took place in the 1980s (Keane and Carr, 2013). The usefulness of UAS for reconnaissance was demonstrated in Vietnam. At the same time, early steps were being taken to use them in active combat at sea and on land. Military drones used solidified in 1982 when the Israeli Air Force used UAVs to wipe out the Syrian fleet with minimal loss of Israeli forces. The Israeli UAVs acted as decoys, jammed communication, and offered real-time video reconnaissance. The Predator RQ-1L UAS from General Atomics was the first UAS deployed in the Balkans war in 1995 (Blom, 2011). This trend continued in modern conflicts.

UAV systems have been used largely in military applications throughout history, as is true of many areas of technology, with civilian applications tending to follow once the development and testing had been accomplished in the military arena (NIAS, 2018). The UAV technology continued to be of interest to the military, but it was often costly and not completely reliable to put into operational use. After concerns about the shooting down of spy planes arose, military revisited the UAVs. Military use of drones soon expanded to play roles in dropping leaflets and acting as spying decoys. Since then, drones have continued to be used in the military, playing critical roles in intelligence, surveillance and force protection, artillery spotting, target following, battle damage assessment and reconnaissance, and weaponry (IWM, 2018). Recognizing the potential of non-military, non-consumer drone applications, the FAA issued the first commercial drone permits in 2006 (FAA, 2010). It opened up new possibilities for companies or professionals who wanted to use drones in business ventures.

The widespread use of the UAV began in 2006 when the US Customs and Border Protection Agency introduced UAVs to monitor the USA and Mexico borders. In late 2012, Chris Anderson, editor-in-chief of Wired magazine, started his own drones company, called 3D Robotics, Inc. (3DR), which started off specializing in hobbyist personal drones, and now marketing its UAVs to aerial photography, film companies, construction, utilities and telecom businesses, and public safety companies, among others (Keane and Carr, 2013). In late 2013, Amazon announced a plan to use commercial drones for goods delivery activities (Kimchi, 2015). In July 2016, Reno-based startup Flirtey successfully delivered a package to a resident in Nevada *via* a commercial drone. Other companies have since followed it. For example, in September 2016, Virginia Polytechnic Institute and State University began a test with Project Wing, a unit of Google-owned Alphabet Inc., to make goods deliveries. In December

2016, Amazon delivered its first Prime Air package in Cambridge, England. In March of 2017, it demonstrated a Prime Air drone delivery in California (Intelligence, 2018).

Table 19.1 presents a summary of the historical developments of UAVs since 1782.

TABLE 19.1 A Summary of the Evolution of UAVs.

Year	Developmental activity
1782	The first unmanned vehicle flight took place by the Montgolfier brothers.
1849	On 22 August 1849, the Austrians used UAV for the first time and attacked Venice using unmanned Balloons loaded with explosives.
1800s	The first drone with a "camera" was deployed at the end of the 19th century.
1894	Unmanned Balloons used by Austrians.
1898	Nikola Tesla invented a small unmanned boat that changes direction on verbal command by using radio frequencies to switch motors on and off and presented it in an exhibition at Madison Square Garden.
1915	During the Battle of Neuve Chapelle, aerial imagery was used by the British Military to capture about 15,000 sky view maps of the German trench fortifications in the region.
1916	During World War I, the first UAV took flight in the United States; though the success of UAV in test flight was erratic the military stamped their potential in the combat.
1917	French artillery officer, Rene' Lorin proposed flying bombs using gyroscopic and barometric stabilization and control.
1918	Charles Kettering (USA) flies Liberty Eagle "Kettering Bug."
1920	Elmer Sperry perfects the gyroscope, and the first enabling technology makes flight control feasible.
1920	Manned quadcopters were first experimented, but their effectiveness was hampered by the technology available at the time.
1920	Etienne Oemichen gave a design that had four rotors and eight propellers; all driven by one motor and recorded over 1000 successful flights. The first recorded distance was 360m in 1924 for a helicopter.

(continued)

Year	Developmental activity
1935	The first used drone appeared as a full-size retooling of the de Havilland DH82B "Queen Bee" biplane.
1940	Reginald Denny sold about 15,000 radio-controlled target drones to the US Military to train the anti-aircraft gunners for World War II.
1941	The first large-scale production of drones by US initiated by Reginald Denny.
1943	Advanced technologies provided control, guidance, and targeting. Gyroscope governed by magnetic compass controlled azimuth—aneroid barometer used for altitude control. Speed was determined by engine performance at max. power and propeller driven "air-log" governed the range.
1943	The German Military inaugurated the FX-1400, or "The Fritz X," a bomb with four small wings and a radio controller weighing 1362 kg. The first remotely controlled munitions were put into operational use and a great breakthrough for guided aerial weapons.
1956	Model A quadcopter, designed by George de Bothezat and Ivan Jerome, was the first successful quadcopter that used varying thrust of all four propellers in order to control pitch, roll, and yaw. However, it was difficult for the pilot to fly because of the workload of controlling all four propellers' thrust simultaneously.
1958	Curtis Wright company designed Curtis Wright VZ-7 as a US Army project that used variable thrust in the four propellers to control flight to simplify the pilot's workload.
1960s	UAVs took a new role during the Vietnam War, that is, stealth surveillance from the early use of target drones and remotely piloted combat vehicles.
1970s	During this time, Israel developed two un-piloted surveillance machines, MASTIFF UAV and the IAA Scout. Also, the success of the Fire Bee continued through the end of the Vietnam War. In the 1970s, the US set its sights on other kinds of UAVs while other countries began to develop their own advanced UAV systems.
1982	Military drone use solidified. Israel outwits Soviet anti-aircraft technology at the outbreak of hostilities with Syria, by revealing its location using a swarm of unmanned aircraft.

Year	Developmental activity
1985	Pioneer UAV program was introduced in the 1980s when a need for an on-call, inexpensive, unmanned, over the horizon targeting, reconnaissance and battle damage assessment (BDA) capability for local commanders was identified by the US military operations in Grenada, Lebanon, and Libya. The Navy started the expeditious acquisition of UAV systems for fleet operations using non-developmental technology in July 1985.
1986	Reconnaissance Drone was a joint US and Israeli Project which produced the "RQ2 Pioneer," a medium-sized reconnaissance drone.
1990	Miniature and Micro UAVs became part of research and soon came into action.
2000	Predator Drone was first deployed in Afghanistan.
2005	The first commercial multicopter was developed by Microdrones in Germany.
2006	The use of drones become widespread.
2010	The Parrot AR Drone, a smartphone-controlled quadcopter for consumers, was introduced at the Consumer Electronics Show in Las Vegas.
2012	Congress required the FAA to integrate small drones into the national airspace by 2015.
2013	The use of commercial drones for delivery activities is planned.
2014	The drone was used for delivery by Amazon for the first time in 2014. Also, the FAA grants an exemption to film and TV production companies for drone use.
2016	Amazon delivered its first Prime Air package in Cambridge, England.
Present	In the past 10 years, many small advanced quadcopters have entered the market, including the DJI Phantom and Parrot AR drone. This new breed of quadcopters is cheap, lightweight, and uses advanced electronics for flight control.

USES OF UAVS

UAVs were most often associated with the military applications in the recent past, where they were used initially for anti-aircraft target practice, intelligence gathering, and as weapons platforms. Still, these days, UAVs are also being employed for the large number of recreational and civilian applications (Sholes, 2007). Table 19.2 lists a few example areas where UAVs are being used for various military and civilian applications (Stoica, 2018). These tasks are performed by a number of different types of UAVs/Drones, from do-it-yourself drones to drones used for the purposes of domestic surveillance to combat drones used in military strikes. There are different models of UAVs; however, some of them are hand-launched and operated remotely or fly autonomously.

TABLE 19.2 A Few Military and Civilian Applications of UAVs/Drones.

Military uses	Civilian uses
Reconnaissance	Aerial photography
Surveillance and Wide Area Aerial Surveillance (WAAS)	Agriculture and forestry
Target Acquisition	Coastguard and conservation
Artillery fire correction	Cities
Battlefield damage assessment (BDA)	Transportation and traffic agencies
Land mine detection	Fire services and forestry
Elimination of unexploded Improvised Explosive Device and land mines	Rivers authorities and fisheries
Battlefield situation awareness and understanding	Gas and oil supply companies
Illumination of targets by laser designations	Information services
Psychological impact on militants	Local authorities/Police authorities
Combat roles	Meteorological services
Swarms and electronic intelligence	Survey organizations/Ordinance survey

Based on the types of the wing, UAVs are broadly classified as fixed-wing or rotary wing types (Cai *et al.*, 2011). The fixed-wing UAVs move horizontally and need a larger open space for take-off and landing. On the other hand, the rotary-wing UAVs move vertically, requiring only a small open space for landing. Table 19.3 presents the basic difference between these two types of UAVs.

TABLE 19.3 Basic Characteristics of Fixed-wing and Rotary-wing UAVs.

	Fixed-Wing	Rotary Wing
Mechanism	The lift generated using wings with forwarding airspeed	The lift generated using blades revolving around a rotor shaft
Advantages	Simpler structure, usually higher payload, higher speed	Can hover, able to move in any direction, vertical take-off and landing
Limitations	Need to maintain forward motion, need a runway for take-off and landing	Usually lower payload, lower speed, shorter range

The UAVs are the most predominant segment of the UAS market. The UAS market includes all unmanned vehicles, such as UAVs, Blimps, and Zeppelins (Freudenrichn, 2010). A Blimp (or a "pressure airship") is a powered, steerable, large-sized balloon with fins and an engine, lighter than air (Figure 19.3a). It has a long endurance but flying at low speeds. It does not have a rigid internal structure; if a Blimp deflates, it loses its shape. Blimps are best known today for their role as advertising and promotional vehicles. Their primary military use is for anti-submarine and reconnaissance roles. They are low-tech and relatively low-cost airships. The Zeppelins are rigid or semi-rigid airships (Figure 19.3b), originally manufactured by the Luftschiffsbau-Zeppelin. They have a rigid metal skeleton consisting of a cigar-shaped, trussed, and covered frame supported by internal gas cells. They are more suitable for long trips in a wider variety of weather conditions, making them expensive (Marsh, 2013). After World War I, Zeppelins were extensively used as bombers and scouts. They are equipped with powerful engines and are capable of lifting heavier loads. The application segments where UAV, Blimps, and Zeppelins may be used are presented in Table 19.4.

TABLE 19.4 Major Segments of UAV, Blimps, and Zeppelins.

Segments	Sub-segments	Intended uses		
		UAV	**Blimps**	**Zeppelins**
Civil	Natural disasters	Low	Medium	Medium
	Humanitarian relief	Low	Least	Least
	Environment	Medium	Medium	Medium
	Precision agriculture	Medium	Low	Low
	Cargo transport	Low	Medium	Medium
Commercial	Weather & storm tracking	Low	Medium to high	Medium
	Advertisement	Low	Medium to high	Medium
Military/ Security	Defense	Medium to high	Medium	Low
	Wireless communications	Medium to high	Medium	Medium

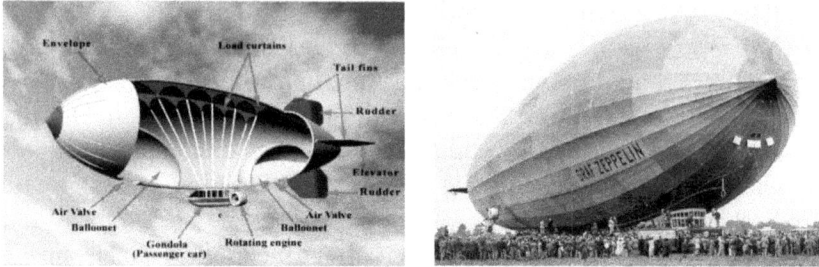

FIGURE 19.3 (a) The Blimp, and (b) The Graf Zeppelin (Source: Encyclopedia Britannica, 1999).

The UAVs have applications in a large number of fields (Figure 19.4), such as aerospace, military, aerial surveying, law enforcement, search and rescue, mining, conservation, pollution monitoring, surveying, oil, gas & mineral exploration & production, disaster relief, archeology, cargo transport, agriculture, precision farming, construction, passenger transport, monitoring criminal & terrorist activities, motion picture film-making, hobby & recreational, journalism, and light show (Ayranci, 2017; Berie, and Buruda, 2018). The uses of UAVs are growing rapidly across many civilian applications, including real-time monitoring, providing wireless coverage, remote sensing, search and rescue, delivery of goods, security and surveillance, precision agriculture, and civil infrastructure inspection (Taladay, 2018). Coupled with the applications from other synergetic technologies, such as 3D modeling, internet of things, artificial intelligence, and augmented reality (AR), as well as Virtual Reality, it has opened up new possibilities for organizations to leverage the use of UAV and its associated technologies across their operations (Dupont *et al.*, 2016).

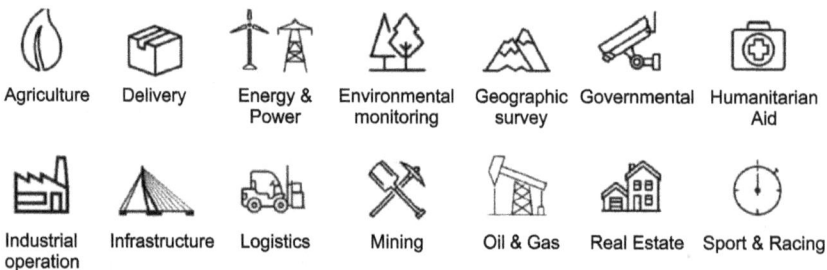

| Agriculture | Delivery | Energy & Power | Environmental monitoring | Geographic survey | Governmental | Humanitarian Aid |
| Industrial operation | Infrastructure | Logistics | Mining | Oil & Gas | Real Estate | Sport & Racing |

FIGURE 19.4 Various uses of UAVs.

Figure 19.5 shows the uses of UAVs to various maturity levels by different sectors of industry (White Paper, 2018). In the lower half of Figure 19.5, humanitarian relief, research & science, safety & security, insurance, and environmental have lower maturity levels due to lack of funds. In contrast, logistics has lower maturity due to regulatory problems (Jha, 2016). The top half of Figure 19.5 shows the potential of full uses of UAVs in these industries. Applications in agricultural mapping usually happen in rural areas; therefore, regulatory hurdles for UAV operations are lower than in urban, densely populated areas.

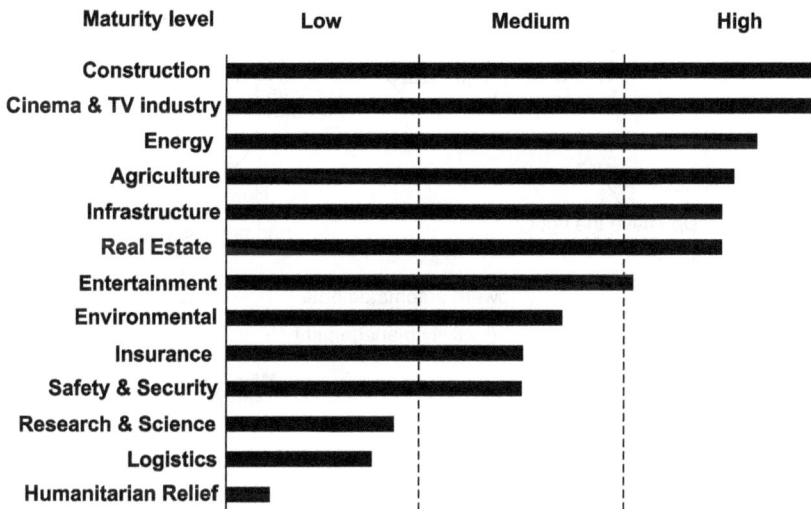

FIGURE 19.5 Uses of UAVs in industries to various maturity levels (White Paper, 2018).

In many business activities, drones can substitute traditional methods of operation. With less human operation and safety infrastructure, drones can reduce the time and costs of work. They can also enhance data analytics, which allows companies to better comprehend and predict operating performance. The drone market is growing steadily in the consumer, commercial, and military sectors. In a 2016 report, Goldman Sachs (2016) estimated that drone technologies would reach a total market size of US $100 billion between 2016 and 2020, though 70% of this figure would be linked to military, 17% to consumer, and 13% to commercial activities. The commercial business represents the fastest growth opportunity, projected to reach US $13 billion between 2016 and 2020.

The abilities of UAV to reduce the cost of compliance and cost of the technology while also enhancing the value of the information gathered through these systems have been the key drivers for the increased adoption of UAV worldwide. Throughout the world, there has been an increasing awareness about UAV in the industry and the government. The stakeholders using the UAV represent the government, UAV developers, and the UAV operators. As we see inFigure 19.6 attempts to describe the interaction of the stakeholders required to realize an automated data collection to improve productivity management through the deployment of UAVs.

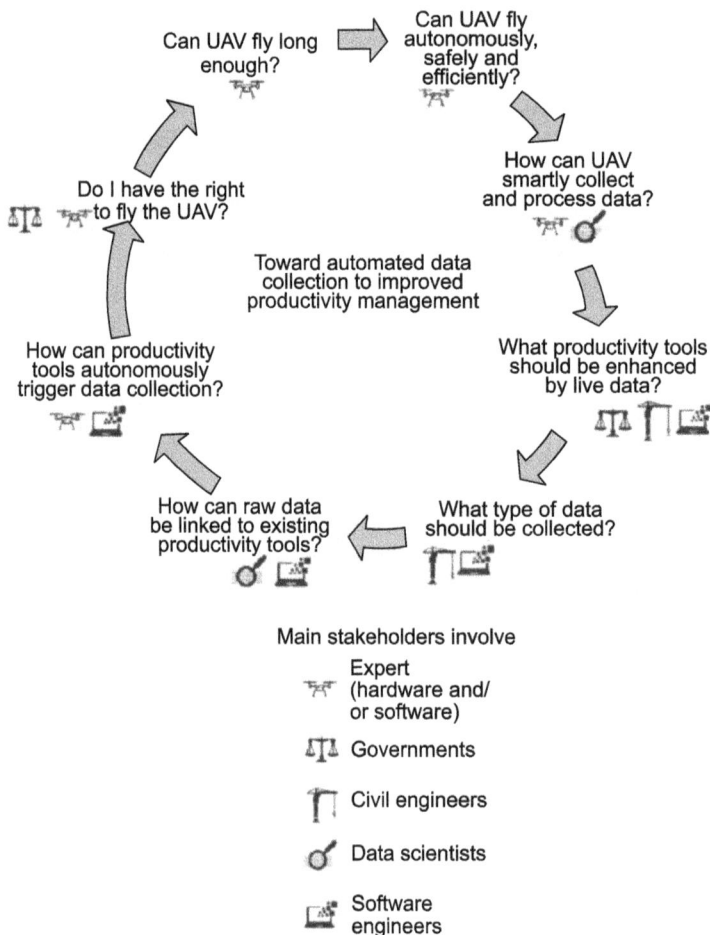

FIGURE 19.6 Cycle of a UAV (Dupont et al., 2016).

Gartner (2018) presented the Hype Cycle for emerging technologies in 2017 (Figure 19.7) which provides a graphical representation of the maturity and adoption of technologies and applications and how they would be potentially relevant to solving real business problems and exploiting new opportunities. Gartner Hype Cycle methodology provides a view of how a technology or application will evolve over time, providing a sound source of insight to manage its deployment within the context of specific business goals. It is a unique graph because it garners insights from more than 2,000 technologies into a small set of 12 emerging technologies and trends. The commercial use of UAVs/drones is approaching the "plateau of productivity" in the Gartner Hype Cycle. This Hype Cycle specifically focuses on the set of technologies that are showing promise in delivering a high degree of competitive advantage over the next 5–10 years.

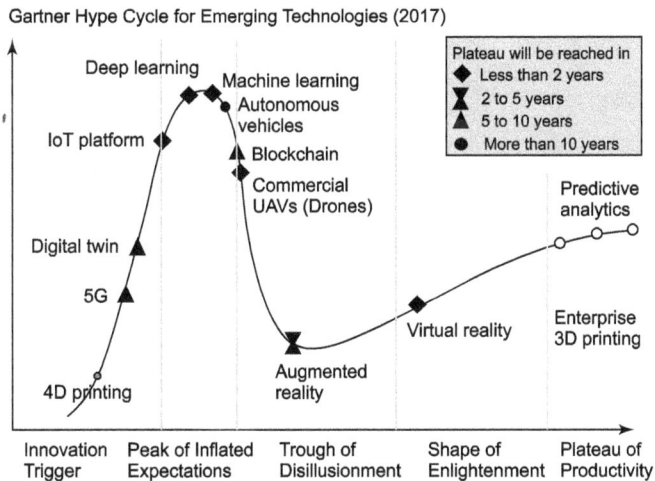

FIGURE 19.7 Hype Cycle for emerging technologies (Culus, 2018).

The UAVs allow users to increase the frequency of data collection and generate accurate and reproducible datasets. The recurring cost for these data sets is going to be very low as compared to other solutions. Thus, a well-designed setup would be required to use UAVs effectively and efficiently (Glover, 2014). All the components of this setup will have a significant role in the derived results. For example, a specific combination of sensors, software, and platform may guarantee accurate data in a particular application. The model in Figure 19.8 explains what it means to choose the right approach to include UAVs in several operations (White paper, 2018). It starts with the client helping users work through requirements and standards to narrow their way down to the required software and hardware. The UAVs application research carried out in many sectors have shown that the emerging interest in UAV-based technologies is related to the reduced size of the vehicles, affordability in terms of cost, low energy consumption, flexibility, minimizing risks, and the resulting high spatial resolution data (Puliti *et al.*, 2015).

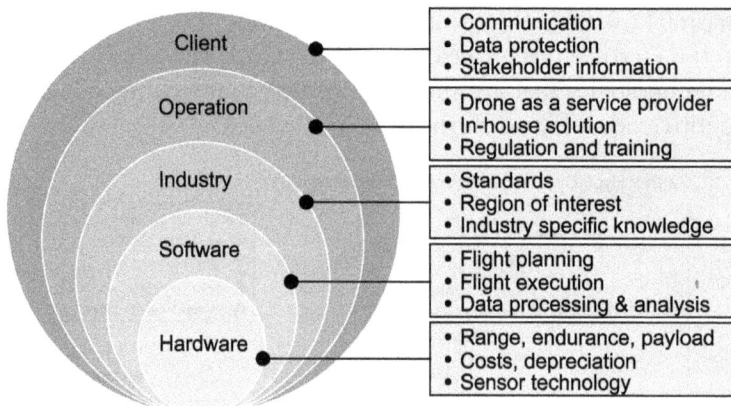

FIGURE 19.8 The onion model of UAV components (White Paper, 2018).

Smart UAVs are the next big revolution in UAV technology, promising to provide new opportunities in different applications, especially in civil infrastructure, to reduce risks and lower cost (Intelligence, 2018). They can be used to inspect nearly any structure, both indoors and outdoors. Still, they are most useful in inspecting structures that are difficult to reach by traditional means, including tall structures, such as buildings, towers. It can also include confined spaces in which space is limited for traditional inspections or structures that are over water, such as bridges or the undersides of oil rig platforms (Jha, 2016).

Globally, some countries have created a conducive environment for organizations to benefit from the applications of UAV and its associated technologies. Hence, these technologies are driving innovation in the UAV market. The UAV market is opening up many new opportunities, from UAV pilots to electronics and cameras. For example, vertical take-off and landing (VTOL) UAVs will be effective and useful in humanitarian aid missions. The relative market of UAVs in different segments in various parts of the world is presented in Table 19.5 (Luncitel, 2011).

TABLE 19.5 Relative Market of UAVs in Different Regions.

Segments	Sub-segments	USA	Europe	Mid-East	Asia-Pacific	Others
Civil	Natural disasters/Humanitarian relief	Medium	Low	Low	Low	Low
	Environment/ Weather & storm tracking	Medium	Low	Low	Low	Low
	Precision agriculture/Cargo transport	Low	Low	Least	Least	Least
Commercial	Advertisement	Least	Least	Least	Least	Least
Military/ Security	Defense	High	Medium to High	Medium to High	Medium to High	Medium
	Wireless communications	High	Medium to High	Medium to High	Medium to High	Medium

As estimated by the US government, 110,000 drones are flying in US airspace, projected to more than quadruple by 2022. According to recent research PwC (2018b), the global UAV market is projected to grow to US $19.85 billion by the end of 2021, at a Compound Annual Growth Rate (CAGR) of nearly 13% over during 1997-2023. Thus, in the future, the drones will be used in a large range of applications, including construction, surveillance, search and rescue operations, logistics, traffic monitoring, exploring hazardous sites, military surveillance, weather and climatic mapping, photography and journalism, delivery services, and sustainable agriculture.

SOME TECHNICAL TERMS

Here some of the basic technical terms used for the operation of UAV have been explained.

Autopilot: A drone/UAV can conduct a flight without real-time human control, for example, following preset GPS coordinates.

Autonomous flight: The flight of a UAV is controlled by internal programming that guides it was to fly without sending radio signals by humans. For example, a UAV may use its onboard GPS system to fly from one predetermined point to another.

Airborne collision avoidance system: An aircraft system based on secondary surveillance radar transponder signals, which operates independently of ground-based equipment to provide advice to the pilot on potential conflicting aircraft that are equipped with Secondary Surveillance Radar transponders.

Waypoints: A waypoint is a set of coordinates that identifies specific points in physical space. Waypoints are used to create a flight path for UAVs. The UAVs with waypoint technology typically utilize GPS and the global navigation satellite system (GLONASS) to create waypoints.

Waypoints navigation: Waypoints allow a UAV/drone to fly on its own to its flying destination or points pre-planned and configured into the remote control navigational software. Waypoints are useful in designing various autonomous missions for UAVs. Mapping would be impossible without the possibility to define these physical locations.

Point of interest: The "point of interest," also known as "region of interest," designates a spot that a UAV must necessarily collect the data of that point.

Line of sight (LOS): Many small UAVs are LOS machines, meaning the operator controlling the device must be in direct sight of the aircraft so that radio signals can be transmitted back and forth. Many larger UAVs are not LOS aircraft because the radio signals that control them are bounced off of satellites or manned aircraft.

Visual line of sight (VLOS): The operation of an unmanned aircraft within the pilot's LOS at all times, without the aid of any device (binoculars) other than corrective lenses (glasses) defines the VLOS (Figure 19.9).

Beyond VLOS (BVLOS): The ability to operate an unmanned aircraft beyond the pilot's LOS is called BVLOS. Flying UAV beyond the VLOS requires a special permit from the regulating authority, for example, FAA in the USA (Figure 19.9).

Extended VLOS (EVLOS): It is a flight from 500 m range to a distance at which the UAV is still within the pilot's sight. It allows remote pilots to extend the proximity of UAV beyond their direct vision (Figure 19.9).

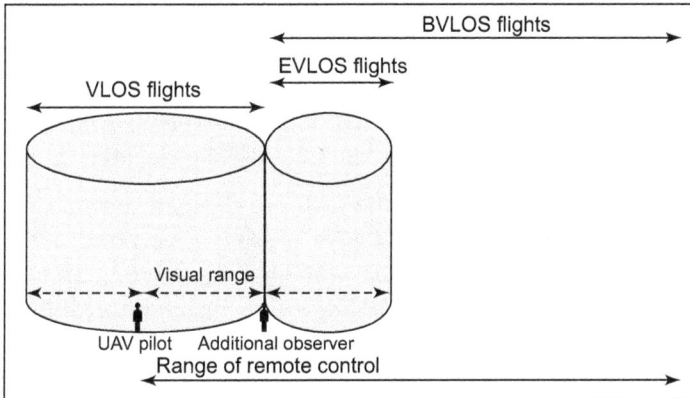

FIGURE 19.9 Distinction between UAV flight ranges-VLOS, EVLOS, and BVLOS.

Geofencing: The use of GPS technology to create a virtual geographic boundary of an area, enabling software to trigger a response when a UAV/drone enters or flies within that particular area, is called geofencing.

First person view (FPV): In FPV, a UAV with a camera transmits video feed wirelessly to goggles, a headset, mobile, or any other display device. The user has a FPV of the UAV/drone flies environment and captures video or still images.

Air traffic control (ATC): A service operated by appropriate authority (such as the FAA in the US) to promote the safe, orderly, and expeditious flow of air traffic.

Ground control station (GCS): It contains software running on a computer that receives telemetry information from a UAV, and displays its progress and status, often including video and other sensor data. It can also be used to transmit in-flight commands to the UAV.

Inertial measurement unit (IMU): The IMU is an electronic device typically used to maneuver aircraft and UAVs (an attitude and heading reference system). It measures the aircraft's specific force, angular rate, and orientation using a combination of accelerometers, gyroscopes, and magnetometers.

Position hold: It is automatic position control of UAV in space using barometric or ground sensors. It is mainly used for inspections and flights to the area where it is not safe to land the UAV, for example, city, forest, etc.

Automatic take-off: It is done with pre-defined height. The runway alignment during automatic take-off is defined by a straight line joining two waypoints surveyed using the navigation system. Take-off sites must be defined as waypoints either inside or outside the mission area. For large sites, it may be important to have the take-off sites located in the middle.

Automatic landing: It is done with pre-defined speed. The automatic landing is initiated from a circuit after passing a waypoint defining where the vehicle exits the circuit and commences the landing approach. Landing sites must be defined as waypoints either inside or outside the mission area. For large sites, it may be important to have landing sites located in the middle.

Return to home: When a UAV/drone automatically returns to the point of take-off or when the operator triggers the function on the remote control to return home. The UAV also returns home if the battery power is low or the signal from the transmitter is lost.

Coming home: It is mainly used if the UAV is flown to longer distances and/or the flight area is not easy to access, for example, mines, open water.

No fly zone: Government regulations restrict areas where flying a UAV. Regions where a UAV could interfere with an airplane or record sensitive information make up most of these areas.

Fail safe: System that helps protect a UAV in case of some error. For example, if a UAV loses control of signals, a fail safe will have the UAV return to the point of take-off (return home).

CHARACTERISTICS OF UAVs

The UAVs are aeronautical platforms that operate without the use of onboard human operators (Garg, 2019). There are broadly two types of UAVs; fixed-wings and rotary-wings (Molina, 2014). The development of new sensors, microprocessors, and imaging systems led to the growth of UAV systems based on fixed and rotatory wings UAVs. Fixed-wing UAVs (Figure 19.10a) have a predetermined airfoil that makes flight possible by generating the lift caused by the UAV forward airspeed. Rotary-wing UAVs (Figure 19.10b) consists of a number of rotor blades that revolve around a fixed mast. The blades themselves are in constant motion, which produces the airflow required to generate the lift. Rotary-wings UAVs are also known as multicopters having more than two motors. Multicopter UAVs are classified depending on the rotor configurations as tricopters, quadcopters, hexacopters, and octocopters.

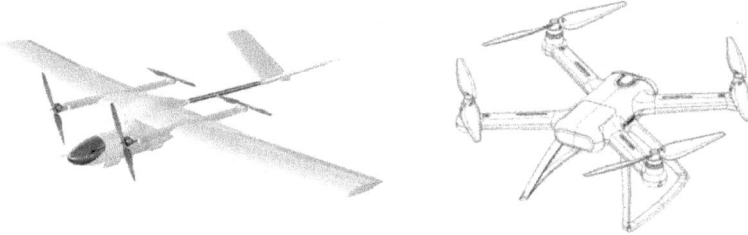

FIGURE 19.10 (a) Fixed-wing type UAV, and (b) Rotary wing UAV (Quadcopter).

The first commercial multicopter appeared in Germany in 2005, developed by Microdrones, and then the industrial sector of multicopters quickly expanded (Glover, 2014). Multicopters control the vehicle motion while varying the relative speed of each rotor to change the thrust and torque with fixed-pitch blades. The great advantage lies in the simplicity of the rotor mechanics required for the flight control, in contrast to conventional single- and double-rotor helicopters, which use complex blade rotations (Vergouw *et al.*, 2016).

The concept of a quadcopter (rotary wing) vehicle is not new; manned quadcopters first experimented in the 1920s, but their effectiveness was hampered by the technology available at the time (Cai *et al.*, 2011). A quadcopter has four propellers (Figure 19.10b) which are fixed and vertically oriented. Each propeller has a variable and independent speed, which allows a full range of movements. Propellers control the conventional helicopters with blades that dynamically pitch around the rotor hub. The components required for blade pitch are expensive; which is one of the reasons why quadcopters are becoming common in many applications (Šustek and Úøedníêek, 2018). Today, UAVs are executing search and rescue missions, tracking cattle rustlers, and monitoring wildfires or landslides with minimal cost and little risk of loss of life (Xu *et al.*, 2014). Other applications of this technology include geomatics, precision agriculture, infrastructure monitoring, and logistics.

The UAVs are generally categorized with reference to several parameters, such as payloads, weight, size, function, payload, geographical range, flight endurance, and altitude (Sholes, 2007). More details are given in Chapter 2. The payload refers to the carrying capacity of the UAV, including the contents. Geographical range, flight endurance, and maximum ceiling are the maximum distance, time, and altitude, respectively, that UAVs can reach during an interrupted mission (Glover, 2014). Size and weight are important characteristics where the size of the UAV (such as nano, mini, regular, large) may vary from that of an insect to the size of a commercial airplane. In contrast, weight may vary from hundreds of grams to hundreds of kilograms (Vergouw *et al.*, 2016). They are also recognized as per their range (such as very close range, close range, short-range, mid-range, etc.). In addition, today, UAVs are characterized as having different energy sources, such as battery cells, solar cells, and traditional airplane fuels. The UAVs are also classified based on autonomy, being fully autonomous to fully control by a remote pilot.

Two primary types of remote sensing systems can be employed in UAVs: active and passive remote sensing systems. The active remote sensing systems include laser altimeter, Light Detection and Ranging (LiDAR), radar, and ranging instrument, while passive remote sensing systems include accelerometer, hyperspectral sensor, and imaging sensor (Wingtra, 2019). In a passive remote sensing system, the sensor detects natural radiation that is emitted or reflected by the object. In contrast, active sensors carry their own light source to illuminate the object and detect natural radiation emitted or reflected by the object. With the advancement of electronic technology in sensors, batteries, cameras, and GPS systems, quadcopters became widely employed over the past decade, both recreationally and commercially (Šustek and Ůøedníêek, 2018; Albeaino *et al.*, 2019). A comparison of UAVs, airborne, and satellite systems is given in Table 19.6.

TABLE 19.6 A Comparison of UAV, Airborne and Satellite Systems (Singhal et al., 2018).

System	System resolution	Degree of availability	Operation mode	Payload capacity	Operating cost
UAV	cm to m	High	Autonomous or remote control	Limited	Low
Helicopters	100 m	High	Human pilot	Limited	Medium
Airborne	Up to 50 m	Moderate	Human pilot	Much more than helicopters	High
Satellite	1 m-1 km	Poor	Autonomous	Limited	Very high

Based on the sensor systems, UAVs are classified as photogrammetry-based UAVs and LiDAR-based UAVs (Austin, 2010). The photogrammetry UAV and LiDAR UAV are actually quite different from each other, even if their three-dimensional (3D) outputs look similar. In survey missions, the choice between photogrammetry UAV and LiDAR UAV depends mainly on the application in hand (Nex and Remondino, 2014; Dering *et al.*, 2019). In addition, operational factors, such as cost and complexity, are considered. UAV photogrammetry indeed opens various new applications in the close range aerial domain, introducing a low-cost alternative to the classical manned aerial photogrammetry for large-scale topographic mapping or detailed 3D recording of ground information and being a valid complementary solution to terrestrial acquisitions (Figure 19.11). UAV images are also often used in combination with terrestrial surveying to complement data for 3D modeling and create orthoimages (Albeaino *et al.*, 2019).

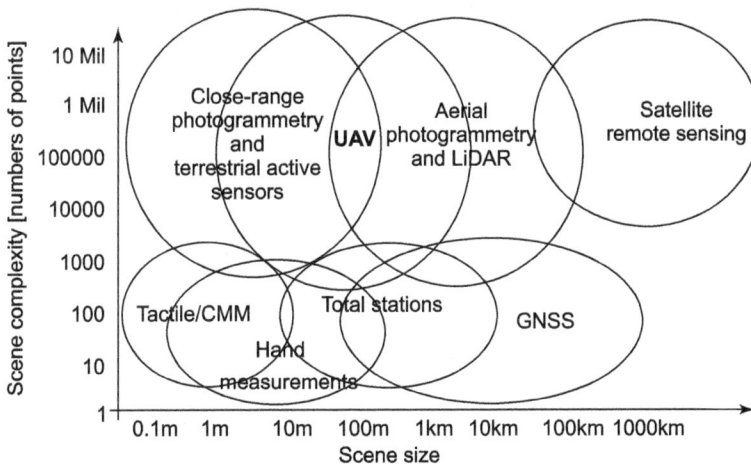

FIGURE 19.11 Available geomatics techniques, sensors, and platforms for 3D recording purposes, according to the scene's dimensions and complexity (Nex and Remondino, 2014).

The latest UAV developments can be explained with low-cost platforms combined with digital cameras and global navigation satellite system (GNSS), Inertial Navigation System (INS), necessary to navigate the platforms, predict the acquisition points, and possibly perform direct georeferencing. Although conventional airborne images still have some advantages, very high-resolution satellite images are filling up the gap between airborne and satellite mapping applications. UAV platforms are an important alternative and solution for studying the environment in 3D (Colomina and Molina, 2014).

Photogrammetry UAVs

UAV photogrammetry is described as "a photogrammetric measurement platform, which operates remotely controlled, semi-autonomously, or autonomously, without a pilot sitting in the vehicle." Photogrammetry UAV captures a large number of high-resolution photos over an area (Wingtra, 2019) with overlap such that the same point on the ground is visible in multiple photos. In a similar way that the human brain uses information from both eyes to have depth perception, photogrammetry uses these multiple images to generate a 3D model. The high-resolution 3D reconstruction contains elevation/height information and texture, shape, and color for every point, enabling easier interpretation of terrain. Photogrammetry UAVs are cost-effective and provide outstanding flexibility regarding where, when, and how 2D or 3D data are captured (Federman *et al.*, 2017).

A UAV coupled with a high-resolution camera can map a large area in a single flight, generating 2D and 3D data (Figure 19.12). Images can be collected from a UAV with a resolution of the order of a few centimeters. Figure 19.13 shows the appropriate camera platform for a photogrammetric survey which depends on both the areal extent and required sampling resolution (Dering *et al.*, 2019). The coverage per flight-hour for a photogrammetry-based UAV exceeds by more than 5x as compared to LiDAR-based UAVs. So, high-quality data can be captured in less time and at an overall lower cost (Buczkowski, 2018). After the acquisition, images can be used for stitching and mosaicking purposes and input in the photogrammetric process. For this purpose, camera calibration and image triangulation are initially performed to create a Digital Terrain Model (DTM) or a Digital Surface Model (DSM), or orthoimages (Colomina and Molina, 2014).

The flight may be undertaken in manual, assisted, or autonomous mode, according to the mission specifications, platform type, and environmental conditions. The onboard GNSS/INS navigation device is used for the autonomous flight (take-off, navigation, and landing) and guide image acquisition. The navigation system, generally called autopilot, allows performing a flight according to the plan and communicates with the platform during the mission. Double-frequency positioning mode or the use of the real-time kinematic method would improve the accuracy of positioning to decimeter level. Still, they are expensive to commonly use on low-cost solutions (Kan *et al.*, 2018). During the flight, the autonomous platform is normally observed with a GCS which shows real-time flight data, such as position, speed, altitude and distances, GNSS observations, battery or fuel status, rotor speed, etc. (Buczkowski, 2018). On the contrary, remotely controlled systems are piloted by the operator from the ground station.

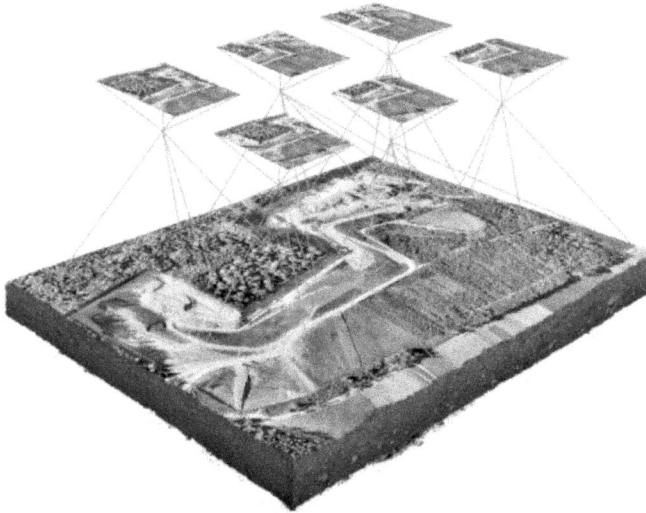

FIGURE 19.12 Photogrammetry UAV (Wingtra, 2019).

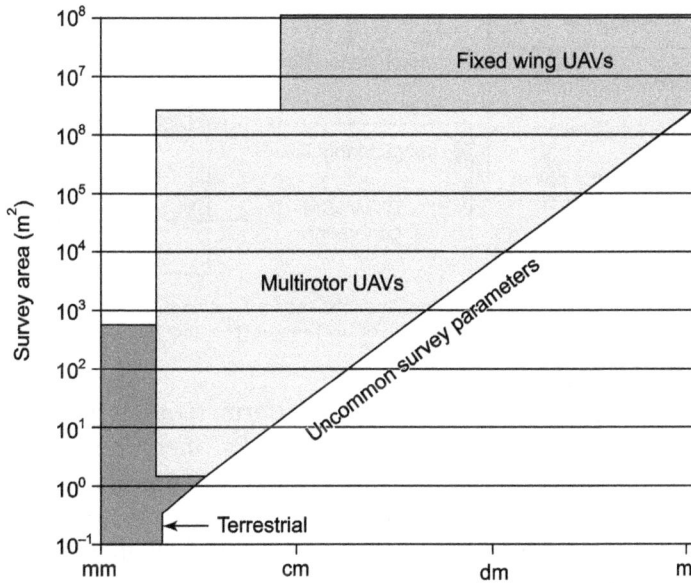

FIGURE 19.13 Sampling resolution with area for the photogrammetric survey (Dering et al., 2019).

The workflow of photogrammetry UAV, presented by Nex and Re-mondino (2014), is shown in Figure 19.14, where the input parameters are flight parameters, available devices, and additional parameters for workflow steps. The flight planning and data acquisition mission is nor-mally planned beforehand using required software, starting from the knowledge of the area of interest, required ground sample distance, and the intrinsic parameters of the onboard digital camera. The desired im-age scale and camera focal length are generally fixed to derive the fly-ing height of the UAV. The camera perspective centers (also called way-points) are computed after deciding the longitudinal and lateral overlap of the strips. All these parameters can vary according to the goal of the flight; for example, missions for a detailed 3D model generation usually require higher overlaps in images.

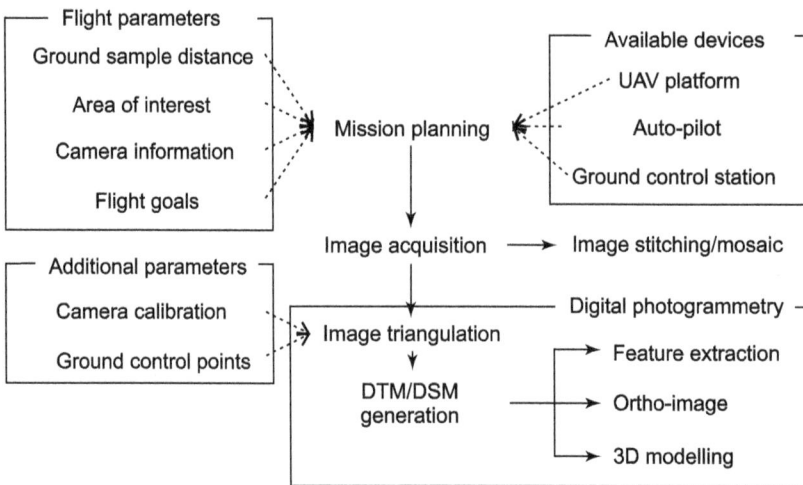

FIGURE 19.14 Typical acquisition and processing of UAV images (Nex and Remondino, 2014).

A typical photogrammetry UAV platform requires flight or mission planning and Ground Control Points (GCPs) measurements for geore-ferencing purposes (Federman *et al.,* 2017). Mission planning and flight altitude are flexible and controlled based on the real needs of users. The flight paths are designed to meet the objectives of the flight mission. Here, mission planning may also have features that contain all safety formalities and applications. Additional information, such as completing a flight log after each flight, can also be attached to this category. The logbook con-tains information about the flight pattern, time-of-flight, battery life, wind speeds, temperature, etc. (Shakhatreh *et al.,* 2018).

Camera calibration and image orientation tasks require the extraction of common features visible in as many images as possible (tie points) followed by a bundle adjustment, that is, a non-linear optimization procedure (Colomina and Molina, 2014). The collected GNSS/INS data can help for automated tie point extraction and allow direct georeferencing of the captured images. Some efficient commercial software that could be used includes, for example, PhotoModeler Scanner, Eos Inc; PhotoScan, Agisoft. Once a set of images has been oriented, 3D reconstruction and modeling steps are surface measurement, orthophoto creation, and feature extraction.

Table 19.7 presents some of the popular software which can be used for photogrammetry UAVs flight planning mission. A good UAV system with professional mission planning and post-processing workflow helps ensure that quality data is captured to get accurate results. Good overlapped images increase the accuracy and provide better error correction as compared to complete reliance on the direct georeferencing method used in the LiDAR survey. Photogrammetry UAV generates not only accurate 3D models but also full-color, high-resolution information for every point on that model (Dering *et al.,* 2019). This provides visual context for 3D data and makes interpretation and analysis of the results much easier than LiDAR point cloud data.

TABLE 19.7 Selected Flight Planning Software/Apps for UAV Photogrammetry (Dering *et al.,* 2019).

Software	*Operating system(s)*	*Cloud processing capabilities*	*Web source*	*Applications*
MapPilot	iOS only	Yes	*https://support.dronesmadeeasy.com*	Flat or hilly landscapes (e.g., coastal platforms, stockpiles, etc.)
FlyLitchi	iOS Android	No	*https://flylitchi.com/*	Complex surveys, such as steep cliffs, pit walls, or tall structures (e.g., towers)

Software	Operating system(s)	Cloud processing capabilities	Web source	Applications
DJI GS Pro	iOS only	No	https://www.dji.com/ground-sta-tion-pro	Flat or hilly ter-rain (e.g., coastal platforms, stock-piles, etc.) and tall structures
UGCS Pro	Android Windows	No	https://www.ugcs.com	All
Pix4D Capture	iOS Android	Yes	https://pix4d.com/produ ct/pix4dcapture	Flat or hilly terrain and tower-shaped structures
DroneDe-ploy	iOS Android	Yes	https://www.dronedeplo y.com	Flat or hilly terrain (e.g., coastal plat-forms, stockpiles, etc.)

Photogrammetry UAV has reached its saturation in terms of cost/accuracy/speed offered. It is preferred in (i) mapping of bare earth sites, (ii) areas not obstructed by trees, buildings, or other objects, (iii) only DSM is to be created, and (iv) when cost consideration is the governing factor.

LiDAR UAVs

The LiDAR technology has been around for many decades, but it is used only recently on UAVs. A LiDAR sensor emits laser pulses and records the time-of-flight and intensity of the reflected energy to be returned to an onboard sensor (Wingtra, 2019). A LiDAR uses oscillating mirrors to send out laser pulses in many directions to generate a "sheet" of light as the UAV moves forward (Figure 19.15). Measuring the timing and intensity of the returning pulses can provide measurements of the ground points. This range is converted to the coordinates of the points (x, y, and z) on the ground using onboard GPS and IMU (Garg, 2019).

FIGURE 19.15 The LiDAR data collection (Wingtra, 2019).

The UAVs which carry these sensors because of their weight and power requirements tend to be significantly larger (typically multicopters). All of these high-end systems carry out direct georeferencing of the raw data. The final output is a large number of cloud points with their known 3D coordinates, which could be used to create DTM or DSM. Laser UAVs can collect data at point densities between 60 and 1500 points/m^2 (Puliti *et al.*, 2015). A high-end LiDAR UAV system can generate results with horizontal (x, y) and elevation (z) accuracies in the range of 1–2 cm (Buczkowski, 2018). To get higher vertical accuracy of 1 cm (0.4 in), the cost of the employed system will go high with an increased level of operational complexity compared to photogrammetry UAV. Recently, efforts to make LiDAR scanners smaller, lighter, and cheaper to be employed in LiDAR UAVs (Barnes, 2018). For example, the Velodyne LiDAR Puck has less than 1 kg weight and is about 100 mm in diameter with a laser pulse range of 100 m.

The LiDAR instrument mounted on UAV collects signals at certain intervals. The instrument sends continuous pulses and can fire a large number of pulses in a second, thus collecting huge point cloud data in a very short time (Dering *et al.*, 2019). The first pulse is used to survey the top of objects, while the last pulse is used to survey the ground below the object. Intermediate pulses convey information about the vertical structure of the object.

In some specific situations, a terrain model below vegetation may be needed as an output. While photogrammetry UAV can effectively create 3D models of the ground in areas with sparse vegetation, the LiDAR UAV is very useful when mapping the areas beneath the dense vegetation, as shown in Figure 19.16 (Dering *et al.*, 2019). This is because LiDAR light pulses can filter through small openings between the leaves and reach the ground below. Due to these advantages, LiDAR UAVs have been successfully used to collect data under trees/forest cover (Corrigan, 2016, Cao *et al.*, 2019). However, the LiDAR systems are expensive for surveying and take a longer time to cover large areas. Some common LiDAR systems and UAV used are given in Table 19.8.

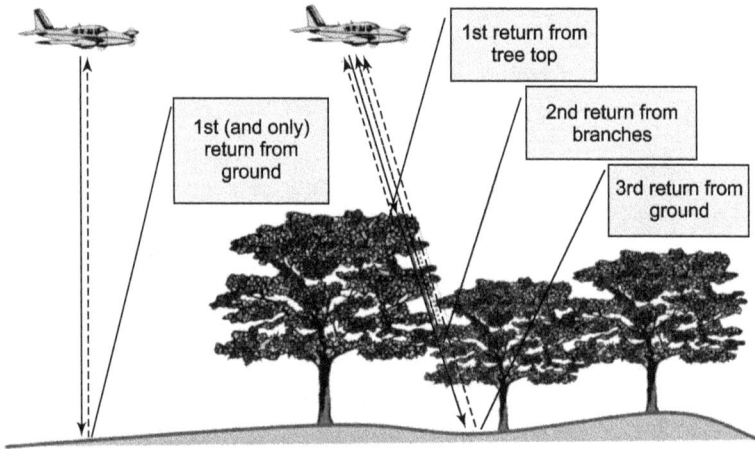

FIGURE 19.16 The LiDAR multiple returns (Ortiz, 2018).

TABLE 19.8 Common LiDAR UAVs and Sensors Used.

S. No.	LiDAR UAVs	LiDAR Sensors
1	DJI M600 Pro LiDAR quadcopter	LeddarTech Vu8
2	Draganflyer Commander	LeddarOne
3	Riegl RiCopter LiDAR UAV quadcopter	Velodyne HDL-32E
4	Harris H4 Hybrid HE UAV quadcopter	Velodyne Puck VLP-16
5	VulcanUAV Harrier Industrial multirotor	Velodyne Puck Lite

S. No.	LiDAR UAVs	LiDAR Sensors
6	VelosUAV helicopter	Velodyne Puck Hi-Res
7	Robota Eclipse fixed-wing drone	VUX-1 UAV
8	DJI Matrice 200 series quadcopter	Routescene—UAV LiDAR Pod
9	OnyxStar Xena drone	YellowScan—3 UAV
10	OnyxStar Fox-C8 HD quadcopter	YellowScan Vx Long Range
11	GeoDrone X4L LiDAR quadcopter	YellowScan Mapper
12	Tron F9 VTOL fixed-wing LiDAR drone	YellowScan Surveyor
13	Boreal long range fixed-wing drone	Leica Geodetics
14	Vapor 55 UAV helicopter	Leica Geo-MMS SAASM

The use of airborne LiDAR has a number of advantages which include:

- Rapid data acquisition is possible.
- The density of points collected enables the determination of man-made objects and infrastructure, such as buildings, bridges, and roads.
- Due to the large number of points collected, a sufficient number of laser pulses will penetrate the gaps between the tree canopy and return back to the sensor, enabling the determination of terrain under the tree canopy.
- Determination of heights of infrastructure, such as power lines or tree canopy, is easy due to a large number of points with recorded coordinates of features. Such information can be very useful for forest management (i.e., tree height, forest density, wood volume, and biomass estimation).
- The generation of digital elevation models is easy and accurate, which is required for many applications.
- LiDAR can be used both as an airborne sensor and as a terrestrial sensor.

Wijeyasinghe and Ghaffarzadeh (2020) have identified four fundamental technology choices (given in Figure 19.17 and explained below) that will strongly impact LiDAR product performance and business growth. The global market for 3D LiDAR in autonomous vehicles is expected to grow to US $5.4 billion by 2030.

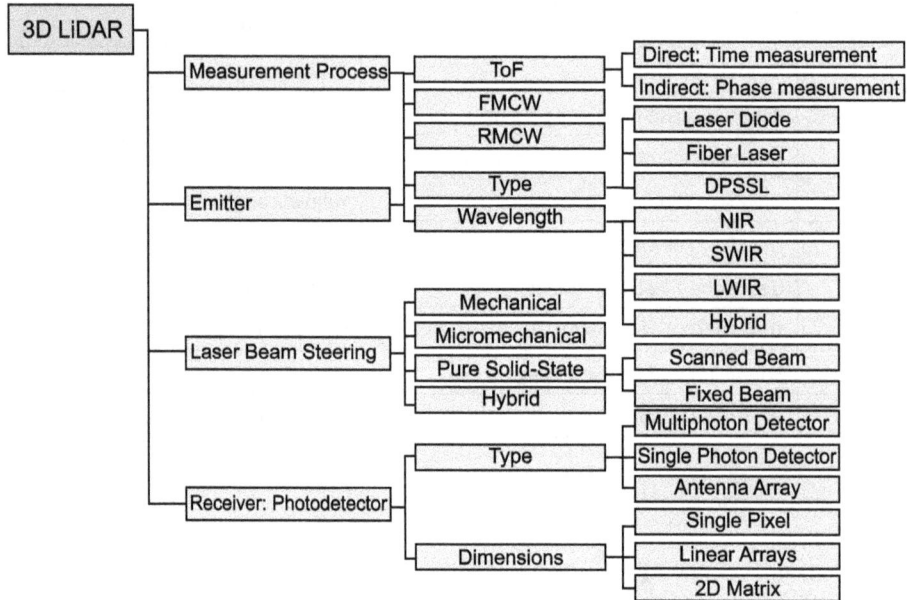

FIGURE 19.17 Four fundamental technologies for the development of LiDAR systems (Wijeyasinghe and Ghaffarzadeh, 2020).

- *Measurement process:* The measurement process determines how the light signal is generated and how it is measured. Most LiDAR companies develop conventional time-of-flight technology, but there is a trend toward emerging frequency-modulated continuous wave and random-modulated continuous wave LiDAR technology.

- *Laser beam:* The short-wave infrared (SWIR) LiDAR is considered safer for human vision than near-infrared (NIR) LiDAR operating at shorter wavelengths. Most LiDAR companies develop conventional NIR technology, but there is a trend toward emerging SWIR LiDAR because it can operate safely at higher power and detect distant objects.

- *Beam steering mechanism:* The laser light must be diffused or scanned by beam steering technology to illuminate the scene. Conventional automotive LiDAR uses a rotating mechanical assembly to scan laser beams by 360°, which enables the vehicle to see around it using a single LiDAR mounted on the roof. They are large, expensive, and prone to mechanical failure. The technology trend is toward small, cheap, and stable module designs. Emerging beam steering options include mechanical systems with limited motion, micro-mechanical systems based on micro-electro-mechanical systems mirrors, and pure solid-state systems. A solid-state LiDAR sensor is a compact and robust device, which is easily embedded into vehicles. For example, 3D flash LiDAR is a type of solid-state LiDAR that usually illuminates the entire scene at once, such as a camera. This enables fast imaging without distortion; however, the typical object detection range is shorter than scanned LiDAR due to laser power limitations, but the technology readiness is too low.

▪ *Photodetector*: The light-absorbing semi-conductor material must match the laser wavelength. Single photon-sensitive detectors are an emerging technology trend, enabling LiDARs to register signals quickly and detect weak signals from distant objects.

A Comparison of Photogrammetry UAVs and LiDAR UAVs

Most of the current photogrammetry UAVs are used for land surveying, which can import geographic information by putting together many aerial photographs. Photogrammetry UAV is good for surveying the sites without any obstacles, but most of the time, lands are not free from obstacles, so laser-based UAV surveying becomes the best choice (Chatzikyriakou and Rickerby, 2019). Photogrammetry UAVs do not provide optimum data when flying through forested areas, as the land features beneath the trees cannot be photographed, while LiDAR surveying gets land information beneath the trees (Cao *et al.*, 2019). In addition, LiDAR systems can take images at night and also work inside the tunnels. Moreover, laser-based surveying is considered to be more accurate than photogrammetry-based (Barnes, 2018).

LiDAR UAVs are preferred (Corrigan, 2019) where (i) mapping below tree canopy or mining sites or bathymetric survey required, (ii) other kinds of obstructions present to sun-light, (iii) need the vertical structure of the object, that is, by using waveform or multiple returns, (iv) mapping the objects, like transmission line, pipelines, roads in a corridor, (v) survey to be done at night or low light condition, (vi) higher accuracy across terrain required, (vii) not possible to establish the GCPs, and (viii) derive the benefits of both LiDAR and photo camera.

Both photogrammetry and LiDAR UAVs can provide a high level of 3D model accuracy, especially compared with terrestrial sampling methods (Corrigan, 2016). In order to generate the best results, a LiDAR system requires all of its components to work perfectly in synchronization.

Small gaps or errors in sensor measurements can lead to significant errors in outputs. The GCPs, which are useful in photogrammetry to correct errors, are difficult to implement with the LiDAR data (Buczkowski, 2018, Chatzikyriakou and Rickerby, 2019). Most of the time, the only solution for erroneous LiDAR data would be to repeat the flights. LiDAR UAV projects require an expert who understands the details of each sub-system and can recognize consistent and accurate data. In contrast, photogrammetry-based workflows are more flexible.

Figure 19.18 illustrates a comparison of UAV-based photogrammetry to other surveying methods in terms of total coverage area and survey error. Arrows demonstrate the UAV photogrammetry region's existing and potentially future expansion through camera hardware, imaging methodologies, and UAV control advancements. It is seen that centimeter-level DTMs developed from UAV-based photogrammetry are a competitive alternative to LiDAR systems. Table 19.9 presents a comparison of UAV-based photogrammetry to LiDAR UAV. Considering the advantages offered and cost coming down, LiDAR UAVs might lead in the future (Dering *et al.*, 2019).

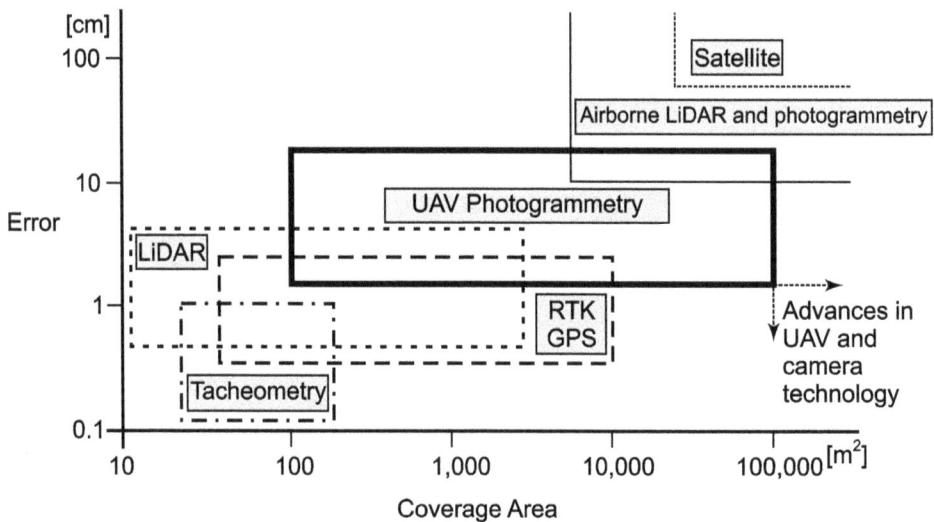

FIGURE 19.18 The UAV application to surveying tasks (Siebert and Teizer, 2014).

TABLE 19.9 A Comparison of Photogrammetry and LiDAR UAVs (Wingtra, 2019).

Parameters	Photogrammetry	LiDAR
Cost	US $20000-30000 for a professional UAV + high-resolution camera system	US $50000 or more for just the sensor. Survey-quality complete systems between US $1,50,000-300,000 range
Operational complexity	No additional sensors required; indirect georeferencing requires longer processing but is resistant to potential workflow errors	LiDAR uses direct georeferencing, which means that multiple components and sensors must work perfectly together in order to gather useful data
Outputs	2D orthomosaic maps, 3D models, point clouds, surface models with visual information as part of the 3D model	3D point clouds, intensity maps with multiple returns, and full-waveform information for classification
Accuracy	1 cm horizontal, 2-3 cm elevation (vertical) over hard surfaces	1-2 cm elevation (vertical) over soft and hard surfaces
Best for	Mapping, surveys, mining, broad-coverage combined with high horizontal and vertical accuracy,	Terrain models below dense vegetation, forestry, 3D modeling of power lines or cables, 3D modeling of complex structures, etc.

WORKING OF A UAV

The most common UAV/drones have two units; receiver and transmitter. The receiver unit consists of the UAV/drone, while the transmitter unit is normally a ground-based controller system to send the commands (by a human) to the UAV/drone. Flight of the UAV may operate with various degrees of autonomy, either under remote control by a human operator or autonomously by an onboard computer (Dering *et al.*, 2019). The UAVs can also be flown autonomously, where modern flight controllers can use software to mark GPS waypoints so that the vehicle will fly to and land or move to a set altitude. This kind of autonomy is becoming increasingly common and contributes to many civilian applications. Although it is possible for the flight controller to control the UAV autonomously, it is generally good to have a remote-controlled transmitter so that the movement of the UAV can be controlled if something goes wrong or the UAV becomes out of control.

A handheld transmitter is adequate enough for larger aerial vehicles, and it is better to have a base station to help with all of the controls. Remote GCS also controls the UAVs, also referred to as a ground cockpit. Transmitters have many channels, and the number of channels of a transmitter is related to the number of separate signals it can send. Usually, a seven-channel radio is adequate enough to operate the UAVs in the beginning. A good transmitter must include a three-position switch or variable knob, as it would be required by most autopilots to switch between various flight modes (Gupta *et al.*, 2018).

The UAV has a flight controller placed at the center and responsible for controlling the spin rate of the motors. Sticks on the controller allow movements in different directions, while trim buttons allow balancing the UAV. The UAV cannot function accurately and efficiently without using external sensors added with the flight controller. So for a stable and precise flight, different sensors, like accelerometer, gyroscope, and IMU, are used to interface with the flight controllers (Gupta et *al.*, 2018). To fly UAV better, some additional sensors, like a barometer, distance measuring sensors, magnetometer, could also be used. The UAVs are also equipped with a different state-of-the-art technology equipment, such as infrared cameras, GPS, and laser devices (Dering *et al.*, 2019). All these sensors and navigational systems are present at the nose of the UAV. The rest of the body is full of UAV technology systems. Computers can also be used to receive live video footage from the onboard camera and to display data being collected by sensors (Šustek and Úøedníèek, 2018).

A typical unmanned aircraft is made of light composite materials to reduce weight and increase maneuverability. The engineering materials used to build the UAV/drone are very lightweight and highly complex composites designed to absorb the vibrations, which decrease the noise produced. The composite material strength allows military drones to cruise at extremely high altitudes. To make the UAV agile, the acceleration and de-acceleration of the UAV must be quick, which means that it should pitch forward in the beginning and when it is time to stop, the UAV should pitch in the opposite direction quickly in an aggressive way. It is necessary to decrease the stopping distance. For the UAV to be agile, it must turn quickly with a small turning radius.

UAVs use a wide range of cameras for their missions. Most cameras used for UAV mapping are lightweight and can be programmed to shoot images at regular intervals or controlled remotely. The cameras required to carry out good mapping work are not the same as those used for professional video or photography work, but they are lightweight, high-resolution cameras designed to be used with UAVs. Some specialized devices that can be mounted on a UAV include LiDAR sensors, infrared cameras equipped for thermal imaging, and air-sampling sensors. Cameras can be mounted to the UAV in various ways. A motorized gimbal holds the camera and provides image stabilization, which can help compensate for turbulence and ultimately produce clear images. Gimbals are also used for changing the angle of the camera from vertical (straight down) to oblique. As UAV mapping is generally performed vertical or at some angle, gimbals may be relatively simple as compared to those used by filmmakers. The internal GPS functionality of some models can program the camera to take pictures at a certain interval or to take a picture based on distance or upon encountering a certain waypoint (Dering *et al.*, 2019).

Fixed-wing UAVs consist of a much simpler structure in comparison to rotary-wing UAVs. The simpler structure provides a less complicated maintenance and repair process, thus allowing the users more operational time at a lower cost. The only disadvantage to a fixed-wing system is the need for a runway or launcher for take-off and landing; however, "vertical take-off and landing" and "short take-off and landing" systems are very popular as they do not require a runway (Gupta *et al.*, 2018). Fixed-wing aircraft also require moving air over their wings to generate enough lift force to stay in a constant forward motion. It means they cannot stay stationary the same way as a rotary-wing UAV can. Therefore, fixed-wing UAVs are not best suited for stationary applications, like inspection work.

The biggest advantage of rotary-wing UAVs is that they can take-off and land vertically. This allows the users to operate within smaller vicinity with no requirement of a landing/take-off area. In addition, their capacity to hover and perform agile maneuvering makes them well suited to applications, like inspections where precision maneuvering for extended periods of time is required (Taladay, 2018).

There are various parameters that are to be considered for flight preparation, as explained below. Preflight checks are very important and help the operator to fly the UAV safely and provide professional results.

Altitude

Altitude is an important consideration when flying a UAV, both for practical purposes and in the interest of flying safely and legally. Legality is an important consideration when deciding an operating altitude. Caution should always be taken when flying UAV at higher altitudes, even if local regulations allow higher-altitude flight, to ensure that flights do not get in the way of manned aircraft. In many countries, it is illegal to fly above 500 feet (or 150 m).

Although higher altitude results in lower resolution, it allows the UAV to cover a larger area. Lower altitude photography gives better image quality but increases the time required to map a certain area. Suppose UAV flies at 100 m height and the area has a maximum elevation of 20 m, so from this height, UAV would cover 80% of the area in one photo compared to the area of one photo with the flat ground shown in Figure 19.19. Therefore, it is important to check the elevation range of the area being covered by the UAV to have proper coverage.

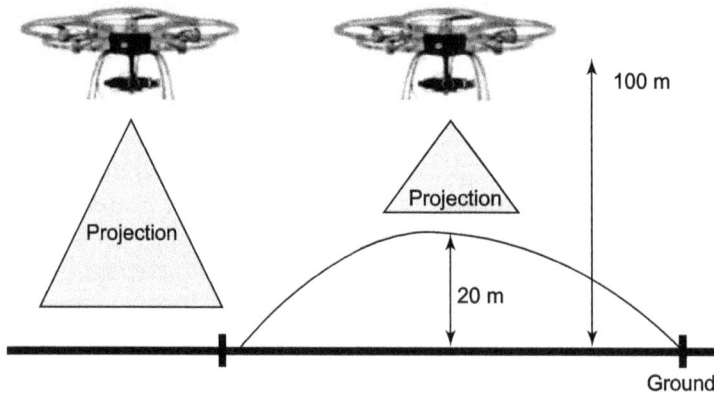

FIGURE 19.19 Terrain area covered by UAV.

Vertical and Oblique Photographs

The two aerial views most commonly used in UAV mapping are vertical (overhead) and oblique photographs (Newcome, 2004). Vertical photographs are taken directly above the object, with the camera looking vertically down. Oblique photographs are taken at a high or low angle to the object below. They can collect the information about the landscape that vertical photos cannot collect. During each flight, the angle of the camera remains fixed; else, this will make the resulting images considerably more difficult to process (Kaminsky *et al.*, 2009). Photos taken from these two views can be combined using photogrammetry software, such as Agisoft PhotoScan or Pix4D, which gives users the ability to view and analyze multiple perspectives in a single computer-generated model.

The 3D models can be created using overlap from vertical photographs. Such 3D models are very useful for many applications, such as post-disaster damage assessment, urban modeling, smart city planning, and creating more accurate flood simulations. The UAV vertical images can be used for the creation of DTM or DSM but cannot capture the details of facades. For façade mapping, oblique images are becoming more popular in the UAV photogrammetric process. Oblique images are easy to capture because it only involves adjusting the built-in camera in the system to the desired angle using the remote control or adjusting the camera before each flight (Wu *et al.*, 2018).

Positioning

Highly accurate UAV navigation is very important when flying for applications, such as creating 3D maps, surveying landscapes, and search & rescue missions. As the UAV is switched on, it searches and detects GNSS satellites. High-end GNSS systems use a satellite constellation, a group of GPS satellites working together, giving coordinates of the point of observation (Garg, 2019). Many of the latest UAVs have dual GNSS, such as GPS and GLONASS, to provide better accuracy. The UAVs can fly in both GNSS and non-satellite modes (Jha, 2016). For example, DJI drones can fly in P-mode (GPS & GLONASS) or Attitude mode, which do not use satellite navigation (Kan *et al.*, 2018).

ADVANTAGES AND DISADVANTAGES OF UAVS

The choice of potential unmanned platforms for low-altitude earth observation is large and continues to grow. In recent years, UAV technologies are evolving rapidly and becoming appropriate for multiple applications. Various reasons make the technologies relatively preferred. Tethered systems navigated manually, such as Kites and Blimps, are ideal for the precise coverage of small sites that requires only a few images (Gupta *et al.*, 2018). However, these systems are hardly used for systematic surveys of larger areas where regular overlaps along evenly spaced flight lines are preferable. The GPS and INS technologies have led to the availability of a range of auto-piloted systems, such as satellites, planes, drones, and multicopters that can autonomously follow the prescribed flight lines (Kan *et al.*, 2018).

It is difficult to generalize the advantages or disadvantages of a particular platform because it's possible applications and working conditions vary greatly around the world (Jha, 2016). However, an attempt has been made to present the performance evaluation of satellites, manned aircraft, and UAVs in Table 19.10.

TABLE 19.10 Performance Evaluation of Satellites, Manned Aircrafts, and UAVs.

S.No.	Attributes	Performance		
		Satellites	Manned aircrafts	UAVs
1	Endurance	High	Medium	Low
2	Payload capacity	Medium	High	Low
3	Cost	Medium	Low	Medium
4	Cloud free images	Low	Medium	High
5	Coverage	Low	Medium	High
6	Maneuverability	Low	Medium	High
7	Deployability	Low	Medium	High
8	Autonomy requirement	Medium	Low	Medium

Advantages of UAV

Any application that could utilize a highly mobile data collection or communication platform can utilize UAV as a primary data collection platform. UAVs have several advantages in military/army over manned aircraft. They provide a persistent presence over a specific area to capture still and video imagery, which provides military intelligence, surveillance, reconnaissance, and weapon delivery capabilities. In politically sensitive areas, UAVs have obvious advantages where the deployment of human soldiers would create too much controversy (Andrews, 2020). Humans on the ground (walking or using ground-based vehicles for data collection) can be challenged by hilly and difficult terrains, physical obstacles, or dangerous site conditions. The UAVs provide a safer method in high-risk situations where the presence of a pilot-driven aircraft is risky. They allow flight in difficult and inaccessible areas without having a risk of causalities since human life is not endangered during the process (Iizuka *et al.*, 2018).

The UAVs are also advantageous in many civilian applications, primarily for accelerating accessibility to remote and dangerous sites, sensor mobility, monitoring, and speed of data collection. The UAVs are important in a monitoring forest fire, especially in areas with rough topography where ground movement is not possible. Hence, UAVs are considered a highly preferred technology for risk management. In particular, UAVs have emerged as an essential data collection tool for applications involving natural hazards (e.g., earthquakes and landslides) where site accessibility is challenging post-event, and the need to collect recent data is urgent. As a result of these challenges, UAVs offer an economic advantage, including reducing costs associated with personnel, travel, and site logistics. Compared to a traditional inspection of structures, UAVs are cheaper, faster, and safer than traditional methods (Gupta et *al.*, 2018).

The demand for UAVs to transport materials and medicines is increasing. As an example, the potential benefits of UAVs in the health sector are extensive. The UAVs can fly over vast distances, enabling the delivery of life-saving medical supplies to those in hard-to-reach communities. The technology could also be used to deliver non-emergency commodities for villages where visits of health workers are not regular, and delivery of commodities is infrequent; thus it greatly reduces the response time. The UAVs could bring meaningful benefits to public health through improved supply chain performance.

The UAV also offers the possibility of making a non-destructive and accurate measurement of many attributes. It enables work measurement and work certification remotely and accurately (e.g., site audits, civil surveys, and measurement) (Taladay, 2018). It can improve measurement accuracy through enhanced visualization of terrain for site surveys and measurement, construction progress, and 3D mapping (Jha, 2016).

UAV-based photogrammetry provides advantages over other remote sensing satellite platforms (Molina, 2014). The UAVs can be a smart and cost-effective complement to traditional photogrammetric data. Satellite images are limited by their revisit time, cloud coverage, and spatial resolution. Advances in UAV platform design have contributed to the innovation of small UAV that fly at low altitudes to provide very fine spatial resolution images (Iizuka *et al.*, 2018). The UAVs can be deployed on demand, and flight parameters can be adjusted to acquire the images at desired high resolution. Hence, this technology provides an efficient method of spatial data collection for multiple applications, such as transportation, urban planning, crop species classification, and yield estimation that require detailed information (Taladay, 2018). The HALE UAV provides a cost-effective and persistent capability to efficiently collect and disseminate high-quality data across large areas than the ground-based systems.

The other advantage of UAV relates to low energy consumption (Banu *et al.*, 2016). Small UAVs use either fuel (gas) or battery as a source of energy. Although the batteries or fuel need to be charged frequently, but the energy required for a mission is small. Therefore, its low energy requirement contributes to the wide acceptance of the technology. The UAVs, being smaller than manned aircraft, can be more easily and cost-effectively stored and transported.

The benefits of the technology are many as UAVs support operations in-

(i) *Safety:* by allowing users to acquire data from the air instead of entering difficult/unstable terrain.

(ii) *Costs:* that are related to reducing operational expenses for data collection, particularly if the area is large.

(iii) *Flexibility:* that is associated with operating missions on schedule or emergency or on demand.

(iv) *Speed:* that enables the collection of real-time data where infrastructure allows.

(v) *Increased productivity:* due to savings in time and a high degree of automation.

(vi) *Frequency:* that opens opportunities for monitoring and tracking of work in progress, and

(vii) *Accuracy:* by generating accurate and reproducible datasets (e.g., 3D models).

Disadvantages of UAV

The disadvantages of UAVs include low maneuverability, low operational speed, vulnerability to attack, cyber or communication link attacks or lost data links, limited area coverage, and the requirement of a sophisticated analysis system to handle Big Data. The UAVs could be used for lethal purposes, criminal activities, like drug smuggling, and terrorist activities when hijacked by terrorist networks under third-party control situations. Another issue relates to safety; when these are not operated by properly trained operators, the chances of UAVs failing are more, leading to fatal accidents

(Iizuka *et al.*, 2018). Therefore, the safety, security, and privacy concerns need to be taken into consideration in technology development, market development, and their use by the public and private parties. In addition, regulations must clarify the limits to use UAVs and ensure that such devices are flown under safe conditions. Some of the advantages and disadvantages of UAVs are presented in Table 19.11.

TABLE 19.11 Advantages and Disadvantages of UAV Types (Heutger and Kückelhaus, 2014).

Types	Advantages	Disadvantages	Photo
Visual Fixed-Wing	Long range High flight time Large coverage area Object resolution (cm/pixel)	Horizontal take-off, requiring substantial space (or support, e.g., catapult) Inferior maneuverability compared to VTOL	
Unmanned Helicopter	VTOL Maneuverability High payloads possible	Expensive Comparably high maintenance requirements	
Multicopter	Inexpensive Easy to launch Low weight The very small take-off area Object resolution (mm/pixel)	Limited payloads Susceptible to wind due to low weight Less flight time Small coverage area	

SUMMARY

UAVs have been widely adopted in the military world, and the success of these military applications is increasingly the driving force to use UAVs in civilian applications. They have recently received a lot of attention, being fairly economical platforms, having navigation/control devices and imaging sensors for quick digital data collection and production. The UAVs provide a real-time capability for fast data acquisition, transmission, and processing. In these conditions, automated and reliable software is required to process the data quickly. Both photogrammetry and LiDAR UAVs can provide images/3D data with high accuracy; each type of data has its own merits and demerits (Corrigan, 2016). The difference between photogrammetry and LiDAR UAV can easily be understood while considering operational and logistical factors.

The biggest advantage of UAV systems is the ability to quickly deliver high temporal and spatial resolution information and allow a rapid response in a number of critical situations where immediate access to this type of 3D geo-information may be difficult. The UAVs can be used in high-risk situations and inaccessible areas, although they still have some limitations, particularly the payload, battery life, and stability. Compared to traditional airborne platforms, UAVs reduce operational costs, providing high accuracy data. The UAVs offer three technical advantages (Andrews, 2020): (i) minimize the risk to the life of the pilot; (ii) not subject to natural human limits (e.g., fatigue); and (iii) data acquisition is generally cheaper than that of manned aircraft.

Besides very dense point cloud generation, the high-resolution UAV images can be used for texture mapping purposes on existing 3D data, orthophoto production, map generation, or 3D building modeling. In addition to large applications in military, UAVs are used for various other applications, such as disaster and healthcare, journalism and photography, early warning system and emergency services, marketing and business, activity monitoring, gaming and sporting, farming and agriculture, military and spy, and many other fields. On the other hand, UAVs might become the most dangerous devices on the earth if the regulations are not followed strictly (Gupta *et al.*, 2018). The same UAV that is used for the delivery of food and products could also be used to transfer bombs and other illegal materials.

The UAVs have been the subject of concerted research over the past few years due to their autonomy, flexibility, and broad application domains. It is predicted that the market for commercial end-use of UAVs might supersede the military market by 2021, cumulatively hitting approximately USD 900 million, while the global market size will touch USD 21.47 billion (Economic Times, 2018). With the implementation of proper monitoring, powerful law and security, and the pros of this technology, UAVs are expected to work for the benefit of society. As the cost will come down, UAVs will become widely available to the general public and used for wider applications. It is expected that in the future, the UAV technology will make the UAVs further smaller, lighter, cheaper, and much more efficient.

EXERCISES

1. Define the terms: UAV, UAS, RPA, and Drone. What is the basic difference between these?

2. Write the historical developments of unmanned aerial vehicles.

3. Discuss the contribution of government and industry to the development of various UAVs.

4. What are the major segments where UAVs, Blimps, and Zeppelins are being used? What is their relative market potential? Also, discuss the market potential of UAVs region-wise.

5. Draw a diagram to show maturity levels of UAVs applications in various sectors.

6. What do you understand by the Hype Cycle for the advancement of technology? Discuss it with reference to emerging technologies.

7. Describe the terms: Waypoints, Point of interest, Line of sight, Position hold, Return to home, and No fly zone. Explain the difference between VLOS, BVOLS, and EVLOS.

8. Discuss the general characteristics of a UAV.

9. Differentiate the working of a Photogrammetry UAV and LiDAR UAV.

10. Discuss the advantages and disadvantages of Photogrammetry UAVs and LiDAR UAVs.

11. Enumerate the advantages and disadvantages of UAVs over satellite images.

THE PROMISE OF ARTIFICIAL INTELLIGENCE (AI) IN INDUSTRIAL AUTOMATION

We are said to be entering a Fourth Industrial Revolution. With the first revolution was in the advent of industrial technology in creating machines to replace or enhance human and animal labor, the second revolution which developed because of the use of infrastructure and systems such as the telegraph, radio, railroads, and electricity, and the third industrial revolution being the advent of digital computing. This Fourth Industrial Revolution is occurring now with the advent of cloud computing, the Industrial Internet of Things (IoT), Big Data and the use of Machine Learning (ML) and Artificial Intelligence (AI). Others prefer the use of the term Industry 4.0 or similar variations.

First Industrial Revolution
• Movement from Human and Animal Power to Mechanical Systems Using Steam and Hydropower • 1760s-1860s
Second Industrial Revolution
• Infrastructure Innovations (Railroads, Telegraph) and Electrification • 1870s-1950s
Third Industrial Revolution
• The Digital Revolution • 1950s-early 2000s
Fourth Industrial Revolution
• Cyber-Physical Systems (CPS), Internet of Things (IoT), Industrial IoT (IIoT), Cloud Computing, Artificial Intelligence • 2000s-present

FIGURE 20.1 The Four Industrial Revolutions

We have explored the use of mechanical robots in Industrial Automation, and the advantages of using digital programming and embedded systems to aid in control and process management. These systems are driven by human-composed computer programs to give step-by-step instructions to the robotic elements of the production process. Sensors provide feedback and the programming tells the robots how to react and adapt to the data derived from the systems. For decades, these systems have grown to be more complex and interdependent, and the human element may seem invisible, but the work of thousands of programmers control each decision that was interlaid in designing the controls.

If we look back to Chapter 18, control was driven through the subsumption architecture, where low level component actuation and sensing is performed by a human operator or through instructions in response to system feedback of embedded systems. Robots could be said to be fully autonomous even if driven by instructions given by human operators via their instructions given in response to expected variables.

AI changes this paradigm in making the leap from human-led development, to allowing the systems to learn information through their performance or by examining sets of data from other systems to suggest and even implement process improvements. For example, what should an automated robot do in response to unexpected variables? In Industrial Automation, we do not want unpredictable reactions when facing unexpected variables. The onus would be on the programmers to build control systems that accounted for as much as possible and for the environment to be tightly controlled to avoid the unexpected. AI might be able to have the systems handle new inputs and variables, learning from its programming and adapting responses that it has modelled from similar systems or from learning its own strengths and weaknesses.

WHAT IS ARTIFICIAL INTELLIGENCE?

We should start with thinking about what is thinking, what is intelligence? This is a philosophical inquiry that has been wrestled with since ancient times. Researchers have tried to distinguish decision making based on instinct or biological urges and a higher level of decision-making that is able to adapt to different environments or inputs. There has been considerable research in artificial intelligence since the middle of the 20th century. We have seen chess-playing computers and chat programs that have attempted to mimic human intelligence so much that we are all familiar with the concept of artificial intelligence on a basic level. The so-called Turing Test, based on the ideas of cryptologist Alan Turing. The test evaluates a computer's AI based on the question that if in a text conversation a human could not distinguish text answers to questions given to both a human and a computer using AI, then the AI would have passed the test.

In modern discussions of AI, the focus is on the following components of intelligence: learning, reasoning, problem-solving, perception, and the ability to use language. While that may seem overly abstract, to put it into context, how would AI be applied in Industrial Automation?

Let us start with the idea of Machine Learning (ML). ML can make use of algorithms to learn patterns from data, to take those insights and apply those insights to recognize patterns and adapt performance to optimize a system. It then could gauge the effects of those applied actions to see if they produced positive or negative outcomes and then learn from those new decisions. As computing processes are not restricted by elements that would hamper human operators (continuous reliability, broad availability, less likely to create computation errors) or where using human operators may be cost-prohibitive to devote the time to make continuous changes and evaluate their effectiveness of each small adaptation.

Deep learning, or more advanced AI may take advantage of larger networks, such as neural networks to model adaptations in more com-

Traditional Linear Processing

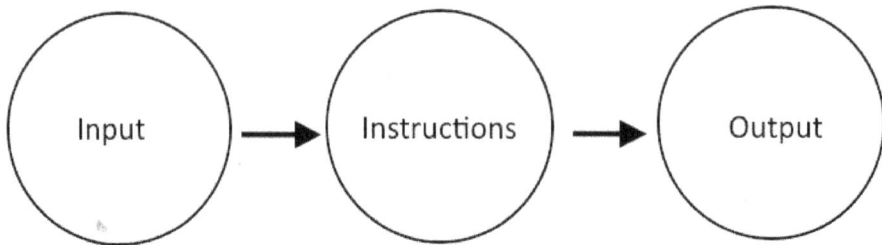

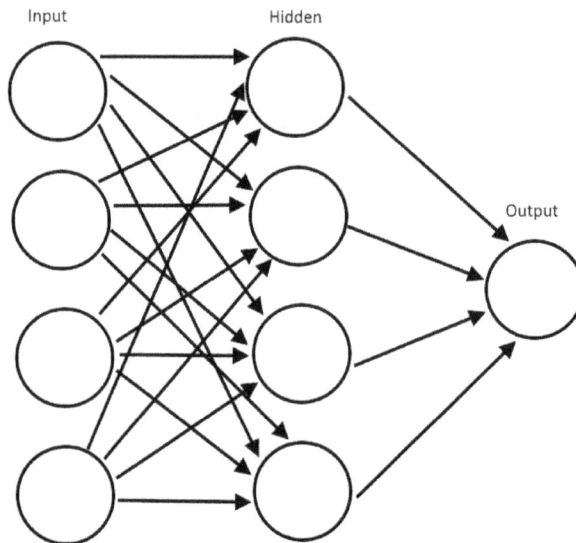

FIGURE 20.2 Linear Systems versus Neural Networks

plex systems. A neural network is an arrangement of computer nodes that attempt to mimic a biological neural network (like a human brain) as opposed to a standard computer architecture that separates memory from processing. The are several approaches to neural networks but the distinction is that computers are connected in a more deeply connected network that goes beyond the more familiar computer processing of just more speed and processing power.

Here is one example of how that might work. As with conventional computer processing, inputs enter the system and outputs are the result. If we have a control system that simply is told, when the system is started (input=start), a system of instructions will be applied, and the system output is for that system to run as instructed. There may be a control parameter that says that if the temperature exceeds a certain threshold, that the system is to shut down safely. The instruction is contained in the programming and the output to a higher than dictated temperature being inputted by a sensor would be for the system to shut down. These are linear processes; if this, then that.

With the addition of ML or AI to a system's controls, it could be a system that is more dynamic and non-linear. If we input that we want the system to operate safely and in the most efficient manner, we could have a neural network adapt to quickly changing conditions and adjust the speed of system operation to maximize output while maintaining safe operations. It could learn how to respond by training on operations in the past, studying data on temperate effects on the materials, studying the ability to manage outputs, adapting to varying costs of energy, reducing workloads when power loads of a plant overall are high, to the human operator staffing levels, Quality Assurance (QA) trends on the outputs and many more. While the training data at first would operator-provided, it would learn on its own and apply those lessons to the system's operations.

APPLICATIONS TO INDUSTRIAL AUTOMATION AND ROBOTICS

There are many examples of how AI could be useful in Industrial Automation and Robotics:

- *Manufacturing Process Improvement and Operation.* Similar to the example given above, AI can analyze and implement constant adjustments based on changing operating and market conditions.

- *Supply Chain.* By analyzing global transport networks, transportation partners, customer sales and demand data, resource availability, can result in more efficient scheduling and adaptive production can reduce waste.
- *Maintenance and Safety.* Inputs from the systems and from operators can be analyzed to make sure that systems are kept in efficient and safe operating order and staffing can be optimized to ensure safe operations.
- *Research and Development.* AI could identify components that are frequent failure points, and systems that could yield better performance. It could identify when expansion of facilities would be optimal or to predict additional resources that would enhance profitability.
- *Quality Inspection and Assurance.* Using computer vision, autonomous camera scanning can inspect components likely to fail, or to check for defects in manufacturing outputs.
- *Automated Security.* AI can provide Cybersecurity by monitoring connections to systems in cyber-physical and IIoT and to operator's computer systems by reporting threats and vulnerabilities. Monitoring physical security of plants can be aided by security robots, access controls, fire/smoke/water monitoring and connection to emergency management systems.
- *Plant design and Optimization.* In planning future expansion, AI can help design layout, resource needs, staffing levels, access to transportation networks and cost of resources more efficiently by learning from operations that it models from existing operations.

FACTORS TO BE CONSIDERED USING ARTIFICIAL INTELLIGENCE IN INDUSTRIAL AUTOMATION

Each of the Industrial Revolutions has faced challenges to achieve great increases in productivity. Profound changes resulted as we moved from one era to another in achieving a world much different from the ones that preceded it.

PLANNING AND IMPLEMENTATION

Early AI adaptations have come from organizations that have invested years in academic research. Training sets of data, the creation of digital models of physical assets, parallel processing for significant stretches of time under multiple operating conditions are all needed in the data-hungry process of training AI. For some processes that are standard across many organizations, such as office automation and customer service, there may be commercially available packages of AI services that will serve as off-the-shelf purposes. With Industrial Automation, there is less likelihood of such solutions as there are competitors who would consider this to be intellectual property not to be shared. There may be solutions for components such as HVAC controls, cyber- or physical-security and supply-chain logistics, but for proprietary processes, the investment would be less likely to be something that could be purchased as a service without significant customization. This requires significant investment in time and computing power.

As each AI module is implemented, there needs to be quality assessment of each new roll-out. Staff would need to be trained, and to have the ability to build-in assessment points where there is monitoring of progress or failure for each stage so that adjustments can be made.

PREPARING THE WORKFORCE FOR AI IMPLEMENTATION

Each period of progress in Industrial Automation has meant significant changes to the workforce. Building an internal culture of innovation that increases opportunity for advancement for the entire organization can help to invest your workforce in working together towards success. If you can educate your employees about the technological changes AI will bring, and to see the possibility of reducing the need for repetitive tasks and career growth can help to build a culture of positivity around changing technology.

Identify AI use cases for automation that have demonstrable worth, where the effort will reduce time waste, mundane, repetitive tasks, and tasks that tend to result in employee error. If new skills are required, training can be parallel to the planning and early implementation phases. This allows for an organization to limit worker-displacement, and to avoid the search for external talent which may be costlier and more difficult based on the market conditions that an organization operates in. These skills will be in high demand, and it is usually less expensive to develop employees with known performance and familiarity with the industry and corporate operations as they are today.

EVALUATING FOR SAFETY AND EFFICIENCY

Implementing AI in Industrial Automation requires that leaders understand the over-arching concern of yielding operator-led or programmer-led system controls to an environment that is more AI-led, is the potential for unforeseen changes that might not be in the best interest of the organization. AI might have greater tolerance for risk than an organization is willing to withstand. Limits must be engineering-informed, to monitor compliance with safety, legal and performance standards. While Industrial Automation is less likely to suffer from racial, gender or other sociological biases, there may be cases where an operator's complexion may affect computer vision decision-making, or where the height or weight of an operator would be a safety issue if it is only looking for typical characteristics of the data sets it was trained on.

EXERCISES

1. What are some of the elements of the so-called Fourth Industrial Revolution?

2. True/False: Artificial Intelligence is a new concept developed since 2010?

3. Provide 5 examples of applications of Artificial Intelligence in Industrial Automation?

4. True/False: An AI system does not require monitoring or evaluation post-implementation.

5. True/False: It is a best practice for organizations to plan to replace their operators with AI experts.

REFERENCES

The following references appear in Chapter 19 "The Use of Unmanned Aerial Vehicles (UAVs) in Industrial Automation"

[Albeaino *et al.*, 2019]. Albeaino, Giles; Gheisari, Masoud; and Franz, Bryan W. (2019), A Systematic Review of Unmanned Aerial Vehicle Application Areas and Technologies in the AEC Domain, Journal of Information Technology in Construction – Vol. 24, ISSN 1874–4753.

[Andrews, 2020]. Andrews, Lena Simone (2020), Attack of the Drones: Ethical, Legal and Strategic Implications of UAV Use, *https://cis. mit.edu/publications/audits/attack-drones-ethical-legal-and-strategic-implications-uav-use*.

[Austin, 2010]. Austin, Reg (2010), Unmanned Aircraft Systems – UAVs Design, Development and Deployment, John Wiley and Sons, Ltd.

[Ayranci, 2017]. Ayranci, Zehra Betul (2017), Use of Drones in Sports Broadcasting, Entertainment and Sports Lawyer, Spring, Volume 33, No. 3.

[Banu *et al.*, 2016]. Banu, Tiberiu Paul; Borlea, Gheorghe Florian; and Banu, Constantin (2016), The Use of Drones in Forestry, Journal of Environmental Science and Engineering B5, 557–562, DOI: 10.17265/2162–5263/2016.11.007.

[Barnes, 2018]. Barnes, Jonathan (2018), Drones (Photogrammetry) vs Terrestrial LiDAR – What Kind of Accuracy Do You Need?, April, *https://www. commercialuavnews.com/infrastructure/drones-vs-terrestrial-lidar-accuracy-need*.

[Berie, and Buruda, 2018]. Berie, Habitamu Taddese; and Buruda, Ingunn (2018), Application of Unmanned Aerial Vehicles in Earth Resources Monitoring: Focus on Evaluating Potentials for Forest Monitoring in Ethiopia, European Journal of Remote Sensing, Vol. 51, No. 1, 326–335, *https://doi.org/10.1080/22797254.2018.1432993*.

[Blom, 2011]. Blom, John David (2011), Unmanned Aerial Systems: A Historical Perspective, Occasional paper-37, Combat Studies Institute Press, US Army Combined Arms Center, Kansas.

[Boucher, 2015]. Boucher, P. (2015), Domesticating the Drone: The Demilitarisation of Unmanned Aircraft for Civil Markets. Sci Eng Ethics, 21, 1393–1412, *https://doi.org/10.1007/s11948-014-9603-3*.

[Buczkowski, 2018]. Buczkowski, Aleks (2018), Drone LiDAR or Photogrammetry? Everything You Need to Know, January 6, *https://geoawesomeness. com/ drone-lidar-or-photogrammetry-everything-your-need-to-know/*.

[Cai *et al.*, 2011]. Cai, G; Chen, B.M.; and Lee, T.H. (2011), Unmanned Rotorcraft Systems. New York: Springer Science & Business Media; pp. 01–267.

[Cao *et al.*, 2019]. Cao, Lin; Liu, Hao; Fu, Xiaoyao; Zhang, Zhengnan; Shen, Xin; and Ruan, Honghua (2019), Comparison of UAV LiDAR and Digital Aerial, Photogrammetry Point Clouds for Estimating Forest Structural Attributes in Subtropical Planted Forests, Forests, Vol. 10, p. 145, DOI: 10.3390/ f10020145.

[Chatzikyriakou and Rickerby, 2019]. Chatzikyriakou, Chara; and Rickerby, Patrick (2019), Comparing Drone LiDAR and Photogrammetry, *https://terra-drone.eu/en/articles-en/comparing-drone-lidar-and-photogrammetry*.

[Colomina and Molina, 2014]. Colomina, I.; and Molina, P. (2014). Unmanned Aerial Systems for Photogrammetry and Remote Sensing: A Review, ISPRS Journal of Photogrammetry and Remote Sensing 92 (June), pp. 79–97, DOI: 10.1016/j.isprsjprs.2014.02.013.

[Corrigan, 2016]. Corrigan, F. (2016), Introduction to UAV Photogrammetry and LiDAR Mapping Basics, *http://www. dronezon.com/learn-about-drones-quadcopters/ introduction-to-uav-photogrammetry-and-lidar-mapping-basics/.*

[Corrigan, 2019]. Corrigan, F. (2019), 12 Top LiDAR Sensors For UAVs, LiDAR Drones And So Many Great Uses, December 31, *https://www. dronezon.com/learn-about-drones-quadcopters/best-lidar-sensors-for-drones-great-uses-for-lidar-sensors/.*

[Culus, 2018]. Culus, Joline; Schellekens, Yves; Agoria; and Smeets, Yannick (2018), A Drone's Eye View, May, PwC, AGORIA, Belgium.

[Dawson, 2018]. Dawson, Philip (2018), Developing a Global Framework for Unmanned Aviation, United Aviation, July 29. *https://www.unitingaviation.com/strategic-objective/safety/developing-a-global-framework-for-unmanned-aviation/Delair.aero.*

[Dering *et al.*, 2019]. Dering, Gregory M.; Micklethwaite, Steven; Thiele, Samuel T.; Vollgger, Stefan A.; and Cruden, Alexander R. (2019), Review of Drones, Photogrammetry and Emerging Sensor Technology for the Study of Dykes: Best Practices and Future Potential, Journal of Volcanology and Geothermal Research, Vol. 373, PP. 148–166, *https:// doi.org/10.1016/j.jvolgeores.2019.01.018.*

[Dronethusiast, 2020]. Dronethusiast (2020), The History of Drones (Drone History Timeline from 1849 to 2019), *https://www.dronethusiast.com/history-of-drones/.*

[Dupont *et al.*, 2016]. Dupont, Quentin F.M.; Chua, David K.H.; Tashrif, Ahmad; and Abbott, Ernest L.S. (2016), Potential Applications of UAV Along the Construction's Value Chain, Science Direct, 7th International Conference on Engineering, Project, and Production Management, DOI: 10.1016/j. proeng.2017.03.155.

[Economic Times, 2018]. Economic Times (2018), India Fastest Growing Market for Unmanned Aerial Vehicles. Retrieved from *https://economictimes.indiatimes.com/news/defence/india-fastest-growing-market-for-unmanned-aerial-vehicles/articleshow/63466658.cms (26 March).*

[Encyclopedia Britannica, 1999]. Encyclopedia Britannica (1999), *https:// www.britannica.com/technology/*.

[FAA, 2010]. FAA. (2010), A Brief History of the FAA, 03 Ocak 2015'de, *http://www.faa.gov/about/history/brief_history*.

[Federman *et al.*, 2017]. Federman, A.; Quintero, M. Santana Quintero; Kretz, S.; Gregg, J.; Lengies, M.; Ouimet, C.; and Laliberte, J. (2017), UAV Photgrammetric Workflows: A Best Practice Guideline, The International Archives of the Photogrammetry, Remote Sensing and Spatial Information Sciences, Volume XLII-2/W5, 2017, 26th International CIPA Symposium 2017, 28 August–01 September, Ottawa, Canada.

[Freudenrichn, 2010]. Freudenrichn Craig (2020), How Blimps Work, *https://science. howstuffworks.com/transport/flight/modern/blimp4.htm*.

[Garg, 2019]. Garg, P.K. (2019), Theory and Principles of Geoinformatics, Khanna Book Publishing House., Delhi.

[Gartner, 2018]. Gartner (2018), Hype Cycle for Emerging Technologies, Gartner Identifies Five Emerging Technology Trends That Will Blur the Lines Between Human and Machine, *https://www.gartner.com/en/ newsroom/press-releases/2018-08-20-gartner-identifies-five-emerging-technology-trends-that-will-blur-the-lines-between-human-and-machine*.

[Glover, 2014]. Glover, John M. (2014), Drone University, DroneUniversity, Edition.

[Gupta *et al.*, 2018]. Gupta, Medha; Bohra, Devender Singh; Raghavan, Raamesh Gowri; and Khurana, Sukant (2018), A beginners' guide to understanding Drones, *https://medium.com/@ sukantkhurana/a-beginners-guide-to-drones-38d215701c4e*.

[Heutger and Kückelhaus, 2014]. Heutger, Matthias; and Kückelhaus, Markus (2014), Unmanned Aerial Vehicle in Logistics – A DHL Perspective on Implications and Use Cases for the Logistics Industry, Published by DHL Customer Solutions & Innovation, Germany.

[ICAO, 2015]. ICAO (2015), International Civil Aviation Organization (ICAO). Manual on Remotely Piloted Aircraft Systems (RPAS); ICAO: Montréal, QC, Canada.

[Iizuka *et al.*, 2018]. Iizuka, Kotaro; Itoh, Masayuki; Shiodera, Satomi; Matsubara, Takashi; Dohar, Mark; and Watanabe, Kazuo (2018), Advantages of Unmanned Aerial Vehicle (UAV) Photogrammetry for

Landscape Analysis Compared with Satellite Data: A Case Study of Postmining Sites in Indonesia, Cogent Geoscience, Vol. 4, No. 1, DOI: 10.1080/23312041.2018.1498180.

[Intelligence, 2018]. Intelligence, S.D. (2018), The Global UAV Payload Market 2017–2027, *https://www.researchandmarkets.com/research/ nfpsbm/the global uav*.

[IWM, 2018]. IWM (2018), A Brief History of Drones, *https://www.iwm. org.uk/ history/a-brief-history-of-drones*.

[Jha, 2016]. Jha, A.R. (2016), Theory, Design, and Applications of Unmanned Aerial Vehicles. New York: CRC Press.

[Kaminsky *et al.*, 2009]. Kaminsky, Ryan S.; Snavely, Noah; Seitz, Steven M.; and Szeliski, Rick (2009), Alignment of 3D Point Clouds to Overhead Images, Second IEEE Workshop on Internet Vision, June.

[Kan *et al.*, 2018]. Kan, M.; Okamoto, S.; and Lee, J.H. (2018), Development of Drone Capable of Autonomous Fight Using GPS. In: Proceedings of the International Multi Conference of Engineers and Computer Scientists, Vol. 2.

[Keane and Carr, 2013]. Keane, John; and Carr, Stephen (2013). A Brief History of Early Unmanned Aircraft. Johns Hopkins Apl Technical Digest, Vol. 32, pp. 558–571.

[Kimchi, 2015]. Kimchi, Gur (2015), Amazon's Top Prime Air Executive Outlines Plans for Delivery Drones to Navigate Skies, *https:// www.supplychain247.com/article/amazon_outlines_plans_for_ delivery_drones_to_navigate_skies/pinc_solutions*.

[Luncitel, 2011]. Lucintel (2011), Growth Opportunity in Global UAV Market, Lucintel – Creating the Equation for Growth, January, *https:// www.lucintel.com/LucintelBrief/UAVMarketOpportunity.pdf*.

[Marsh, 2013]. Marsh, Randall (2013), Goodyear Replacing Its Current Blimp Fleet With Zeppelins, Aircraft, *https://newatlas.com/ goodyear-blimp-replacement-zeppelins/28335/*.

[Molina, 2014]. Molina, Colomina P. (2014), Unmanned Aerial Systems for Photogrammetry and Remote Sensing: A Review, ISPRS Journal of Photogrammetry and Remote Sensing, Vol. 92, June, pp. 79–97, *https:// doi.org/10.1016/j.isprsjprs.2014.02.013*.

[Newcome, 2004]. Newcome, L.R. (2004), Unmanned Aviation: A Brief History of Unmanned Aerial Vehicles, AIAA.

[Nex and Remondino, 2014]. Nex, Francesco; and Remondino, Fabio (2014), UAV for 3D Mapping Applications: A Review, Appl Geomat., Vol. 6, Nos. 1–15, DOI 10.1007/ s12518-013-0120-x.

[NIAS, 2018]. NIAS (2018), The Evolution of Commercial Drone Technology, *https://nias-uas.com/evolution-commercial-drone-technology/.*

[Ortiz, 2018]. Ortiz, Nora Raboso (2018), LiDAR Technology, *https://www.cursosteledeteccion.com/donde-descargar-datos-lidar/lidar-technology/.*

[PwC, 2018]. PwC (2018a), Flying High, PwC-Engineering Council of India, November, *https://www.pwc.in/assets/pdfs/publications/2018/flying-high.pdf.*

[PwC, 2018b]. PwC (2018b), Clarity from Above, PwC Global Report on the Commercial Applications of Drone Technology, *https://www.pwc.pl/pl/pdf/clarity-from-above-pwc.pdf.*

[Puliti *et al.*, 2015]. Puliti, Stefano; Ørka, Hans; Gobakken, Terje; and Næsset, Erik (2015), Inventory of Small Forest Areas Using an Unmanned Aerial System. Remote Sensing, Vol. 7, pp. 9632–9654. DOI: 10.3390/rs70809632.

[Shakhatreh *et al.*, 2018]. Shakhatreh, Hazim; Sawalmeh, Ahmad; Al-Fuqaha, Ala; Dou, Zuochao Dou; Almaita, Eyad; Khalil, Issa; Othman, Noor Shamsiah; Khreishah, Abdallah; and Guizani, Mohsen (2018), Unmanned Aerial Vehicles: A Survey on Civil Applications and Key Research Challenges, arXiv:1805.00881v1 [cs. RO] 19 Apr 2018.

[Sholes, 2007]. Sholes, E. (2007), Evolution of a UAV Autonomy Classification Taxonomy. In Proceedings of the IEEE Aerospace Conference, Big Sky, MT, USA, 3–10 March.

[Siebert and Teizer, 2014]. Siebert, Sebastian; and Teizer, Jochen (2014), Mobile 3D Mapping for Surveying Earthwork Projects Using an Unmanned Aerial Vehicle (UAV) System. Automation in Construction, Vol. 41, pp. 1–14, DOI: 10.1016/j.autcon.2014.01.004.

[Singhal *et al.*, 2018]. Singhal, Gaurav; Bansod, Babankumar; and Mathew, Lini (2018), Unmanned Aerial Vehicle Classification, Applications and Challenges: A Review, *http://dx.doi.org/10.20944/preprints201811.0601.v1.*

[Stoica, 2018]. Stoica, A.A. (2018), Emerging Legal Issues Regarding Civilian Drone Usage, Challenges Knowl. Soc., Vol. 2018, No. 12, pp. 692–699.

[Šustek and Úøedníêek, 2018]. Šustek, Michal; and Úøedníêek, Zdenìk (2018), The Basics of Quadcopter Anatomy, EDP Sciences, CSCC 2018, MATEC Web of Conferences 210, 01001 (2018), *https://doi. org/10.1051/matecconf/201821001001.*

[Taladay, 2018]. Taladay, Katie (2018), Pros and Cons of Using Unmanned Aircraft Systems (UAS) for Generating Geospatial Data: Case Studies from Hawai'i. DOI: 10.13140/RG.2.2.10224.35847.

[Vergouw *et al.*, 2016]. Vergouw, B.; Nagel, H.; Bondt, G.; and Custers, B. (2016), Drone Technology: Types, Payloads, Applications, Frequency Spectrum Issues and Future Developments, in the Future of Drone Use. Springer, pp. 21–45.

[White Paper, 2018]. White Paper (2018), UAV Workflow Integration, Drone Industry Insights, May. DroneII.com.

[Wijeyasinghe and Ghaffarzadeh, 2020]. Wijeyasinghe, Nilushi; and Ghaffarzadeh, Khasa (2020), LiDAR 2020–2030: Technologies, Players, Markets & Forecasts, IDTechEx, *https://www.idtechex.com/en/research-report/ lidar-2020- 2030-technologies-players-markets-and-forecasts/694.*

[Wingtra, 2019]. Wingtra (2019), Drone Photogrammetry vs. LiDAR: What Sensor to Choose for a Given Application, *https://wingtra.com/ drone-photogrammetry-vs-lidar/.*

[Wu *et al.*, 2018]. Wu, B.; Xie, L.; Hu, H.; Zhu, Q.; and Yau, E. (2018), Integration of Aerial Oblique Imagery and Terrestrial Imagery for Optimized 3D Modeling in Urban Areas, ISPRS J. Photogramm. Remote Sens., Vol. 139, pp. 119–132.

[Xu *et al.*, 2014]. Xu, Zhiqiang; Yang, Jiansi; Peng, Chaoyong; Wu, Ying; Jiang, Xudong; Li, Rui; Zheng, Yu; Gao, Yu; Liu, Sha; and Tin, Baofeng (2014), Development of an UAS for post-earthquake Disaster Surveying and Its Application in MS7.0 Lushan Earthquake, Sichuan, China, Computers & Geosciences, Vol. 68, pp. 22–30.

INDEX